AF352851

Eco-Friendly Materials for Civil Construction: Utilization and Advantages

Eco-Friendly Materials for Civil Construction: Utilization and Advantages

Editors

Carlos Maurício Fontes Vieira
Gustavo de Castro Xavier
Henry Alonso Colorado Lopera
Sergio Neves Monteiro

MDPI • Basel • Beijing • Wuhan • Barcelona • Belgrade • Manchester • Tokyo • Cluj • Tianjin

Editors

Carlos Maurício Fontes Vieira
Advanced Materials
Laboratory
State University of the North
Fluminense Darcy Ribeiro
Campos dos Goytacazes
Brazil

Gustavo de Castro Xavier
Civil Engineering Laboratory
State University of the North
Fluminense Darcy Ribeiro
Campos dos Goytacazes
Brazil

Henry Alonso Colorado
Lopera
Composites Lab
University of Antioquia
Medellin
Colombia

Sergio Neves Monteiro
Department of Materials
Science
Military Institute of
Engineering
Rio de Janeiro
Brazil

Editorial Office
MDPI
St. Alban-Anlage 66
4052 Basel, Switzerland

This is a reprint of articles from the Special Issue published online in the open access journal *Sustainability* (ISSN 2071-1050) (available at: www.mdpi.com/journal/sustainability/special_issues/ Eco_Friendly_Materials).

For citation purposes, cite each article independently as indicated on the article page online and as indicated below:

LastName, A.A.; LastName, B.B.; LastName, C.C. Article Title. *Journal Name* **Year**, *Volume Number*, Page Range.

ISBN 978-3-0365-7087-7 (Hbk)
ISBN 978-3-0365-7086-0 (PDF)

Contents

About the Editors

Carlos Maurício Fontes Vieira

Carlos Maurício Fontes Vieira graduated as a mechanical engineer (1992) at the Santa Úrsula University (USU). He received his M.Sc. (1997) and Ph.D. (2001) from the State University of the North Fluminense (UENF), followed by a post doctorate (2004). In 2005, he joined the Advanced Material Laboratory of UENF as associated professor and was Under-Rector for Research and Postgraduation (2014). Since 2019, he is Titular Professor and Coordinator of Postgraduate Program in Materials Science and Engineering. Dr. Vieira has published more than 600 articles in journals, conference proceedings and book chapters and author of more than 10 patents. He is researcher (1C) of the Brazilian Council for Scientific and Technological Development (CNPq) and Scientist of Our State of the Superior Council of the State of Rio de Janeiro (FAPERJ). He leads a research group in sustainable materials for civil construction, including topics of heavy clay ceramics, engineered rocks, natural fibers, alkali-activated materials and additive manufacturing.

Gustavo de Castro Xavier

Gustavo de Castro Xavier is associate professor Laboratory of Civil Engineering at the State University of North Fluminense, professor at the Postgraduation Programs in Civil Engineering (PPGEC/UENF) and Materials Science Engineering (PPGECM/UENF). He holds a degree in Civil Engineering (1999), M.Sc. (2001) and Ph.D. (2006) from State University of North Fluminense. In 2007, he joined the UENF to Laboratory of Civil Engineering as a permanent professor at PPGEC/UENF, later (2020) he joined the PPGECM/UENF. He was advisor and Under-Rector of the Under-Rectory of Extension and Community Affairs (2007/2011) and was Mayor of UENF (2011/2015). He was elected as Head of Laboratory (2017/2020). Reviewer for several national and international journals, he has published more than 200 articles in journals, book chapters and conference proceedings. He was Young Scientist of Our State of the Superior Council of the State of Rio de Janeiro (2014/2019). He is a researcher at the Brazilian Council for Scientific and Technological Development (CNPq) and Ad Hoc consultant at the same institution. He works by developing alternative materials for Civil Engineering/Materials, with emphasis on the durability of construction materials and the use of industrial waste in mortars and red ceramics, in addition to researching ornamental rocks and ballast for railway purposes.

Henry Alonso Colorado Lopera

Henry A. Colorado L. obtained his Ph.D. and M.Sc. in Materials Science and Engineering at University of California Los Angeles (UCLA) in 2013; and M.Sc. and B.Sc. in Mechanical Engineering at Universidad Nacional de Colombia in 2005. He is a Full Professor at Universidad de Antioquia in Medellin, Colombia, and leads a research group in sustainable materials and manufacturing, including topics of solid waste management, composites, natural fibers, additive manufacturing, and construction materials.

He is a Full Professor associated to Mechanical and Materials Engineering graduate schools at the Universidad de Antioquia, and an active collaborator with several institutions worldwide. He is also very active member of American Ceramics Society (Acers), where he is a Global Ambassador.

Sergio Neves Monteiro

Sergio Neves Monteiro graduated as a metallurgical engineer (1966) at the Federal University of Rio de Janeiro (UFRJ). He received his M.Sc. (1967) and Ph.D. (1972) from the University of Florida, followed by a 1975 course in energy at the Brazilian War College, and a post doctorate (1976) at the University of Stuttgart. In 1968, he joined the Metallurgy Department of UFRJ as full professor of the postgraduation program in engineering (COPPE). He was elected as head of department (1978), coordinator of COPPE (1982), Under-Rector for Research (1983), and was invited as Under-Secretary of Science for the State of Rio de Janeiro (1985) and Under-Secretary of the College Education for the Federal Government (1989). He retired in 1993 from the UFRJ and joined the State University of North Rio de Janeiro (UENF), where he retired in 2012. He is now a professor at the Military Institute of Engineering (IME), Rio de Janeiro. Dr. Monteiro has published more than 2,000 articles in journals and conference proceedings and has been honored with several awards including the ASM Fellowship and several TMS awards. He is the top researcher (1A) of the Brazilian Council for Scientific and Technological Development (CNPq) and Emeritus Scientist of State of Rio de Janeiro (FAPERJ). He was president of the Superior Council of the State of Rio de Janeiro Research Foundation, FAPERJ (2012), and currently is coordinator of the Engineering Area of this foundation. He has also served as president of the Brazilian Association for Metallurgy, Materials and Mining (ABM, 2017–2019), as a consultant for the main Brazilian R&D agencies, and as a member of the editorial board of five international journals as well as associate editor-in-chief of the Journal of Materials Research and Technology. He is the author of 130 patents and a top world researcher in "Natural Fiber Composites" and "Ballistic Armor", Scopus 2020.

Preface to "Eco-Friendly Materials for Civil Construction: Utilization and Advantages"

It is unsustainable for Planet Earth to possess societies with a lifestyle that is increasingly consumerist and predatory to the environment. The linear economy has failed as a viable form of human development, as it entails a huge generation of waste and the consequential negative impact on the environment. We currently consume poorly designed and energy-inefficient products that do not have adequate reusability, and therefore the development of sustainable materials is mandatory as one of the pillars for the transition from a linear to a circular economy.

This Special Issue is dedicated to papers from academic researchers working on multidisciplinary experiments, as well as modeling related to environmental sustainability. Papers on construction materials used worldwide are encouraged aiming at improving efficiency, productivity, and competitiveness in world markets.

For this Special Issue, both scientific and technical papers advanced in understanding on eco-friendly building materials, which might promote a saving in energy, as well as pollutants reduction in the extraction of raw materials, manufacturing process, or in the final application. These building materials can be more durable and also incorporated with urban, agricultural, or industrial wastes.

The editors of this book express their thanks and gratitude to Sustainability staff for giving the Special Issue editors the opportunity to publish this book. The editors also thank the publisher, MDPI, who produced the book, and the authors.

Carlos Maurício Fontes Vieira, Gustavo de Castro Xavier, Henry Alonso Colorado Lopera, and Sergio Neves Monteiro
Editors

sustainability

Article

Application of Glass Waste on Red Ceramic to Improve Sintering

Geovana Delaqua [1],*, Juan Magalhães [1], Markssuel Marvila [1], Fernando Vernilli, Jr. [2], Sérgio Monteiro [3], Henry Colorado [4] and Carlos Vieira [1]

1   LAMAV—Advanced Materials Laboratory, UENF—State University of the Northern Rio de Janeiro, Av. Alberto Lamego, 2000, Campos dos Goytacazes 28013-602, RJ, Brazil
2   DEMAR—Materials Engineering Department, Lorena Engineering School—EEL, Area I—Campinho Municipal Road, s/n°, Lorena 12602-810, SP, Brazil
3   Department of Materials Science, Instituto Militar de Engenharia—IME, Praça General Tibúrcio 80, Praia Vermelha, Urca, Rio de Janeiro 22290-270, RJ, Brazil
4   CCComposites Laboratory, Universidad de Antioquia (UdeA), Medellín 050010, Colombia
*   Correspondence: geovanagirondi@pq.uenf.br

Abstract: Given the current huge generation of solid waste worldwide, alternative and innovative methodologies for incorporating these materials should be encouraged elsewhere. In this context, the objective of this research is to evaluate the use of glass waste as a substitute for sand as raw material in ceramics. Formulations containing from 0% to 20% of glass waste were produced, thus replacing natural sand. Extruded and calcined specimens were produced at temperatures of 800, 900 and 1000 °C. The characterization results demonstrated the compatibility and their potential for the glass waste for improving the properties of ceramics. Results of density, water absorption and flexural strength improved when 20% of glass waste was added due to the porosity reduction, provided by the formation of a liquid phase and then by a sintering, promoted by the glass waste. This resulted in coherent properties with ceramic applications in the form of tiles and blocks, at a calcining temperature of 800 °C. On the contrary, results without glass did not reach the necessary parameters even at 1000 °C. In conclusion, the feasibility of using glass waste has been proven, which, in addition to improving the material's properties, provides economy benefits for the ceramic industry, with the calcination process at milder temperatures.

Keywords: glass waste; sand; liquid phase; reuse

Citation: Delaqua, G.; Magalhães, J.; Marvila, M.; Vernilli, F., Jr.; Monteiro, S.; Colorado, H.; Vieira, C. Application of Glass Waste on Red Ceramic to Improve Sintering. *Sustainability* **2022**, *14*, 10454. https://doi.org/10.3390/su141610454

Academic Editor: Cinzia Buratti

Received: 13 July 2022
Accepted: 18 August 2022
Published: 22 August 2022

Publisher's Note: MDPI stays neutral with regard to jurisdictional claims in published maps and institutional affiliations.

1. Introduction

In recent years, the generation of industrial solid waste has increased exponentially around the world. This increase has as main reasons the population increase and the consequent increase in waste production [1]; urbanization processes, generating an increase in the number of inhabitants in large cities and the generation of waste [2]; the industrialization process [3]; and the increase in inappropriate places for the final disposal of waste, through dumps [1,2], which further aggravates this environmental problem.

Given this, the need to find ways to reuse waste is increasingly urgent. An alternative, which will be studied in this work, is the reuse of these wastes in the red ceramic industry, due to the diversity of functions that wastes can play in these types of materials [4]. It is known that the clays used for the production of ceramic artifacts are essentially plastic in nature, which enables, for example, use of waste that introduces non-plastic elements into the clayey mass to control this property [5,6]. As an example, the work of Moreira et al. (2008) [7] evaluated the use of ornamental rock waste in ceramics, proving the technological feasibility and economic advantages obtained with the use of this waste. Similar results were obtained by Amaral et al. (2019) [8]. In this research, the authors evaluated the use of ornamental rock residue in ceramic materials produced by uniaxial pressing. The authors' objective was to evaluate the possibility of applying this waste in paving blocks. The results obtained highlight the feasibility of using waste in the proposed application.

Another possibility is the use of waste that helps to reduce the energy expenditure of the material in the calcination stage. This action can be direct through the supply of energy through exothermic reactions, or indirect through the reduction of the melting temperature of the ceramic material. The work by Delaqua et al. (2020) [9] evaluated the use of powdered cigarette waste in red ceramics with satisfactory results in the evaluated properties and proving the energy gain of the ceramic material during burning. Another relevant work with similar characteristics was developed by Delaqua et al. (2020) [10], where the authors evaluated the use of macrophyte biomass in ceramic materials, with positive conclusions.

A waste of great importance from an industrial point of view is glass waste. Only in Brazil, all products made with glass correspond on average to 3% of urban waste, and about 47% of glass packaging is recycled annually in the country, totaling 470 thousand tons per year [11]. Moreover, in general, this waste is 100% recyclable, that is, it can be indefinitely recycled [12]. However, when this waste is wrongly disposed of, it can be harmful, for reasons such that it is not biodegradable, and in most cases, it has a high cutting power [13]. In this article, the use of wastes from the manufacturing process of tempered glass as a melting element in red ceramics was studied, with the waste being removed from the initial manufacturing stage, that is, from the cutting of flat glass.

Other relevant information on glass waste, especially its application in ceramics, is highlighted by Silva et al. (2017) [14]. In this research, the authors developed a literature review on the use of glass waste in ceramic materials, highlighting several advantages obtained with the use of waste. An advantage is that the use of glass waste can reduce the environmental impact of the ceramic industry. Another issue addressed is the diversity of glass waste, which can be obtained from different sources. This information highlights the importance of conducting research with this material.

Another point that deserves to be highlighted is the need to use a typical ternary composition containing clay (30–60% by weight%), feldspar (15–40% by weight%) and quartz (5–30% by weight%) to produce ceramic materials [15,16]. This composition was studied such as in Luo et al. (2021) [17], where the authors evaluated the effect of replacing quartz and clay with fly ash to produce ceramics. In this study, the authors proved the need to use the typical ternary composition to obtain ceramic materials with adequate properties [18]. Thus, it is expected that the use of glass waste will contribute to the ternary composition, adding feldspar to the mix and providing adequate technological properties.

The main innovations of this work are to propose the use of glass waste to replace the sand used as a substitute for clayey masses. Although there are other published works with glass wastes [14,19], the analysis of new methodologies that enable the real application of the waste in ceramic materials is necessary.

2. Materials and Methods

The materials used in the research were commercial clayey mass, composed of two clays typical of the region used by the industries of Campos dos Goytacazes, RJ, Brazil. The mass is predominantly kaolinitic and has been used in other studies [5,8,9]. Moreover, natural quartz sand and glass waste were used in the formulation's soda–calcium plan. The waste was obtained from a company located in Rio das Ostras, RJ, Brazil.

The raw materials used in the research were characterized by X-ray fluorescence (XRF), using an S2 POLAR equipment, from BRUKER supplier. In addition, they were characterized through physical particle size tests, using the procedure of NBR 7181 [20], and through scanning electron microscopy (SEM) using a Shimadzu SSX-550 Superscan equipment. The samples used in SEM were prepared in advance using sanding and polishing. The samples were tested immediately after the metallization process, not requiring a conditioning step.

The formulations were produced using 60% clay 1, 20% clay 2 and 20% natural sand, as shown in Table 1. In addition, it was proposed to replace 0–20% glass waste, as a substitute for natural sand.

Table 1. Formulations used for production of red ceramics.

Formulation	Clay 1 (%)	Clay 2 (%)	Natural Sand (%)	Glass Waste (%)
0%	60	20	20	0
5%	60	20	15	5
10%	60	20	10	10
20%	60	20	0	20

The formulations defined in Table 1 were initially evaluated for linear and optical dilatometry using a Netzsch brand DIL 402 pc equipment with a heating rate of 5 °C/min and a maximum temperature of 1050 °C. The rate used in this assay can vary in the range of 3–5 °C/min (slow burning), as the dilatometry results are close and do not affect the discussion of the results [21,22]. The masses of approximately 2 g were previously sieved in a sieve with 42 μm opening and prepared with 8% moisture, for subsequent compaction in a cylindrical shape with a force of 2 tons, with an application time of 5 s. In the formulations before firing, plasticity tests were also carried out, through the Atterberg limits defined by NBR 6459 [23] and NBR 7180 [24]. In this way, it was possible to obtain the prognosis of extrusion.

Afterward, the specimens were produced by extrusion, using a prismatic geometry of 15 × 25 × 115 mm. These specimens were evaluated for linear shrinkage by drying and dry density before the calcination process. The firing was carried out in a laboratory Muffle furnace by Maitec model FL 1300, at 800, 900 and 1000 °C. In the burning, a heating rate of 3 °C/min was used until reaching the burning temperature, with a plateau of 240 min. Cooling was by natural convection. After firing, density after firing, linear retraction after firing, water absorption, and flexural strength, were all tested. The procedure was adapted from NBR 15310 [25] and NBR 15270 [26]. Finally, microstructural analysis was performed using SEM.

3. Results and Discussion

3.1. Characterization of Raw Materials

Table 2 presents the XRF results of the raw materials. It is observed that the chemical composition of clays is predominantly based on SiO_2 and Al_2O_3, which is typical of clay minerals. The presence of Fe_2O_3 is important, as it offers the reddish after-firing color typical of ceramic artifacts [27]. The presence of TiO_2 helps in the surface hardness of the material, which can improve its mechanical properties, but it makes the material more fragile [28,29]. It can also act as a coloring oxide, clearing the ceramic piece after firing. The presence of K_2O, MgO, Na_2O and CaO is important because these alkaline and earthy alkaline elements act in the formation of the liquid phase, which is the main sintering mechanism of ceramic materials [6,30]. The loss of ignition, above 10%, is problematic because it is an indication of the presence of organic matter that can increase the material's shrinkage and porosity, hence the importance of using corrective materials such as sand and glass waste. The use of two types of clay is justified to minimize dependence on a single deposit and to make the material less heterogeneous.

The chemical composition of natural sand is predominantly based on SiO_2, which is compatible with its mineralogical composition based on quartz. The glass waste, however, has predominantly SiO_2 in its composition, which is typical of glass. This material is probably in an amorphous form, contributing to the reduction of the material's porosity. There is also a high content of Na_2O (6.86%) and CaO (15.54%). These components aid in the formation of a liquid phase and improve strength, as they reduce the porosity of the material. The high content of CaO can cause bleaching in ceramic pieces, producing other static patterns for the pieces [31,32].

Figure 1 presents the results of the granulometry of the raw materials, while Figure 2 presents the SEM of the glass waste. It is easy to observe that the particle size of glass waste is close to the particle size of natural sand, which is another result that indicates the feasibility of using glass waste as a substitute for sand in ceramic materials. Regarding the morphology of the waste, an irregular pattern can be seen in Figure 2, due to the amorphism of the material, and particles with a high specific surface area, which is important to increase the resistance of the ceramic materials after firing.

Table 2. Chemical composition by XRF of raw materials.

Composition	Clay 1 (wt%)	Clay 2 (wt%)	Natural Sand (wt%)	Glass Waste (wt%)
SiO_2	47.04	49.34	81.10	72.58
Al_2O_3	32.56	30.71	11.90	1.82
Fe_2O_3	3.48	3.66	1.20	0.55
TiO_2	1.29	1.21	0.47	0.01
K_2O	1.01	0.99	1.50	0.36
MgO	0.55	0.61	0.62	2.53
Na_2O	0.34	0.24	0.84	6.86
CaO	0.24	0.22	0.51	15.54
Others	0.10	0.11	0.16	0.02
L.O.I.	13.39	12.91	1.70	0.00

L.O.I., loss of ignition.

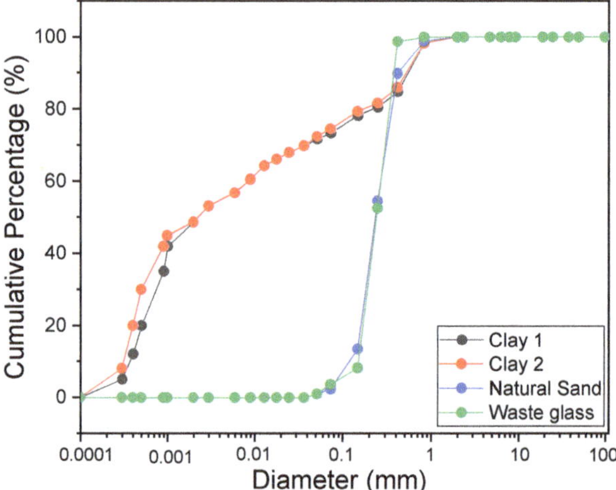

Figure 1. Granulometry of raw materials.

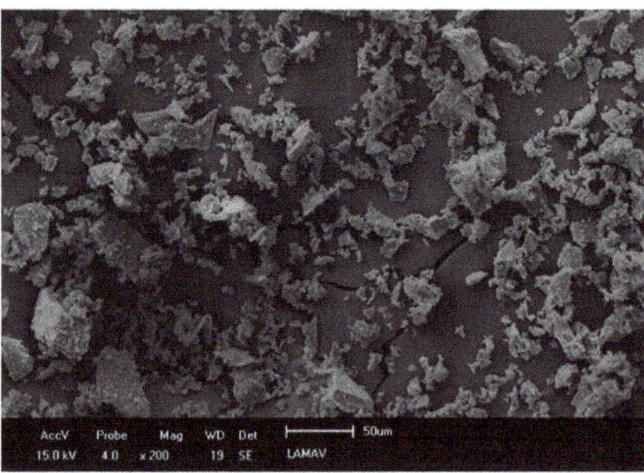

Figure 2. SEM of the glass waste.

3.2. Characterization of Ceramic Masses before Firing

Figure 3 presents the results of the dilatometry of the ceramic masses, revealing that the use of glass waste promotes the formation of larger amounts of liquid phase than natural sand, due to the presence of amorphous silica. Around 600 °C, the first relevant event occurs, which is when the glass enters to a softening point, showing a high reduction in its viscosity [33,34]. This is also evident by the optical dilatometry shown in Figure 3. The ceramic mass with 0% glass waste, and which has only natural sand, suffers an adverse event at this same temperature due to the allotropic transformation of the quartz, which can cause defects in ceramic materials [35]. The use of an amorphous material that can fulfill the same roles as natural sand is an advantage in this regard.

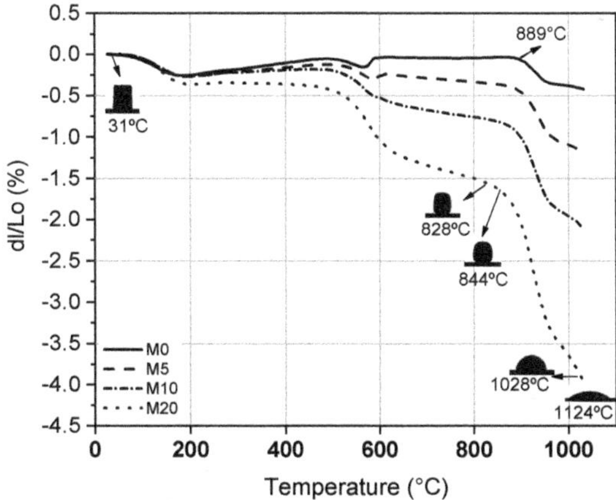

Figure 3. Dilatometry results.

Another important event takes place between 800 and 1000 °C, where the glass fusion occurs. These temperatures promote greater formation of the liquid phase, which can help the ceramic properties due to the reduction of porosity [33]. The excess of this phase can be harmful, due to the high shrinkage that can cause excessive defects in the manufacturing stage [36]. Thus, it is necessary to evaluate the results obtained after firing, especially for the mass containing 20% of glass waste, to verify that the retraction obtained was not too excessive. This will be discussed further later in this text.

Figure 4 presents the extrusion prognostic results. It is observed that all the studied ceramic masses are within the acceptable extrusion region. This point is important for using the glass waste in place of the natural sand because problems during the extrusion step will impact the properties of the ceramic material before and after firing [8]. This indicates that the behavior of the waste is non-plastic, compatible with natural sand, which also behaves this way.

Figure 5 shows the results of apparent density during the drying and linear drying. Note that the apparent density of the specimens increased with the use of the glass protector. This characteristic is because the greater the density, the greater the mass of the material occupying the same volume. This information can help to reduce the porosity of ceramic pieces after firing [37]. Regarding linear drying, note that glass reduces shrinkage. This is also positive due to the formation of particles, which will probably increase after the retraction of ceramic pieces and the firing due to the liquid phase. The reduction in drying shrinkage can compensate the high shrinkage after firing [38].

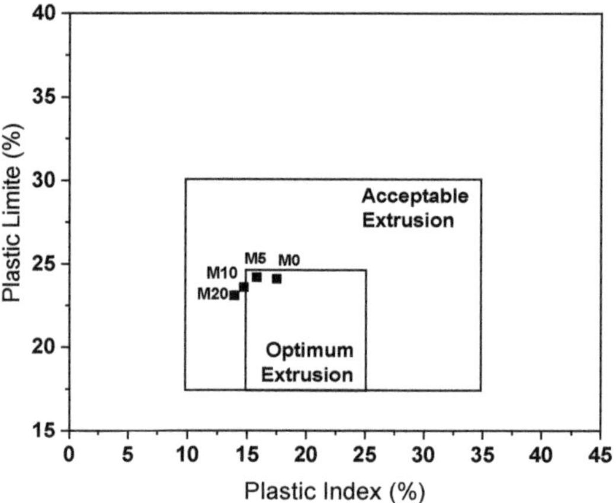

Figure 4. Extrusion prognosis.

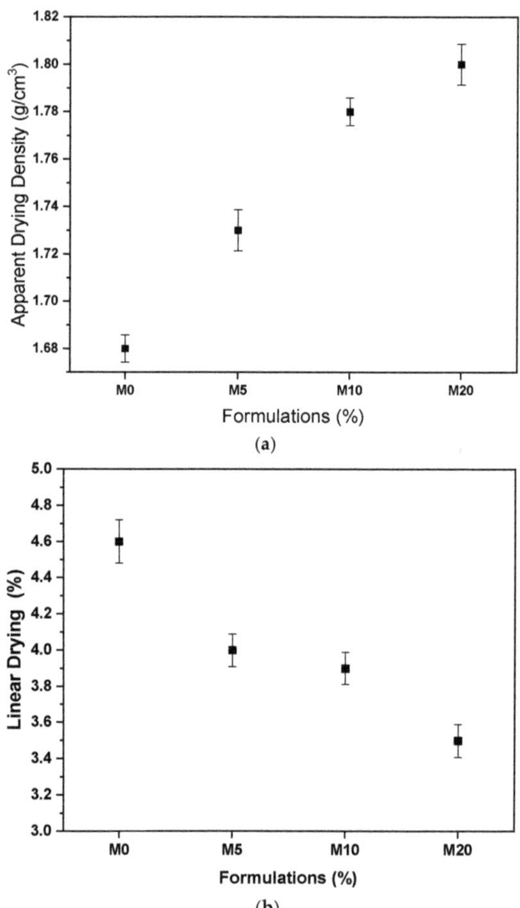

Figure 5. Results of (**a**) apparent drying density; (**b**) linear drying.

3.3. Characterization of Ceramic Masses after Firing

Figure 6 shows the post-fire density and post-fire linear shrinkage results. It is observed that as the calcination temperature increases, the apparent density of the ceramic material increases, as does the linear shrinkage. This behavior is the pattern of this type of material and has already been reported by other authors [7,39]. However, the glass waste promoted an increase in the densification of the ceramic material, which is directly related to the formation of a liquid phase and greater sintering of resistant phases in the material. A negative characteristic is that the use of the waste promoted an increase in the firing shrinkage, which was already expected based on the other results. Compositions 10% and 20% show shrinkage above 4% at a firing temperature of 1000 °C. This can be a problem during the manufacture of ceramic parts due to defects in the developed products. The other retraction values, however, are within the standards established by other authors, such as Girondi et al. (2020) [9], Delaqua et al. (2020b) [10], and Delaqua, et al. (2022) [22].

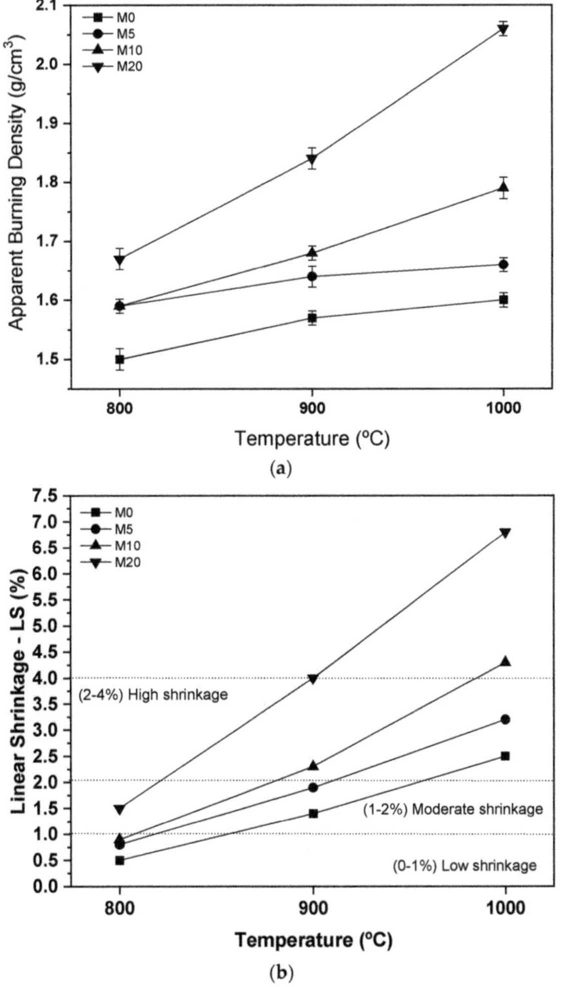

Figure 6. Results of (**a**) apparent burning density; (**b**) linear shrinkage.

Figure 7 presents the water absorption results, in which the mass containing only natural sand (0%) does not meet the limits for tiles, since at all temperatures, it presents water absorption greater than 20%. The masses containing glass waste, especially the one with 20% of the material, meet the water absorption requirements for both tiles and blocks, already in the calcination at 800 °C [7]. This is a great advantage because it provides savings in the manufacture of ceramics. The reduction of water absorption is related to the reduction of porosity and to the formation of liquid phase. As observed in Figure 3, from 600 °C onward, the glass already reaches its softening point, and therefore, it already has a functional liquid phase mechanism. At 1100 °C, glass still meets its operating point, updated, even more with the interconnection mechanisms of the materials [13,33]. These results do not prove the feasibility of using the glass waste, as they are used to use this material.

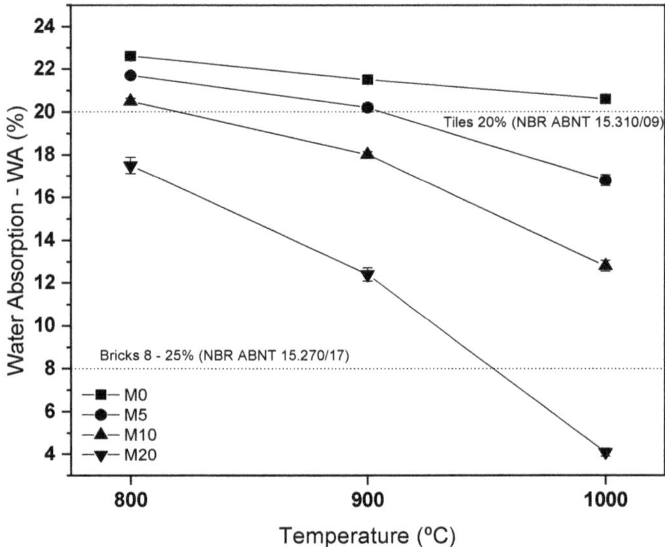

Figure 7. Water absorption results.

Figure 8 presents the flexural strength results. It is observed that the 20% composition meets the requirements of blocks and tiles already at a temperature of 800 °C, proving the savings highlighted in Figure 7. That is, it does not make much sense to propose calcination at higher temperatures if in this calcination range the composition of 20% already meets the requirements for water absorption and flexural strength, promoting savings in the firing stage. The same results are not possible without the use of glass waste, since the 0% composition does not meet the 6.5 MPa limit for tiles nor at the highest calcination temperature [8,10]. The results obtained are consistent with those highlighted by other authors and can be attributed to the formation of a liquid phase, as previously discussed [32].

Based on the discussion of the results, the effect of glass waste on ceramic materials is clear. It is observed that the role of natural sand is to act as a source of quartz for ceramic materials. In ceramic materials, quartz has the function of controlling shrinkage and reducing defects in the material. Quartz replacement, therefore, should be limited so as not to exacerbate these problems. However, the use of glass waste contributes feldspar to the ceramic material. This component, as highlighted in the introduction, contributes to the typical ternary composition of ceramic materials, containing clay, feldspar, and quartz [15,16]. The presence of glass waste contributes to a decrease in the firing temperature and an increase in the apparent density, resulting in a reduction in porosity and an increase in flexural rupture strength. Conversely, glass waste causes defects in ceramic materials, which is why its content should be limited to 20%, as proven in this research. At 1000 °C, the introduction of 20% of the glass waste caused an increase in flexural rupture strength from approximately 5 to almost 20 MPa. Conversely, it caused an increase in linear shrinkage from 2.5% to approximately 7.0%, which leads to an increase in material defects, as will be discussed below. Figure 9 shows a picture of the 20% specimens calcined at 1000 °C, where the formation of defects in the material, which, although they did not affect the strength gain, make the material of low aesthetic value. Therefore, the use of lower temperatures is more advantageous, since at 800 °C, it is possible to obtain values compatible with the main applications of ceramic materials.

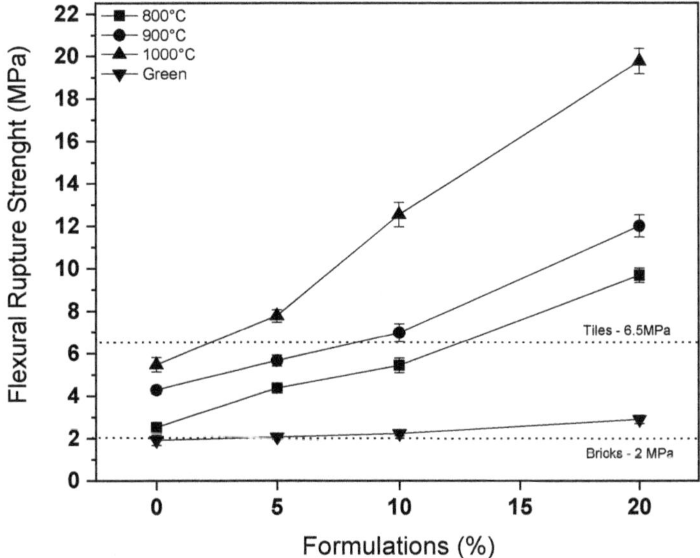

Figure 8. Flexural rupture strength results.

Figure 9a shows a typical defect of ceramics produced with clays containing high levels of Fe_2O_3 called black heart. This defect is essentially due to the high iron content present in the clay, which occurred in all compositions calcined at 1000 °C in this research. It is observed that the thermal effect is a catalyst for the black heart. Thus, an essential action to reduce these defects is to use lower calcination temperatures [40]. It can be seen from the results presented above that the use of glass waste makes it possible to reduce the calcination temperature without affecting the properties evaluated. Thus, an efficient action to reduce these defects is the use of glass waste to improve sintering in ceramic materials.

The formation of surface defects, highlighted in Figure 9b, is related to the high linear shrinkage values of this composition, shown in Figure 6b, and to the sintering mechanism promoted by glass waste. This sintering mechanism is potentiated at a temperature of 1000 °C, because as illustrated in Figure 3, at this temperature range, the glass waste melts, causing the surface defects observed in the material [4].

(a)

(b)

Figure 9. Macroscopic analysis of the 20% composition: (**a**) inner surface; (**b**) external surface.

Figure 10 presents the SEM results of the ceramic masses. Figure 10a,b shows the difference between the 0% and 20% compositions calcined at 800 °C, where in Figure 10b, the presence of a more compact and dense structure is visible. This can be attributed to the sintering promoted by the glass waste. Figure 10c,d shows the effects of calcination at 1000 °C in the 0% and 20% formulations, where the formation of a glassy phase in the 20% composition is clear. This is directly related to the use of glass, which promotes the formation of a liquid phase and reduces the porosity of the material. This happens due to the melting process of the glass waste, observed in Figure 3. In this temperature range, the glass becomes liquid, densifying the matrix and filling the pores not accessible by the sand that the glass waste replaces. Thus, the mechanism involved is liquid phase diffusion, which fills the accessible pores of the ceramic material and contributes to the flexural rupture strength [33]. The images are compatible with the results obtained by Figures 7 and 8.

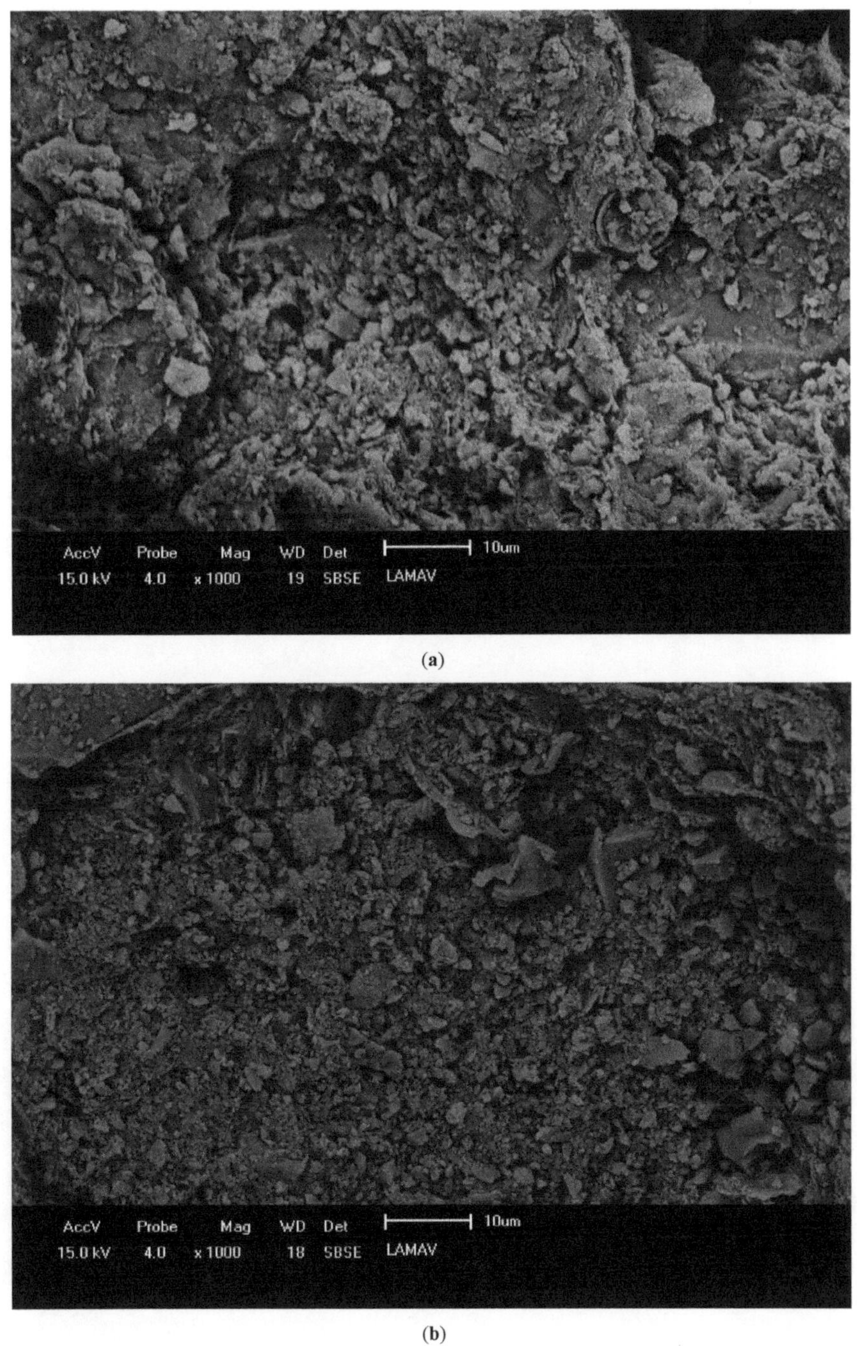

(**a**)

(**b**)

Figure 10. *Cont.*

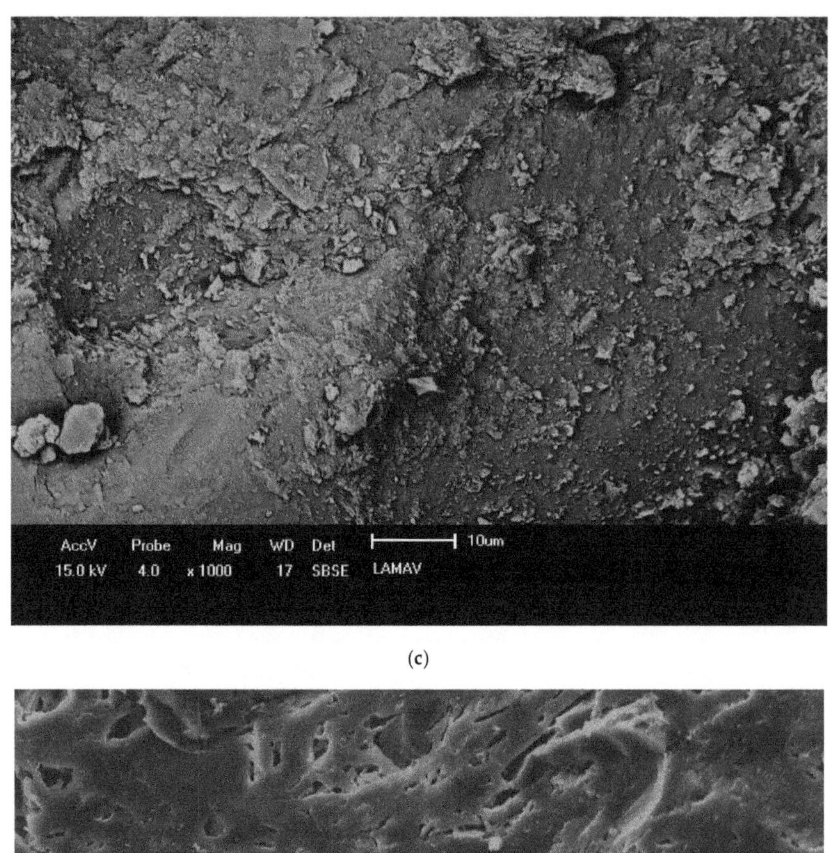

(c)

(d)

Figure 10. SEM of specimens: (**a**) 0% composition calcined at 800 °C; (**b**) 20% composition calcined at 800 °C; (**c**) 0% composition calcined at 1000 °C; (**d**) 20% composition calcined at 1000 °C.

4. Conclusions

Based on the results obtained, the following conclusions were summarized:

- The characterization of the glass waste shows the material's compatibility with natural sand, enabling its use, in addition to indicating a potential for improving flexural rupture strength, due to the alkali levels, which increases the sintering of the ceramic.
- The masses containing glass waste did not present problems in the extrusion prognosis, in addition to improving the densification of the ceramic material after drying and after firing.
- After firing, the use of 20% glass waste promoted the sintering of the ceramic mass, due to the formation of a liquid phase, reducing porosity and improving the properties of water absorption and flexural strength. There were problems with burn retraction, but within tolerated limits.
- The masses containing 20% of glass waste obtained parameters compatible with industrial applications at a calcination temperature of 800 °C, while the composition containing only natural sand did not present the same behavior even at temperatures of 1000 °C. This indicates the feasibility of using glass waste.

As perspectives for future work, the following stand out: (i) evaluate the influence of the granulometry of the glass waste on the properties of red ceramics; (ii) evaluate the effect of using 15% and 25% of glass waste as a substitute for sand; (iii) carry out durability and degradation tests on the red ceramic containing glass waste.

Author Contributions: Methodology, C.V., G.D. and J.M.; formal analysis G.D. and J.M.; resources S.M. and C.V.; formal analysis, H.C. and F.V.J.; writing—original draft preparation, G.D., J.M. and M.M.; writing—review and editing, C.V., G.D., M.M. and H.C.; supervision, C.V.; project administration, C.V.; funding acquisition, C.V. All authors have read and agreed to the published version of the manuscript.

Funding: This research was funded by Fundação Carlos Chagas Filho de Amparo à Pesquisa do Estado do Rio de Janeiro: E-26/200.847/2021 and Conselho Nacional de Desenvolvimento Científico e Tecnológico: 301634/2018.1.

Institutional Review Board Statement: Not applicable.

Informed Consent Statement: Not applicable.

Data Availability Statement: Not applicable.

Acknowledgments: The authors would like to thank FAPERJ and CNPQ as well as NewTemper for the waste and the Laboratory of Advanced Materials—LAMAV/UENF for the support.

Conflicts of Interest: The authors declare no conflict of interest.

References

1. Parthan, S.R.; Milke, M.W.; Wilson, D.C.; Cocks, J.H. Cost Estimation for Solid Waste Management in Industrialising Regions—Precedents, Problems and Prospects. *Waste Manag.* **2012**, *32*, 584–594. [CrossRef] [PubMed]
2. Mrayyan, B.; Hamdi, M.R. Management Approaches to Integrated Solid Waste in Industrialized Zones in Jordan: A Case of Zarqa City. *Waste Manag.* **2006**, *26*, 195–205. [CrossRef]
3. Hu, D.; Wang, R.; Yan, J.; Xu, C.; Wang, Y. A Pilot Ecological Engineering Project for Municipal Solid Waste Reduction, Disinfection, Regeneration and Industrialization in Guanghan City, China. *Ecol. Eng.* **1998**, *11*, 129–138. [CrossRef]
4. Dondi, M.; Raimondo, M.; Zanelli, C. Clays and Bodies for Ceramic Tiles: Reappraisal and Technological Classification. *Appl. Clay Sci.* **2014**, *96*, 91–109. [CrossRef]
5. Hossain, S.; Roy, K. Cerâmica sustentável derivada de resíduos sólidos: Uma revisão. *J. Asian Ceram. Soc.* **2020**, *8*, 984–1009. [CrossRef]
6. Mota, L.; Toledo, R.; Machado, F.; Holanda, J.N.F.; Vargas, H.; Fariajr, R. Thermal characterisation of red clay from the Northern Region of Rio de Janeiro State, Brazil using an open photoacoustic cell, in relation to structural changes on firing. *Appl. Clay Sci.* **2008**, *42*, 168–174. [CrossRef]
7. Moreira, J.; Manhães, J.P.; Holanda, J.N. Processing of red ceramic using ornamental rock powder waste. *J. Mater. Process. Technol.* **2008**, *196*, 88–93. [CrossRef]
8. Amaral, L.F.; Carvalho, J.P.R.G.d.; da Silva, B.M.; Delaqua, G.C.G.; Monteiro, S.N.; Vieira, C.M.F. Development of Ceramic Paver with Ornamental Rock Waste. *J. Mater. Res. Technol.* **2019**, *8*, 599–608. [CrossRef]
9. Girondi, G.D.; Marvila, M.M.; de Azevedo, A.R.G.; de Souza, C.C.; Souza, D.; de Brito, J.; Vieira, C.M.F. Recycling Potential of Powdered Cigarette Waste in the Development of Ceramic Materials. *J. Mater. Cycles Waste Manag.* **2020**, *22*, 1672–1681. [CrossRef]

10. Girondi Delaqua, G.C.; Marvila, M.T.; Souza, D.; Sanchez Rodriguez, R.J.; Colorado, H.A.; Fontes Vieira, C.M. Evaluation of the Application of Macrophyte Biomass Salvinia Auriculata Aublet in Red Ceramics. *J. Environ. Manag.* **2020**, *275*, 111253. [CrossRef]
11. Vieira, C.M.F.; Morais, A.S.C.; Monteiro, S.N.; Delaqua, G.C.G. Teste Industrial de Cerâmica Vermelha Incorporada Com Resíduo de Vidro de Lâmpada Fluorescente. *Cerâmica* **2016**, *62*, 376–385. [CrossRef]
12. Zhang, J.; Liu, B.; Zhang, S. A Review of Glass Ceramic Foams Prepared from Solid Wastes: Processing, Heavy-Metal Solidification and Volatilization, Applications. *Sci. Total Environ.* **2021**, *781*, 146727. [CrossRef] [PubMed]
13. Li, B.; Guo, Y.; Fang, J. Effect of MgO Addition on Crystallization, Microstructure and Properties of Glass-Ceramics Prepared from Solid Wastes. *J. Alloy. Compd.* **2021**, *881*, 159821. [CrossRef]
14. Silva, R.V.; de Brito, J.; Lye, C.Q.; Dhir, R.K. The Role of Glass Waste in the Production of Ceramic-Based Products and Other Applications: A Review. *J. Clean. Prod.* **2017**, *167*, 346–364. [CrossRef]
15. Luo, Y.; Zheng, S.; Ma, S.; Liu, C.; Wang, X. Ceramic Tiles Derived from Coal Fly Ash: Preparation and Mechanical Characterization. *Ceram. Int.* **2017**, *43*, 11953–11966. [CrossRef]
16. Menezes, R.R.; Neto, H.G.M.; Santana, L.N.L.; Lira, H.L.; Ferreira, H.S.; Neves, G.A. Optimization of Wastes Content in Ceramic Tiles Using Statistical Design of Mixture Experiments. *J. Eur. Ceram. Soc.* **2008**, *28*, 3027–3039. [CrossRef]
17. Luo, Y.; Wang, J.; Wu, Y.; Li, X.; Chu, P.K.; Qi, T. Substitution of Quartz and Clay with Fly Ash in the Production of Architectural Ceramics: A Mechanistic Study. *Ceram. Int.* **2021**, *47*, 12514–12525. [CrossRef]
18. Sasui, S.; Kim, G.; Nam, J.; van Riessen, A.; Hadzima-Nyarko, M. Effects of Waste Glass as a Sand Replacement on the Strength and Durability of Fly Ash/GGBS Based Alkali Activated Mortar. *Ceram. Int.* **2021**, *47*, 21175–21196. [CrossRef]
19. Yang, H.; Chen, C.; Pan, L.; Lu, H.; Sun, H.; Hu, X. Preparation of Double-Layer Glass-Ceramic/Ceramic Tile from Bauxite Tailings and Red Mud. *J. Eur. Ceram. Soc.* **2009**, *29*, 1887–1894. [CrossRef]
20. *NBR 7181*; Solo-Análise Granulométrica. Associação Brasileira de Normas Técnicas-ABNT: São Paulo, Brasil, 2016.
21. de Strieder, M.C.; Ratusznei, F.; Pereira, M.; Montedo, O.R.K. Laser-Assisted Glass-Based Sealing of Polished Porcelain Stoneware Tile Surface to Increase Stain Resistance. *J. Eur. Ceram. Soc.* **2020**, *40*, 3478–3488. [CrossRef]
22. Girondi Delaqua, G.C.; Ferreira, M.d.N.; Amaral, L.F.; Sánchez Rodríguez, R.J.; Atem de Carvalho, E.; Fontes Vieira, C.M. Incorporation of Sludge from Effluent Treatment Plant of an Industrial Laundry into Heavy Clay Ceramics. *J. Build. Eng.* **2022**, *47*, 103451. [CrossRef]
23. *NBR 6459:2016*; Solo—Determinação Do Limite de Liquidez. Associação Brasileira de Normas Técnicas-ABNT: São Paulo, Brasil, 2016.
24. *NBR 7180:2016*; Solo—Determinação Do Limite de Plasticidade. Associação Brasileira de Normas Técnicas-ABNT: São Paulo, Brasil, 2016.
25. *NBR 15310*; Componentes Cerâmicos—Telhas—Terminologia, Requisitos e Métodos de Ensaio. ABNT: São Paulo, Brasil, 2005.
26. *NBR 15270-1*; Componentes Cerâmicos—Blocos e Tijolos Para Alvenaria Parte 1: Requisitos. ABNT: São Paulo, Brasil, 2017.
27. Gencel, O.; Sutcu, M.; Erdogmus, E.; Koc, V.; Cay, V.V.; Gok, M.S. Properties of Bricks with Waste Ferrochromium Slag and Zeolite. *J. Clean. Prod.* **2013**, *59*, 111–119. [CrossRef]
28. Gülen, Ş.M.; Çöpoğlu, N.; Yılmaz, Y.B.; Karaahmet, O.; Cengiz, T.; Gökdemir, H.; Çiçek, B. TiO$_2$-Based Glass-Ceramic Coatings: An Innovative Approach to Architectural Panel Applications. *Case Stud. Constr. Mater.* **2022**, *16*, e00805. [CrossRef]
29. Özcan, M.; Birol, B.; Kaya, F. Investigation of Photocatalytic Properties of TiO$_2$ Nanoparticle Coating on Fly Ash and Red Mud Based Porous Ceramic Substrate. *Ceram. Int.* **2021**, *47*, 24270–24280. [CrossRef]
30. Conconi, M.S.; Morosi, M.; Maggi, J.; Zalba, P.E.; Cravero, F.; Rendtorff, N.M. Thermal Behavior (TG-DTA-TMA), Sintering and Properties of a Kaolinitic Clay from Buenos Aires Province, Argentina. *Cerâmica* **2019**, *65*, 227–235. [CrossRef]
31. Pardo, F.; Jordan, M.M.; Montero, M.A. Ceramic behaviour of clays in Central Chile. *Appl. Clay Sci.* **2018**, *157*, 158–164. [CrossRef]
32. Dondi, M.; Guarini, G.; Raimondo, M.; Zanelli, C. Recycling PC and TV Waste Glass in Clay Bricks and Roof Tiles. *Waste Manag.* **2009**, *29*, 1945–1951. [CrossRef]
33. Karamanov, A.; Dzhantov, B.; Paganelli, M.; Sighinolfi, D. Glass Transition Temperature and Activation Energy of Sintering by Optical Dilatometry. *Thermochim. Acta* **2013**, *553*, 1–7. [CrossRef]
34. Khoeini, M.; Hesaraki, S.; Kolahi, A. Effect of BaO Substitution for CaO on the Structural and Thermal Properties of SiO$_2$–B$_2$O$_3$–Al$_2$O$_3$–CaO–Na$_2$O–P$_2$O$_5$ Bioactive Glass System Used for Implant Coating Applications. *Ceram. Int.* **2021**, *47*, 31666–31680. [CrossRef]
35. Fontes Vieira, C.M.; Monteiro, S.N. Firing Behavior of the Clay Fraction of a Natural Kaolinitic Clay: Are They Different? *Mater. Res.* **2019**. [CrossRef]
36. Assías, S.G.; Pabón, F.; Cala, N.; Delvasto, P. Incorporation of Fluorescent Lamp Waste into Red-Clay Bricks: Defect Formation, Physical and Mechanical. Properties. *Waste Biomass Valorization* **2021**, *12*, 1621–1632. [CrossRef]
37. Eliche-Quesada, D.; Martínez-Martínez, S.; Pérez-Villarejo, L.; Iglesias-Godino, F.J.; Martínez-García, C.; Corpas-Iglesias, F.A. Valorization of Biodiesel Production Residues in Making Porous Clay Brick. *Fuel Proces. Technol.* **2012**, *103*, 166–173. [CrossRef]
38. Oummadi, S.; Nait-Ali, B.; Alzina, A.; Paya, M.-C.; Gaillard, J.-M.; Smith, D.S. Optical Method for Evaluation of Shrinkage in Two Dimensions during Drying of Ceramic Green Bodies. *Open Ceram.* **2020**, *2*, 100016. [CrossRef]

39. Kagonbé, B.P.; Tsozué, D.; Nzeukou, A.N.; Ngos, S. Mineralogical, physico-chemical and ceramic properties of clay materials from Sekandé and Gashiga (North, Cameroon) and their suitability in earthenware production. *Heliyon* **2021**, *7*, e07608. [CrossRef]
40. Izzo, F.; Guarino, V.; Ciotola, A.; Verde, M.; de Bonis, A.; Capaldi, C.; Morra, V. An Archaeometric Investigation in a Consumption Context: Exotic, Imitation and Traditional Ceramic Productions from the Forum of Cumae (Southern Italy). *J. Archaeol. Sci. Rep.* **2021**, *35*, 102768. [CrossRef]

sustainability

MDPI

Article

Circular Economy of Construction and Demolition Waste: A Case Study of Colombia

Henry A. Colorado [1,*], **Andrea Muñoz** [1] **and Sergio Neves Monteiro** [2]

[1] CCComposites Laboratory, Universidad de Antioquia UdeA, Calle 70 N°. 52–21, Medellín 050010, Colombia; andreamuzzap@gmail.com

[2] Departamento de Ciência dos Materiais, Instituto Militar de Engenharia, Rio de Janeiro 22290-270, Brazil; snevesmonteiro@gmail.com

* Correspondence: henry.colorado@udea.edu.co

Abstract: This paper presents the results of research into construction and demolition (C&D) waste in Colombia. The data and analyses are shown in a local and Latin American context. As the situation in Colombia is quite similar to that in many developing countries worldwide, this research and its findings are potentially applicable to similar economies. Several factors were calculated and compared in order to evaluate which best fit the data from Colombia. We also included an experimental characterization and analysis of several key types of C&D waste from important infrastructure projects in Colombia, specifically by using the X-ray diffraction and scanning electron microscopy techniques. For the quantification of CDW, a calculation was performed based on the area and four factors of volume and density, followed by an econometric analysis of the detailed information using the Hodrick–Prescott filter, which revealed the CDW trends. Our results revealed that there are limitations regarding the availability of information and effective treatments for this waste, as well as shortcomings in education and other issues, not only for Colombia but also for other countries in Latin America.

Keywords: solid waste; construction and demolition; circular economy

Citation: Colorado, H.A.; Muñoz, A.; Neves Monteiro, S. Circular Economy of Construction and Demolition Waste: A Case Study of Colombia. *Sustainability* **2022**, *14*, 7225. https://doi.org/10.3390/su14127225

Academic Editor: Ming-Lang Tseng

Received: 23 April 2022
Accepted: 30 May 2022
Published: 13 June 2022

Publisher's Note: MDPI stays neutral with regard to jurisdictional claims in published maps and institutional affiliations.

1. Introduction

Global warming is a very serious threat that could cause a devastating worldwide disaster if actions are not taken by all countries to reduce the excessive consumption of materials [1], develop sustainable agriculture [2], and, most importantly, apply politics to find real solutions [3]. Due to the urgency of this situation, it is necessary to increase corrective measures in all areas of solid waste management and recycle as much as possible, particularly in sectors that produce large amounts of waste that consists of metallurgical slags [4,5], rubber tires [6–8], organics [9], and construction and demolition waste (CDW) [10]. Moreover, preventive strategies such as improved economic models must be developed. The popular circular economy model [11,12] has been implemented in many sectors, e.g., by modifying manufacturing and design processes so that products last longer and can be used in other applications after their lifecycles (upcycling). For this reason, many sectors that involve materials and electronic devices are now experiencing a revolution, with processes being redesigned to fit the circular economy model [13], for example, in additive manufacturing [14], construction and building materials [15], smart materials [16], thermoelectric modules [17], and the transportation industry [18]. Furthermore, models have been developed for the sustainable design of solid-waste management on a city [19] and country scale [20].

Construction, demolition, and renovation activities generate a large amount of waste which harms the environment and can have a significant impact on global warming if it is not managed properly [21]. It is estimated that around 35% of construction and demolition waste (CDW) ends up in landfills without any type of treatment [22]. Therefore, new solutions, technology, and approaches are required for the management of CDW. The European Commission proposed that by 2020, a minimum of 70% of construction and demolition waste should be recycled. However, there were some member states of the European Union in which recycling rates already exceeded 70% [23]. This has been achieved through good waste management practices that essentially implement circular economy principles in the construction and demolition sector and beyond. The most effective practices are aimed at maximizing the reuse of items by facilitating recycling, material recovery, and the secondary use of materials through quality assurance schemes for waste-derived materials. This involves considering the entire value chain of the construction sector and implementing these principles throughout [24].

To establish a construction and demolition waste management system, an appropriate CDW quantification must be determined that can help the government make realistic decisions and policies and determine the places wherein disposal can be granted (e.g., end of waste disposal [25]). The globally adopted methodologies for quantifying CDW can be summarized into six categories: the method of visiting the site, the method of calculating the generation rate, the method of life analysis, the method of classifying the accumulation system, the method of modeling variables, and other assorted methods [26]. The site visit method (SV) requires that investigators go to the construction or demolition site. The generation rate calculation method (GRC) seeks to obtain the waste generation rate and estimate certain alternative parameters such as the financial value based on the area. The life analysis method is mainly used to quantify demolition waste based on mass balance, assuming that every building will at some point be demolished and therefore the amount of demolition residue must be equal to the mass of the built structure. The accumulation system classification method (CSA) is based on the generation rate calculation method but involves a quantification system for different special materials. The variable modeling method uses the modeling of variables to simulate the generation of waste, taking into account five factors (specific activity, type of work and equipment, type of material and storage, site and weather conditions, and company policy). The other methods include estimating waste production as a percentage of the purchase of materials and calculating the amount of residue for some materials based on their chemical characteristics.

In this article, the amount of construction and demolition waste generated in Colombia was quantified using innovative methods involving the area of construction and the volume and density factors of the materials. Additionally, a case study from Colombia was considered and the composition of the waste was characterized. Subsequently, an econometric study was carried out to better understand the behavior and factors that influence the generation of CDW in Colombia. Finally, the CDW situation in Colombia and other countries was evaluated and compared.

2. Materials and Methods

In this work, a review was carried out to quantify construction and demolition waste in Colombia. The aim was to obtain the first quantified values of this kind in the country and generate relationships and useful data for future applications and regulations, all with an eye on implementing a circular-economy strategy.

First, an information search was conducted within the National Administrative Department of Statistics (DANE, according to its Spanish acronym), which is the entity responsible for the planning, collection, processing, analysis, and dissemination of official statistics in Colombia [27]. DANE produces a census of buildings in urban and metropolitan areas, reporting the licensed sites, completed projects, projects in progress, and discontinued projects. In this article, all of these areas were considered since it is certain that they all generate waste.

Due to the information available in Colombia, the best strategy for quantifying CDW is to calculate the rate of waste generation based on the area. After obtaining the area of completed construction projects, a volume factor is obtained for transforming units from m^2 to m^3. A density factor is then obtained to transform m^3 into tons. In this work, several factors from different sources were used in order to obtain the one that best adapts to the available information. The factors that were used to obtain the amount of waste generated were:

- Factor 1: based on the quantitative data of the rubble generated by the private sector which has been obtained from the building censuses available in DANE and in the Colombian Chamber of Construction (CAMACOL) [28].
- Factor 2: obtained from an article published by the University of Eafit [29].
- Factor 3: based on the Fatta model, in which the amount of CDW is calculated from the density and the factor presented by the author [30].
- Factor 4: based on the law of mixtures according to the composition of waste in Bogotá.
- The value of the factors used in the quantification of CDW in Colombia used in this research is summarized in Table 1.

Table 1. Volume and density factors.

Name	Volume (m^3/m^2)	Density (ton/m^3)
Factor 1	1.42	1.48
Factor 2	1.35	1.40
Factor 3	0.6	1.6
Factor 4	1.12	1.48

For the experimental phase, sampling was carried out at the construction and demolition landfill facility "La Escombrera" in Medellín, Colombia. The X-ray diffraction (XRD) experiments were conducted using a PANanalytical X'Pert PRO diffractometer (Cu Kα radiation of 1.5406 Å) run at 45 kV, with scanning between 8° and 55°. Thirty subsamples from different parts of the demolition landfill facility were collected in order to obtain a composite sample of 10 Kg, which was representative of all building debris.

Econometric Analysis

The area below the curve gave the amount of CDW produced over time. The CDW flow per year was obtained as well. The gross domestic product from construction gave the cash flow per year. The statgraphics software was used to find the correlation between the flow of CDW and the flow of GDP from construction.

The Hodrick–Prescott and Baxter–King filters were used to compare the statistical cyclical component properties of the time series for both GDP and CDW generated from construction, showing how the properties of the cyclical component vary depending on the filter used in a cycle-trend decomposition. After quantifying CDW in Colombia, the data were compared with respect to information available in Latin America and the rest of the world.

3. Results and Analysis

3.1. Construction Waste Management in Colombia

The management of waste generated by construction activity in Colombia is shown in Figure 1 (solid lines). Note that this type of waste does not undergo any type of treatment and is typically sent to a final disposal site. This has created environmental and logistical problems due to the large volume of waste, which reduces the lifespan of landfills.

Some of the main problems in the management of CDW in Colombia [31] are: construction starting without having prepared the integral management of the CDW, no waste separation at the source, few real waste treatment technologies available, low demand for CDW processed materials, poor logistics in CDW management, and a lack of education and awareness around the proper management of CDW. Because of this, in 2017, the standards for the management of CDW and public clearance services in Colombia were updated (resolution 472) to regulate CDW use and recovery. Since CDW is classified as special, waste disposal companies do not collect it, which means that it is often disposed of in a non-proper way [32]. The law also establishes that the public administration is responsible for generating the required mechanisms and having spaces for their execution. Each city must therefore carry out studies and plans to educate the community and manage CDW (PGIRS). Such plans are expected to control and monitor CDW, generate strategies, and implement actions that lead to good waste management. Unfortunately, so far, the reality is quite far from the plans of PGIRS. The evolution of CDW management is shown in Figure 1 (dashed line). In Colombia, the comprehensive management of construction and demolition waste is governed by several standards [31]: Resolution 541 of 1994 "By means of which the loading, unloading, transport, storage and final disposal of rubble, materials, elements, concrete and loose aggregates of construction, demolition, organic layer, soil and excavation subsoil is regulated" [33]; Decree 948 of 1995 "Regulations in relation to the prevention and control of atmospheric pollution and protection of air quality" [34]; National Decree 1713 of 2002 Article 44 "Collection of rubble. It is the responsibility of the producers of rubble to arrange its collection, transport, and disposal in authorized dumps. The Municipality or District and the company that provides the cleaning service are responsible for coordinating these activities within the framework of the programs established in the PGIRS plans [35], which is also regulated by Law 1259 of 2008 by means of which the application of the environmental directive is established in the national territory against the violators of the rules of cleanliness, cleaning and collection of rubble" [36]; Decree 2981 of 2013 "By which the provision of the public cleaning service is regulated" [37]; and Resolution 472 of 2017 "By which the comprehensive management of waste generated in construction and demolition (CDW) activities is regulated and other provisions are issued" [38].

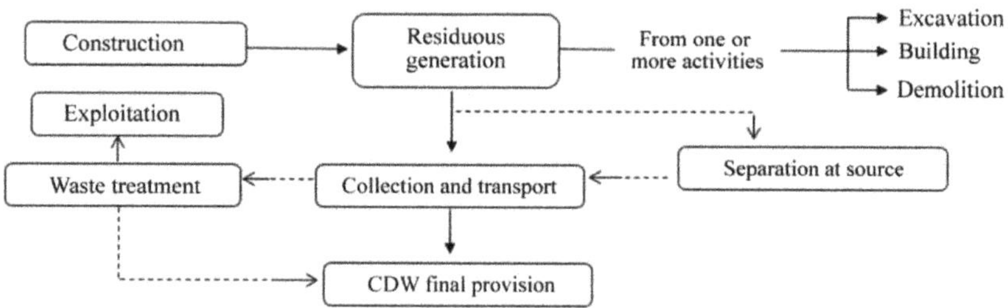

Figure 1. CDW cycle in Colombia.

3.2. CDW Quantification

The process to select the appropriate methodology is shown in Figure 2. Figure 3 shows results from the quantification method used to calculate the waste generation rate based on the construction area and the aforementioned factors. The amount of construction and demolition waste generated was recorded from 2008 to 2017 for the cities of Bogotá and Medellín in Colombia, as well as Colombia as a whole.

Considering the information available from the Ministry of Environment and Sustainable Development and the Institute of Environmental Studies, it was found that by 2011, the main cities in the country (12 of Colombia's 32 cities) had generated 22 million tons of CDW. In the year 2015, the city of Bogota generated approximately 8 million tons and the city of Medellin generated approximately 2.2 million tons [31]. These data serve as a reference value to compare against the new data obtained in this investigation.

Figure 3a shows the amount of CDW generated in the city of Medellín. It can be seen that when using Factors 1, 2, and 4, it is above the reference value, which is 2.2 million tons for the year 2015. In contrast, when using Factor 3, the CDW generated is close to the reference value.

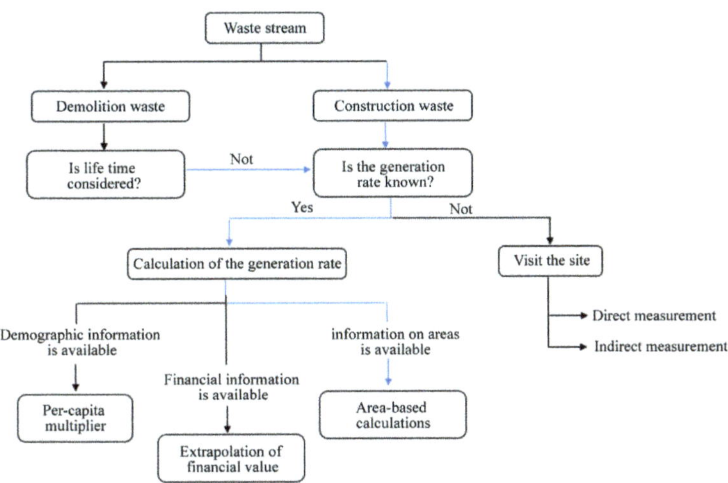

Figure 2. Selection of the quantification methodology.

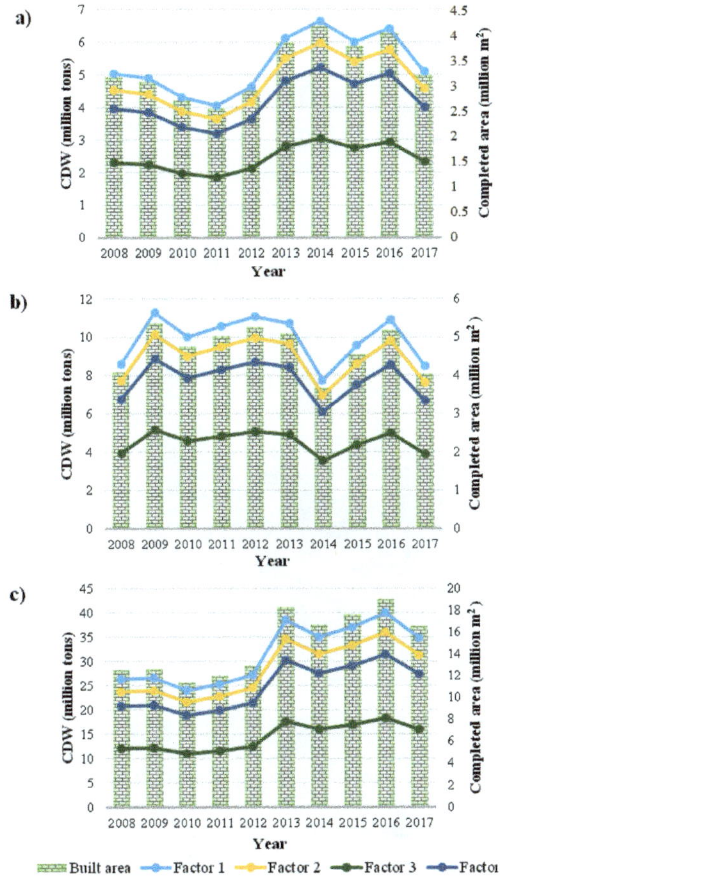

Figure 3. Amount of CDW generated in (**a**) Medellín, (**b**) Bogotá, and (**c**) Colombia as a whole.

The CDW generated in the city of Bogota is shown in Figure 3b. It can be observed that when using Factor 4, the value of CDW generated is the closest to the reference value, which is 8 million tons for the year 2015. When using Factors 1 and 2, the CDW generated is above the reference value, and when using Factor 3, it is below the reference value.

This behavior also happens for the CDW generated in Colombia, which can be evidenced in Figure 3c. It is clear that when using Factor 4, the CDW is the closest to the reference value, which is 22 million tons for the year 2015. When using Factors 1 and 2, the CDW generated is above the reference value, and when using Factor 3, it is below the reference value.

In general, it should be noted that the reference values were only obtained from legally reported CDW, which is certainly below the real data as illegal CDW disposal is not registered. In addition, the reference value does not include all 32 Colombian cities due to the lack of information in some parts of the country. Furthermore, Factor 3, which is based on the Fatta model, assumes that for every 1000 m^2 of construction, 50 m^3 of CDW is generated; in reality, more m^3 of CDW is generated, meaning that the results obtained are below the reference values. As a result of the aforementioned issues, this research focuses on Factor 1.

3.3. CDW Characterization in Colombia
3.3.1. X-ray Diffraction (XRD)

Figure 4 shows the morphologies of some CDW samples obtained from La Pradera landfill without any type of treatment. Figure 4a shows CDW with combined powders of concrete and red clay products, Figure 4b shows mostly bricks, Figure 4c shows CDW mainly composed of concrete but with some red clay products, and Figure 4d shows mostly clean concrete waste.

Figure 4. Samples obtained from the La Pradera waste dump.

The results of the XRD analysis obtained from the construction and demolition waste samples are shown in Figure 5. The main phases found are calcium oxide, iron oxide, magnesium oxide, feldspar (AlKSi$_3$O$_8$), Albite, anorthite, ettringite, portlandite, silicate and hydrated calcium calcite and gypsum. These phases are consistent with those found in bricks, tiles, ceramics, and concrete. It can also be seen that the same phases are found in the four diffractograms. Only the intensity of the peaks changes.

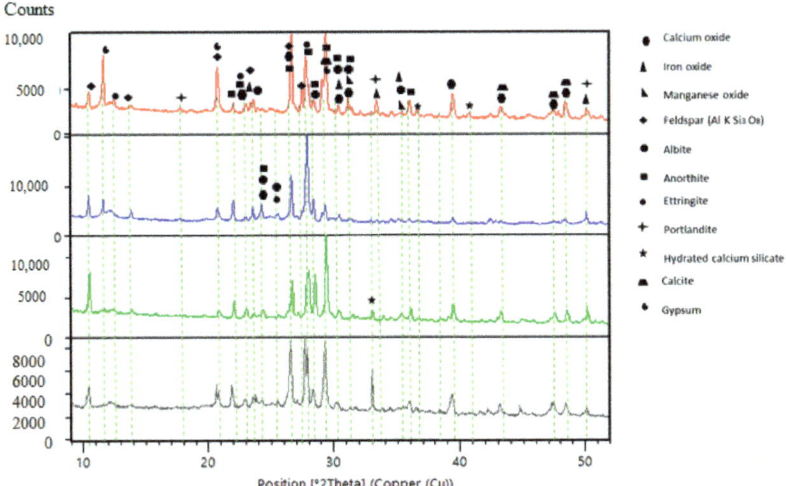

Figure 5. X-ray diffractogram of CDW in Colombia.

3.3.2. Comparison of CDW Composition

Considering the results of X-ray diffraction data and the information available in the literature, the composition of the CDW generated in Colombia was compared with that generated in other countries. Figure 6 shows the composition of the CDW generated in Colombia. Waste classified as others contributes 38%, concrete contributes 28%, brick 22%, ceramic 8%, and mortar 4% [39]. In addition, it can be seen that China has the highest percentage of metallic materials, which are insignificant in the United States, Norway, and Colombia. In all countries, inert waste, which includes concrete, mortar, brick, and ceramics, makes the following contributions: 85% in Kuwait, 84% in Norway, 77% in the United States, 91% in Spain, 96% in Portugal, 87% in China [40], and 60% in Colombia. The difference in these values is due to the variety of construction structures and materials used in the different countries, mainly due to the difference in climates, cultures, and types of construction.

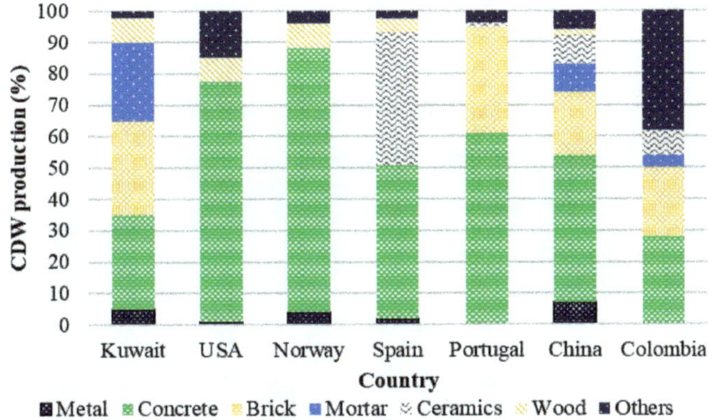

Figure 6. Composition of CDW production in different countries.

3.4. Econometric Analysis

3.4.1. Correlation Analysis

In order to carry out this study, it had to be discovered which factors were related and which were the most appropriate to estimate the generation of CDW in Colombia. Figure 7 compares the construction and demolition waste with the population, gray cement production, and gross domestic product (GDP) from construction in Colombia. In Figure 7a, it can be seen that the population has an increasing linear growth through the years, while the amount of CDW increases or decreases. This shows that, in general, the population growth does not significantly affect the amount of CDW generated annually. Figure 7b shows that cement production varies over the years, but this variation does not have the same trend as the CDW generation. Figure 7c reveals the GDP from construction, which shows that GDP decreases or increases with a similar trend to the generation of CDW. The flow of CDW and GDP from construction was determined through time and is presented in Figure 8a. The flow of GDP from construction can be seen to increase with time, while the flow of CDW (waste generated) decreases in the periods 2008–2009, 2009–2010, and 2013–2014, which is due to a decrease in areas being built on during this time. Figure 8b shows the correlation between the flow of CDW and the flow of GDP from construction, giving a value of 0.7104, indicating that the model estimates fit quite well.

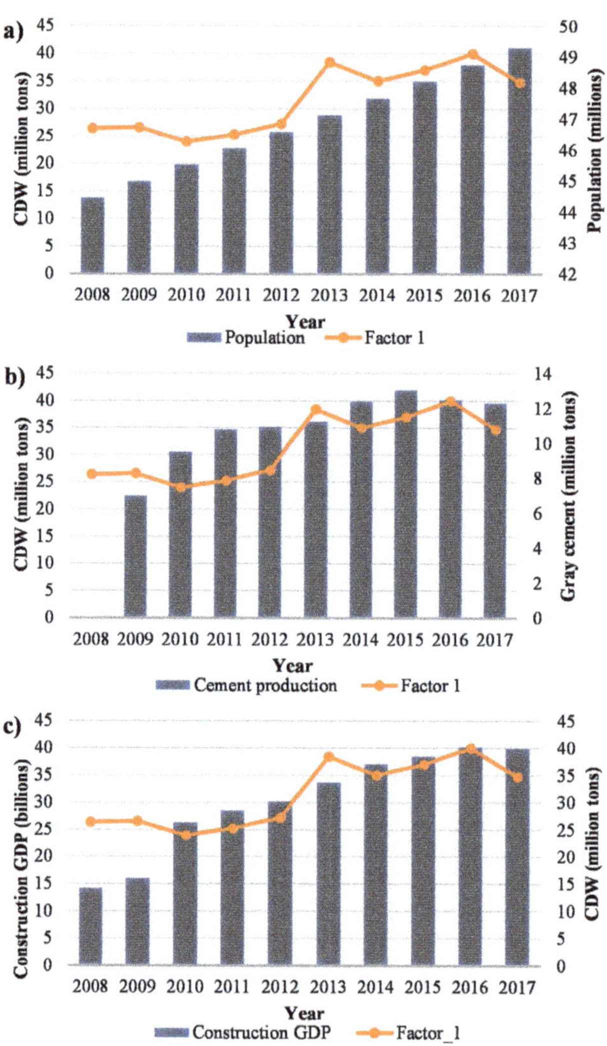

Figure 7. Comparison of the amount of CDW generated with (**a**) population, (**b**) gray cement production, and (**c**) GDP from construction in Colombia.

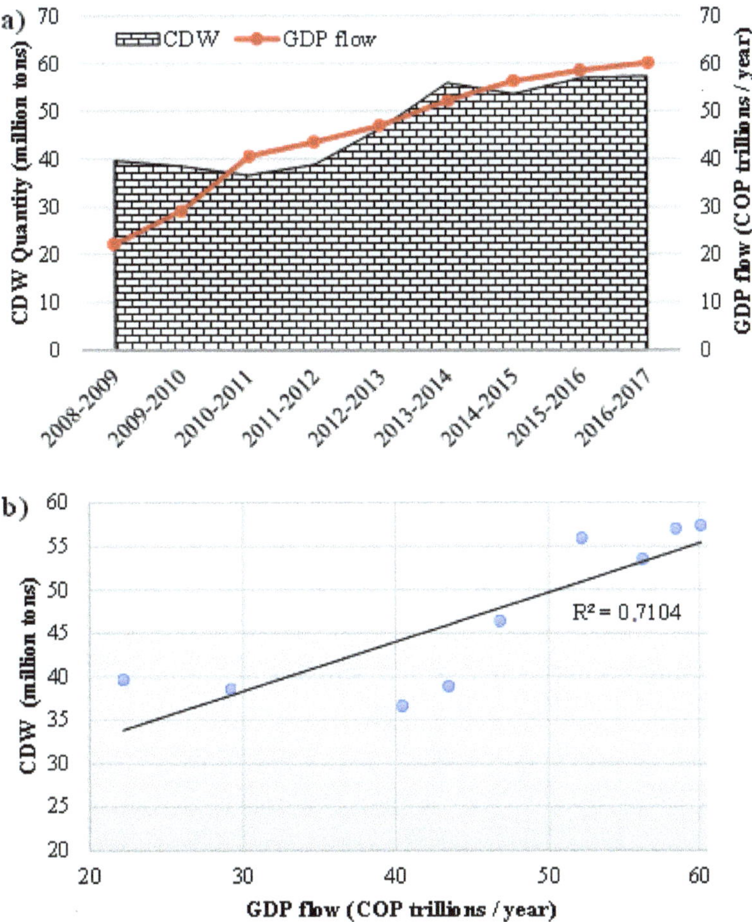

Figure 8. Flows of CDW and GDP from construction in Colombia: (**a**) evolution over time and (**b**) correlation between flows.

3.4.2. Filters for Cyclic Components

The Hodrick–Prescott filter was originally designed to decompose the series of GDP values into long-term growth and a cyclical component. In the context of CDWs, the soft part can be interpreted as the long-term seasonal component or trend and the volatile part as the stochastic component [41]. The Baxter–King filter consists of a linear filter that eliminates very slow or low frequency movements (trend) and high frequency components (irregular), while retaining the intermediate components (cycle) [42].

The linear component of the Hodrick–Prescott filter for the CDWs generated in Colombia and the GDP from construction are shown in Figure 9a,b, respectively, while the cyclical component obtained by the Hodrick–Prescott and Baxter–King filter is shown in Figure 10a,b, respectively, for both the CDW generated and the GDP from construction.

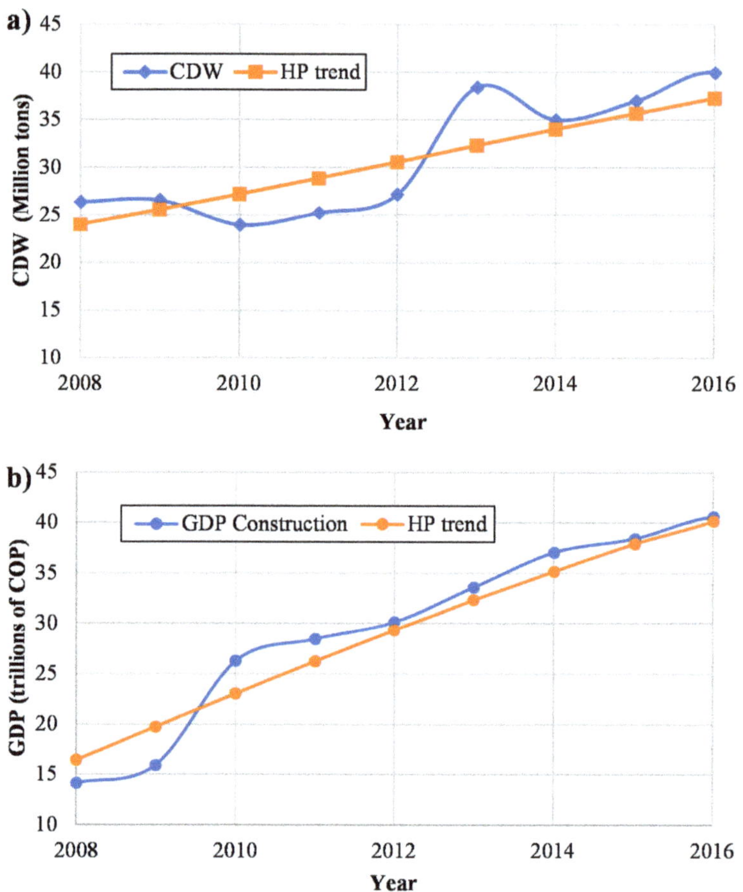

Figure 9. Hodrick–Prescott filter: (**a**) amount of CDW and (**b**) GDP from construction.

In both Figure 10a,b, it can be observed that the CDW cycle has three changes over time. The first corresponds to the period from 2008 to 2012. The reason this period is below the CWD trend is due to a decrease in planning permission granted for construction, in addition to the implementation of laws 1295 and 1333 that refer to environmental penalty fees (comparendos) and environmental sanctioning, respectively. The second change corresponds to the 2012–2016 period. This is above the trend due to the increase in the number of areas granted planning permission for housing construction. The decrease in 2013 was mainly due to the number of areas granted permission for buildings and housing of social interest. The third change corresponds to the 2016–2017 period. The CDW is below the trend line due to the implementation of Decree 1077, which regulates the housing, city, and land sectors with respect to special waste, and due to resolution 472, that requires the comprehensive management of waste generated by construction and demolition activities.

The cycle of GDP from construction shows three changes over time. The first corresponds to the period from 2008 to 2009, which is below the trend. Here, the decrease is due to the financial crisis that caused a collapse in the United States economy. The resulting increase in inflation and rise in interest rates reduced the purchasing power and decelerated loans in the construction sector. Planning approval for construction therefore decreased. It can also be seen that in 2009, there is growth, which is due to the increase in public works and a reduced inflation rate. The second change corresponds to the period from 2009 to

2015. Although the CDW is above the trendline, a decrease is observed due to the delay in the approval times of environmental licenses, which were required before any building work could begin. In 2014, another decrease is seen due to the fall in housing construction. The third change comprises the period from 2016 to 2017, which is below the trendline.

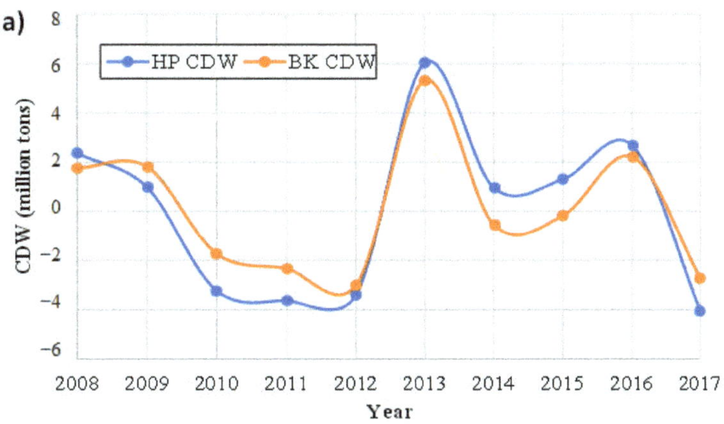

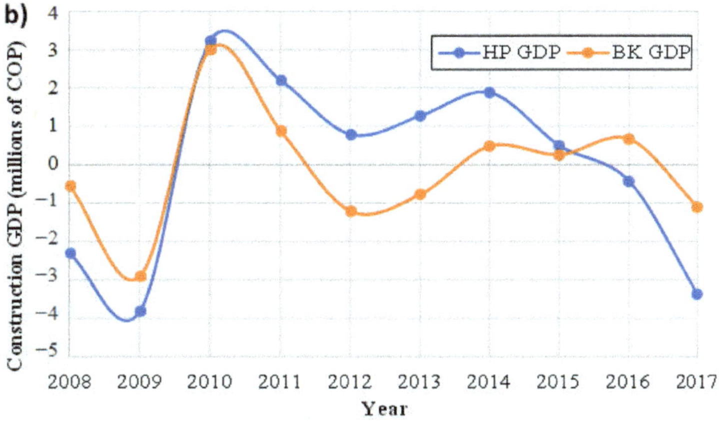

Figure 10. Cyclic components obtained by Hodrick–Prescott and Baxter–King filters from (**a**) CDW and (**b**) GDP from construction.

3.5. Information Available in other Countries

3.5.1. CDW in the Latin American framework

The information available on the management and generation of CDW in some Latin American countries is summarized below:

- Argentina

In general, there are problems with waste management. The Secretariat of Environment and Sustainable Development (SA and DS), dependent on the Ministry of Health and Environment, has designed the National Strategy for Urban Solid Waste (ENGIRSU). The mission of this strategy is to comprehensively manage waste, reduce the disposal of waste in open dumps, and increase its disposal in landfills designed, built, and operated in an appropriate way in a program that transfers, recovers, recycles, and finally disposes of urban solid waste and CDW [43]. To promote better management of CDW in Argentina, incentives are granted to companies who recover or reuse materials in order to mitigate the impact on the environment [44].

- Bolivia

The 2011 Solid Waste Management Diagnosis established that at a national level, there is no adequate management of this type of waste. CDW is deposited in public areas, rivers, streams, vacant lots or along the entrance roads to dumps. Despite the existence of local ordinances and regulations, the corresponding actions are not carried out correctly. Recently, in the main cities of the country, service companies, through municipal ordinances, have been regulating the collection and final disposal of construction waste [45].

- Brazil

The Brazilian model establishes in its Resolution CONAMA No. 307/2002 [46] all actions that must be carried out to minimize the generation of waste, and the technical aspects of its treatment, transport, and final disposal. Waste material is classified to obtain the maximum possible separation, in order to minimize transportation costs and to contribute to recycling and reuse of materials, thus reducing the burden on the environment.

Due to this resolution, Brazil became the first country in Latin America to have a recycling plant for CDW, through Resolution 307 created by CONAMA in 2002, which specifically establishes the guidelines for the management of construction waste; from there, municipalities such as Sao Pablo, Salvador, and others have taken better actions to maximize the recycling of CDW.

- Chile

The Environment Committee of the Chilean Chamber of Construction, the company signatories of the APL (Clean Production Agreement), together with the support of a Development Project (PROFO) of the Production Development Corporation (CORFO) [47], created the company Regeneradora de Materiales de la Construcción S.A (REGEMAC) [48] in the year 2000, which developed a monitoring system for 100% of waste collected. This describes the life cycle of the waste, from its origin to its final disposal. In the metropolitan region, there are places designated by SEREMI for the disposal of rubble and inert materials to ensure this waste is properly disposed of.

- Costa Rica

Costa Rica is developing a comprehensive solution for the management of CDWs, guiding the government and private sectors to promote the reduction of waste, the recovery of materials, the use of energy, and the treatment of waste, as well as promoting competitiveness and environmentally friendly behavior in the private sector [49].

- Ecuador

In Ecuador, it has been established that the Municipal Decentralized Autonomous Governments are directly responsible for the management of their solid waste. However, very little management actually takes place since most municipalities have created units to provide the service under the hierarchical dependency of the hygiene directorates, while other municipalities manage waste through the municipal police stations, which have a weak institutional image and do not have administrative or financial autonomy [50].

- Mexico

CDW management presents serious deficiencies because few entities have adequate infrastructure for waste management. Only in the Federal District were recycling plants identified as being in operation. Although most of the CDW generated in small construction projects is removed by private cargo vehicles, just over 5% is transported by municipal solid waste (MSW) collection vehicles, and it is estimated that about 10% of waste is disposed of on conservation land and public roads. In the case of public and private works, it is estimated that 67% of their waste is transported by private cargo vehicles. Only 20% is disposed of in authorized sites, and only 3% is recycled. The rest is used for land releveling and taken to landfills [51]. Most of the CDW is deposited in abandoned properties and public roads; another part is deposited in landfills, where, due to its characteristics and volume, its useful life is short. Only a small part of the waste is deposited in sites specifically designed for this purpose. Just four states in the republic have authorized final disposal sites, including Mexico City, the State of Mexico, Guanajuato, and Baja California [52].

- Panama

There is no specific regulation for the management of CDW; therefore, its management is at the discretion of the project owner or contractor. The final disposal of CDW is carried out in public roads, public drainage systems (sewerage), vacant lots, and riverbanks and streams, and close to the limits of protected areas, natural parks, and mangroves. It is also disposed of in clandestine dumps and landfills for neighborhoods and road infrastructure projects. The National Council of Private Companies and the National Cleaner Production Center in Panama proposes regulations for the proper management and use of CDWs, promoting a culture wherein waste is separated from the source and classified as well as other important initiatives for the proper management of CDW [53].

- Peru

It is the obligation and responsibility of the related institutions to coordinate ways to reduce CDW, and its reuse, storage, collection, commercialization, transport, treatment, transfer, and final disposal. The projects included in the National System for Environmental Impact Assessment (SEIA) must formulate a waste management plan including technical and administrative procedures [54]. Because in Peru there are few waste dumps for the final disposal of CDW, a large amount of construction waste is dumped into the sea and onto riverbanks without prior treatment [55].

- Dominican Republic

The Dominican Republic does not have specific regulations for the treatment of CDW. This waste is generally managed irregularly and deposited in makeshift places, including sidewalks, wetlands, parks, and riverbanks, without permits. Since there is no coherent national policy for the management of CDW, companies who generate it are responsible for its disposal. The legislation issued by the Ministry of Environment and Natural Resources regulates the operation of landfills [56].

3.5.2. CDW in the World

Construction and Demolition Waste Management

Around the world, CDW policies and legislation adopt the 3R or 4R waste minimization system, i.e., reduce, reuse, recycle, and recover [57,58]. This system focuses on four categories: strategies for the management of construction and demolition waste, prevention techniques, and collection, reuse, and recovery practices [59]. The first category includes planning for the management of CDW, taking into account all the areas involved in the construction process. The CDW is identified and quantified in order to decide what treatments are necessary for its separation and recycling. Economic tools that encourage waste management systems and maximize the environmental performance of waste are also taken into account. For prevention and collection, a construction life-cycle study is carried out in the design phase, which demonstrates opportunities for the use of prefabricated and recycled materials and modern construction methods. In addition, success in managing CDW is achieved via adequate logistics and innovation in the practices of handling and storage of the materials. Recovery practices are used in both stationary and mobile plants to maximize the production of high-quality recycled material.

Legislative Framework for the Management of Construction and Demolition Waste

The European Union's waste framework directive (2008/98/EC) proposed that by 2020 the recycling of non-hazardous CDW should be at least 70% of its weight [23,60]. In Taiwan, CDW must be transported to waste processing facilities [61], while in China and Hong Kong, regulations have been created to encourage the use of recycled construction materials in foundation works and roads, among other areas. In Australia, some policies promote the reuse of CDW at the construction site and the implementation of sustainable construction practices. In the United States, there are laws that specify the use of recycled materials in several infrastructure applications [62].

3.5.3. Comparison of CDW in Colombia and Other Countries

Figure 11 shows the amount of waste generated in 2012 in millions of tons for several countries, among which is Colombia. In this graph, the main generator of construction and demolition waste is China, with 1.02 billion tons, followed by India with 530 million, the United States with 519 million, France with 246.7 million, and Germany with 201.3 million [63]. Colombia ranks 12th with 24.45 million tons of CDW.

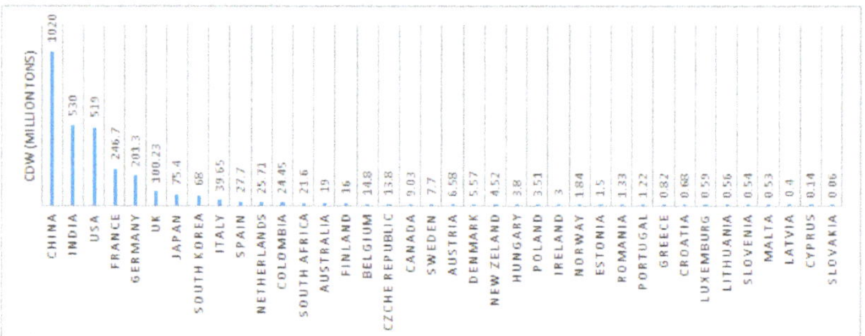

Figure 11. Amount (millions of tons) of CDW generated around the world in the year 2012.

Figure 12 summarizes the CDW generated and the gross domestic product (GDP) from construction for several countries. On average, 400 tons of CDW are generated per 1 million USD of GDP from construction [22]. Taking this into account, countries such as Colombia, Hong Kong, the United States, and China generate more waste than the average, although Colombia is above these countries with 1,697,141 per 1 million USD of GDP from construction. This is because in Colombia, CDW management is inefficient. In contrast, countries like Slovenia, Slovakia, Spain, Poland, and Portugal are well below the average, generating less than 100 tons per million USD GDP.

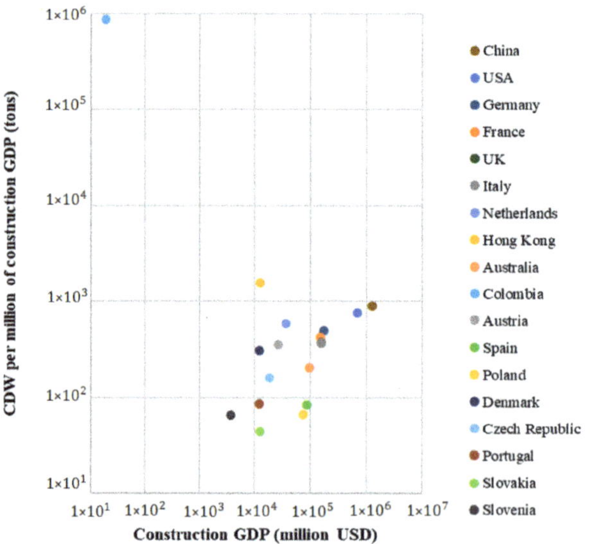

Figure 12. Comparison of CDW and GDP of various countries in 2014.

4. Conclusions

- The information available on the management of construction and demolition waste in Colombia is very scarce. There is no reliable data on the amount of CDW generated annually. However, taking into account the data that are available, it was concluded that the method used for the quantification of CDW was correct. In addition, when comparing the results of the different factors used for quantification, it was found that Factor 1 best fits with the government data.

- From the case study, the X-ray diffraction results agree with the information found on the composition of CDW in Colombia, where 62% of the residue is made up of brick, concrete, and ceramic. Subsequently, comparing the composition of CDW in Colombia with that of other countries, little similarity was found in composition. This is because the structures and construction materials used in different countries vary from country to country due to the different construction materials, motivated by variations in climates, cultures, and types of construction.

- The cycles obtained by the Hodrick–Prescott and Baxter–King filters show similar patterns when describing the movement of the GDP from construction and the CDW generated. However, the Hodrick–Prescott filter gives a smoother graph than the Baxter–King filter. In addition, this research showed several factors that affect both the GDP from construction and the CDW generated.

- When searching for information on CDW in Latin America, it should be noted that most countries do not have solid data on its generation and quantification. However, with the little information found, it can be concluded that the situation in the region is similar to Colombia in terms of CDW management, which is very limited. It should also be noted that Brazil is the Latin American country that has the best management of this type of waste.

- Education is one of the most important aspects for the implementation of a sustainable economy, but education is poorly funded in developing countries. Perhaps active learning methods [64] and technologies such as additive manufacturing [65] could contribute to the rapid implementation of circular economy models.

- It was found that both the management and the legislation of CDW is much more advanced in countries such as the United States, Germany, Portugal, and Spain because they use the 4R system, which minimizes the generation of CDW and reuses the greatest amount of waste possible. In Latin America, CDW management is just starting; thus, there is a lack of any real plans in this area.

- The differences among the different countries shown in this research revealed that many aspects have to be considered to effectively manage CDW. It is not only the economic aspects of different regions that have a profound impact on waste, but also climate and culture. Thus, the problem should not be left solely to experts and politicians; the real solution will require a multidisciplinary team.

- The uniformity of materials found in CDW is certainly one of the biggest concerns for its use on a large scale, as the variability from building to building can be high. Using CDW directly from the construction site can optimize transportation but limits its application due to the high variability. Therefore, it is more strategic to process CDW in a large facility in which thousands of tons of materials are mixed and the uniformity can be improved. The experimental case study shown in this research confirms the particularly of the phases found, which may vary significantly not only from region to region but also from country to country.

Author Contributions: Conceptualization, H.A.C.; methodology, H.A.C.; software, A.M.; validation, H.A.C. and A.M.; formal analysis, H.A.C. and S.N.M.; investigation, H.A.C.; resources, H.A.C.; writing—original draft preparation, A.M.; writing—review and editing, H.A.C. and S.N.M.; project administration, H.A.C.; funding acquisition, H.A.C. and S.N.M. All authors have read and agreed to the published version of the manuscript.

Funding: This research received no external funding.

Institutional Review Board Statement: Not applicable.

Informed Consent Statement: Not applicable.

Data Availability Statement: Not applicable.

Conflicts of Interest: The authors declare no conflict of interest.

References

1. Barbarossa, V.; Bosmans, J.; Wanders, N.; King, H.; Bierkens, M.F.P.; Huijbregts, M.A.J.; Schipper, A.M. Threats of Global Warming to the World's Freshwater Fishes. *Nat. Commun.* **2021**, *12*, 1701. [CrossRef] [PubMed]
2. Magomedov, I.A.; Dzhabrailov, Z.A.; Bagov, A.M. Subsistence Agriculture l and Global Warming. *IOP Conf. Ser. Earth Environ. Sci.* **2021**, *677*, 032109. [CrossRef]
3. Nukusheva, A.; Ilyassova, G.; Rustembekova, D.; Zhamiyeva, R.; Arenova, L. Global Warming Problem Faced by the International Community: International Legal Aspect. *Int. Environ. Agreem. Politics Law Econ.* **2020**, *21*, 219–233. [CrossRef]
4. Shang, W.; Peng, Z.; Huang, Y.; Gu, F.; Zhang, J.; Tang, H.; Yang, L.; Tian, W.; Rao, M.; Li, G.; et al. Production of Glass-Ceramics from Metallurgical Slags. *J. Clean. Prod.* **2021**, *317*, 128220. [CrossRef]
5. Colorado, H.A.; Garcia, E.; Buchely, M.F. White Ordinary Portland Cement Blended with Superfine Steel Dust with High Zinc Oxide Contents. *Constr. Build. Mater.* **2016**, *112*, 816–824. [CrossRef]
6. Hittini, W.; Mourad, A.H.I.; Abu-Jdayil, B. Utilization of Devulcanized Waste Rubber Tire in Development of Heat Insulation Composite. *J. Clean. Prod.* **2021**, *280*, 124492. [CrossRef]
7. Revelo, C.F.; Colorado, H.A. A Green Composite Material of Calcium Phosphate Cement Matrix with Additions of Car Tire Waste Particles. *Int. J. Appl. Ceram. Technol.* **2021**, *18*, 182–191. [CrossRef]
8. Agudelo, G.; Cifuentes, S.; Colorado, H.A. Ground Tire Rubber and Bitumen with Wax and Its Application in a Real Highway. *J. Clean. Prod.* **2019**, *228*, 1048–1061. [CrossRef]
9. Dhanya, B.S.; Mishra, A.; Chandel, A.K.; Verma, M.L. Development of Sustainable Approaches for Converting the Organic Waste to Bioenergy. *Sci. Total Environ.* **2020**, *723*, 138109. [CrossRef]
10. Aslam, M.S.; Huang, B.; Cui, L. Review of Construction and Demolition Waste Management in China and USA. *J. Environ. Manag.* **2020**, *264*, 110445. [CrossRef]
11. Munir, M.T.; Mohaddespour, A.; Nasr, A.T.; Carter, S. Municipal Solid Waste-to-Energy Processing for a Circular Economy in New Zealand. *Renew. Sustain. Energy Rev.* **2021**, *145*, 111080. [CrossRef]
12. Colorado-Lopera, H.A.; Echeverri-Lopera, G.I. The Solid Waste in Colombia Analyzed via Gross Domestic Product: Towards a Sustainable Economy. *Rev. Fac. De Ing. Univ. De Antioq.* **2020**, *96*, 51–63. [CrossRef]
13. Sumter, D.; de Koning, J.; Bakker, C.; Balkenende, R. Circular Economy Competencies for Design. *Sustainability* **2020**, *12*, 1561. [CrossRef]
14. Colorado, H.A.; Gutiérrez-Velásquez, E.I.; Monteiro, S.N. Sustainability of Additive Manufacturing: The Circular Economy of Materials and Environmental Perspectives. *J. Mater. Res. Technol.* **2020**, *9*, 8221–8234. [CrossRef]
15. Zapata, J.F.; Azevedo, A.; Fontes, C.; Monteiro, S.N.; Colorado, H.A. Environmental Impact and Sustainability of Calcium Aluminate Cements. *Sustainability* **2022**, *14*, 2751. [CrossRef]
16. Villegas, J.E.; Camilo, J.; Gutierrez, R.; Colorado, H.A. Active Materials for Adaptive Building Envelopes: A Review. *J. Mater. Environ. Sci.* **2020**, *2020*, 988–1009.
17. Castañeda, M.; Gutiérrez-Velásquez, E.I.; Aguilar, C.E.; Monteiro, S.N.; Amell, A.A.; Colorado, H.A. Sustainability and Circular Economy Perspectives of Materials for Thermoelectric Modules. *Sustainability* **2022**, *14*, 5987. [CrossRef]
18. Miller, P.; de Barros, A.G.; Kattan, L.; Wirasinghe, S.C. Public Transportation and Sustainability: A Review. *KSCE J. Civ. Eng.* **2016**, *20*, 1076–1083. [CrossRef]
19. Heidari, R.; Yazdanparast, R.; Jabbarzadeh, A. Sustainable Design of a Municipal Solid Waste Management System Considering Waste Separators: A Real-World Application. *Sustain. Cities Soc.* **2019**, *47*, 101457. [CrossRef]
20. Shekdar, A.V. Sustainable Solid Waste Management: An Integrated Approach for Asian Countries. *Waste Manag.* **2009**, *29*, 1438–1448. [CrossRef]
21. Alsheyab, M.A.T. Recycling of Construction and Demolition Waste and Its Impact on Climate Change and Sustainable Development. *Int. J. Environ. Sci. Technol.* **2021**, *19*, 2129–2138. [CrossRef]
22. Menegaki, M.; Damigos, D. A Review on Current Situation and Challenges of Construction and Demolition Waste Management. *Curr. Opin. Green Sustain. Chem.* **2018**, *13*, 8–15. [CrossRef]
23. Dahlbo, H.; Bachér, J.; Lähtinen, K.; Jouttijärvi, T.; Suoheimo, P.; Mattila, T.; Sironen, S.; Myllymaa, T.; Saramäki, K. Construction and Demolition Waste Management—A Holistic Evaluation of Environmental Performance. *J. Clean. Prod.* **2015**, *107*, 333–341. [CrossRef]
24. Galvez-Martos, J.L.; Styles, D.; Shoenberger, H. Best Environmental Management Practice in the Building and Construction Sector. Available online: https://susproc.jrc.ec.europa.eu/product-bureau/sites/default/files/inline-files/ConstructionSector.pdf (accessed on 22 April 2022).

25. Bergsdal, H.; Bohne, R.A.; Brattebø, H. Projection of Construction and Demolition Waste in Norway. *J. Ind. Ecol.* **2007**, *11*, 27–39. [CrossRef]
26. Wu, Z.; Yu, A.T.W.; Shen, L.; Liu, G. Quantifying Construction and Demolition Waste: An Analytical Review. *Waste Manag.* **2014**, *34*, 1683–1692. [CrossRef]
27. DANE Departamento Administrativo Nacional de Estadística (National Administrative Department of Statistics). Available online: https://www.dane.gov.co/ (accessed on 25 March 2020).
28. Cámara Colombiana de La Construcción-Camacol. Available online: https://camacol.co/ (accessed on 6 October 2021).
29. Vásquez Hernández, A.; Botero Botero, L.F.; Carvajal Arango, D. Fabricación de Bloques de Tierra Comprimida Con Adición de Residuos de Construcción y Demolición Como Reemplazo Del Agregado Pétreo Convencional. *Ing. Y. Cienc-Ing. Cienc.* **2014**, *11*, 197–220. [CrossRef]
30. Fatta, D.; Papadopoulos, A.; Avramikos, E.; Sgourou, E.; Moustakas, K.; Kourmoussis, F.; Mentzis, A.; Loizidou, M. Generation and Management of Construction and Demolition Waste in Greece—An Existing Challenge. *Resour. Conserv. Recycl.* **2003**, *40*, 81–91. [CrossRef]
31. Minambiente Colombia Ministerio de Ambiente y Desarrollo Sostenible. Available online: https://www.minambiente.gov.co/ (accessed on 6 February 2020).
32. Orozco Gutiérrez, C.J.; Gómez Rodríguez, F.S.; Severiche Ramírez, J.A.; Rico Gómez, K.J.; Pinto Fajardo, N.A.; Zambrano Echeverría, V.A.; Alarcón, W.A.; Elorza López, Y.A.; Figueroa García, Y.X. *Guía Para la Elaboración del Plan de Gestión Integral de Residuos de Construcción y Demolición (RCD) en Obra*; Secretaría Distrital de Ambiente: Bogota, Colombia, 2014.
33. Ministerio del Medio Ambiente. *Resolución 541*; Ministerio del Medio Ambiente: Bogota, Colombia, 1994.
34. Minambiente Colombia. *Decreto 948 de 1995*; Minambiente: Bogota, Colombia, 1995.
35. Pastrana Arango, A.; Pizano de Narváez, E.; Mayr Maldonado, J. *Decreto 1713 de 2002*; El Ministro de Desarrollo Económico: Bogota, Colombia, 2002.
36. Congreso de Colombia. *Ley 1259 de 2008*; Congreso de Colombia: Bogota, Colombia, 2008.
37. Presidencia de la Republica de Colombia. *Decreto 2981 de 2013*; Presidencia de la Republica de Colombia: Bogota, Colombia, 2013.
38. Minambiente Colombia. *Resolución 472 de 2017–Manejo y Disposición de Residuos de Construcción–Solames*; Minambiente: Bogota, Colombia, 2017.
39. Porras, Á.C.; Cortes, N.L.G.; Duarte, M.C.C. Determinación de Propiedades Físico-Químicas de Los Materiales Agregados En Muestra de Escombros En La Ciudad de Bogotá D. C. *Rev. Ing.* **2013**, *12*, 45–58. [CrossRef]
40. Zheng, L.; Wu, H.; Zhang, H.; Duan, H.; Wang, J.; Jiang, W.; Dong, B.; Liu, G.; Zuo, J.; Song, Q. Characterizing the Generation and Flows of Construction and Demolition Waste in China. *Constr. Build. Mater.* **2017**, *136*, 405–413. [CrossRef]
41. Hodrick, R.J.; Prescott, E.C. Postwar U.S. Business Cycles: An Empirical Investigation. *J. Money Credit Bank.* **1997**, *29*, 1–16. [CrossRef]
42. Baxter, M.; King, R.G. *Measuring Business Cycles Approximate Band-Pass Filters for Economic Time Series*; National Bureau of Economic Reserch: Cambridge, MA, USA, 1995. [CrossRef]
43. Gisela, I.; González, L. *Residuos Solidos Urbanos Argentina Tratamiento y Disposicion Final Situacion Actual y Alternativas Futuras*; Camara Argentina de la Construcción: Buenos Aires, Argentina, 2010.
44. Ferré Olive, E.H. *Foro Virtual de Contabilidad Ambiental y Social*; Facultad de Ciencias Económicas de la Universidad de Buenos Aires: Buenos Aires, Argentina, 2010.
45. Vaca Zelaya, J.L.; Sánchez, M.J.C. *Manual Educativo Manejo de Residuos Sólidos*; Abbase Ltda: La Paz, Bolivia, 2007.
46. Conselho Nacional do Meio Ambiente. *Resolução Conama No 307 de 05/07/2002*; Conselho Nacional do Meio Ambiente: Brasilia, Brazil, 2002.
47. Gobierno de Chile. CORFO. Available online: https://www.corfo.cl/sites/cpp/quienessomos (accessed on 19 October 2021).
48. REGEMAC Gestión Sustentable de Residuos Sólidos Inertes. Available online: https://regemac.cl/home/ (accessed on 19 October 2021).
49. Morales Alpízar, M.; Flórez-Estrada, M.V. *Guía de Manejo de Escombros y Otros Residuos de La Construcción*; Oficina Regional para Mesoamérica y la Iniciativa Caribe, Unión Internacional para la Conservación de la Naturaleza UICN: San José, Costa Rica, 2011.
50. Ministerio del Ambiente Gestión Integral de Desechos Sólidos (PNGIDS). Available online: http://inabio.biodiversidad.gob.ec/2019/01/30/26-gestion-integral-de-desechos-solidos-pngids/# (accessed on 28 July 2021).
51. Cámara Mexicana de la Industria de la Construcción. *Plan de Manejo de Residuos de La Construcción y La Demolición*; Cámara Mexicana de la Industria de la Construcción: Mexico City, Mexico, 2014.
52. Gobierno Del Estado De México. Infraestructura Para El Manejo de Residuos Peligrosos. In *Bases Conceptuales y de Diagnóstico del Programa Para la Prevención y Manejo Integral de Residuos Peligrosos en la Zona Metropolitana del Valle de México*; Gobierno del Estado de México: Toluca, Mexico, 2013.
53. Consejo Municipal de Panama. *Acuerdo N° 124*; Consejo Municipal de Panama: Panama City, Panama, 2015.
54. Ministerio de Vivienda, Construcción y Saneamiento VIVIENDA. Decreto Supremo N° 003-2013-VIVIENDA. Available online: http://sial.segat.gob.pe/normas/aprueba-reglamento-gestion-manejo-residuos-las-actividades-las (accessed on 28 July 2021).
55. Organismo de Evaluación y Fiscalización Ambiental–OEFA. *La Fiscalización Ambiental en Residuos Sólidos*; Organismo de Evaluación y Fiscalización Ambiental–OEFA: Chachapoyas, Peru, 2014.

56. Gobierno de la Republica Dominicana Ministerio de Medio Ambiente y Recursos Naturales. Available online: https://ambiente.gob.do/transparencia/ (accessed on 28 July 2021).
57. Esa, M.R.; Halog, A.; Rigamonti, L. Strategies for Minimizing Construction and Demolition Wastes in Malaysia. *Resour. Conserv. Recycl.* **2017**, *120*, 219–229. [CrossRef]
58. Yang, H.; Xia, J.; Thompson, J.R.; Flower, R.J. Urban Construction and Demolition Waste and Landfill Failure in Shenzhen, China. *Waste Manag.* **2017**, *63*, 393–396. [CrossRef] [PubMed]
59. Gálvez-Martos, J.L.; Styles, D.; Schoenberger, H.; Zeschmar-Lahl, B. Construction and Demolition Waste Best Management Practice in Europe. *Resour. Conserv. Recycl.* **2018**, *136*, 166–178. [CrossRef]
60. Alexandridou, C.; Angelopoulos, G.N.; Coutelieris, F.A. Mechanical and Durability Performance of Concrete Produced with Recycled Aggregates from Greek Construction and Demolition Waste Plants. *J. Clean. Prod.* **2018**, *176*, 745–757. [CrossRef]
61. Lai, Y.-Y.; Yeh, L.-H.; Chen, P.-F.; Sung, P.-H.; Lee, Y.-M. Management and Recycling of Construction Waste in Taiwan. *Proc. Environ. Sci.* **2016**, *35*, 723–730. [CrossRef]
62. U.S. Environmental Erotection Agency. Reduce, Reuse, Recycle. Available online: https://www.epa.gov/recycle (accessed on 19 October 2021).
63. Akhtar, A.; Sarmah, A.K. Construction and Demolition Waste Generation and Properties of Recycled Aggregate Concrete: A Global Perspective. *J. Clean. Prod.* **2018**, *186*, 262–281. [CrossRef]
64. Lopera, H.A.C.; Gutierrez-Velasquez, E.; Ballesteros, N. Bridging the Gap between Theory and Active Learning: A Case Study of Project-Based Learning in Introduction to Materials Science and Engineering. *IEEE Rev. Iberoam. de Tecnol. del Aprendiz.* **2022**, *17*, 160–169. [CrossRef]
65. Colorado, H.A.; Mendoza, D.E.; Valencia, F.L. A Combined Strategy of Additive Manufacturing to Support Multidisciplinary Education in Arts, Biology, and Engineering. *J. Sci. Educ. Technol.* **2020**, *30*, 58–73. [CrossRef]

Article

Structural Behavior of LC-GFRP Confined Waste Aggregate Concrete Square Columns with Sharp and Round Corners

Rattapoohm Parichatprecha [1], Kittipoom Rodsin [2,*], Krisada Chaiyasarn [3], Nazam Ali [4], Songsak Suthasupradit [1], Qudeer Hussain [5] and Kaffayatullah Khan [6]

[1] Department of Civil Engineering, School of Engineering, King Mongkut's Institute of Technology Ladkrabang, Bangkok 10520, Thailand

[2] Center of Excellence in Structural Dynamics and Urban Management, Department of Civil and Environmental Engineering Technology, College of Industrial Technology, King Mongkut's University of Technology North Bangkok, Bangkok 10800, Thailand

[3] Thammasat Research Unit in Infrastructure Inspection and Monitoring, Repair and Strengthening (IIMRS), Thammasat School of Engineering, Faculty of Engineering, Thammasat University Rangsit, Khlongluang, Pathumthani 12000, Thailand

[4] Department of Civil Engineering, School of Engineering, University of Management and Technology, Lahore 54770, Pakistan

[5] Center of Excellence in Earthquake Engineering and Vibration, Department of Civil Engineering, Chulalongkorn University, Bangkok 10330, Thailand

[6] Department of Civil and Environmental Engineering, College of Engineering, King Faisal University, Al-Hofuf 31982, Al-Ahsa, Saudi Arabia

* Correspondence: kittipoom.r@cit.kmutnb.ac.th

Citation: Parichatprecha, R.; Rodsin, K.; Chaiyasarn, K.; Ali, N.; Suthasupradit, S.; Hussain, Q.; Khan, K. Structural Behavior of LC-GFRP Confined Waste Aggregate Concrete Square Columns with Sharp and Round Corners. *Sustainability* **2022**, *14*, 11221. https://doi.org/10.3390/su141811221

Academic Editors: Carlos Maurício Fontes Vieira, Gustavo de Castro Xavier, Henry Alonso Colorado Lopera and Sergio Neves Monteiro

Received: 28 July 2022
Accepted: 2 September 2022
Published: 7 September 2022

Publisher's Note: MDPI stays neutral with regard to jurisdictional claims in published maps and institutional affiliations.

Abstract: Reusing construction brick waste to fabricate new concrete is an economical and sustainable solution for the ever-increasing quantity of construction waste. However, the substandard mechanical properties of the concrete made using recycled crushed brick aggregates (RBAC) have limited its use mainly to non-structural applications. Several studies have shown that the axial compressive performance of the concrete is a function of the lateral confining pressure. Therefore, this study proposes to use low-cost glass fiber-reinforced polymer (LC-GFRP) wraps to improve the substandard compressive strength and ductility of RBAC. Thirty-two rectilinear RBAC specimens were constructed in this study and tested in two groups. The specimens in Group 1 were tested without the provision of a corner radius, whereas a corner radius of 26 mm was provided in the Group 2 specimens. Specimens in both groups demonstrated improved compressive behavior. However, the premature failure of LC-GFRP wraps near the sharp corners in Group 1 specimens undermined its efficacy. On the contrary, the stress concentrations were neutralized in almost all Group 2 specimens with a 26 mm corner radius, except low-strength specimen with six layers of LC-GFRP. As a result, Group 2 specimens demonstrated a more significant improvement in peak compressive strength and ultimate strain than Group 1 specimens. An analytical investigation was carried out to assess the efficiency of existing compressive stress–strain models to predict the peak compressive stress and ultimate of LC-GFRP-confined RBAC. Existing FRP models were found unreliable in predicting the key parameters in the stress–strain curves of LC-GFRP-confined RBAC. Equations were proposed by using nonlinear regression analysis, and the predicted values of the key parameters were found in good agreement with the corresponding experimental values.

Keywords: low-cost GFRP; square; recycled brick aggregate; regression

1. Introduction

The rapid growth of the construction industry has put a great demand on the natural resources that are used for construction practices. It has been suggested that the global demand for concrete will increase to approximately 18 billion tons per year by 2050 [1] and an estimated yearly consumption approaching approximately 30 billion tones [2]. This suggests that there exists an enormous usage of natural resources, mainly coarse and fine aggregates resulting in their rapid depletion [3]. From the view of sustainability, this rapid depletion of natural resources must be tackled in an effective way.

The demolition of existing buildings produces a considerable quantity of waste that demands proper disposal. Roughly 700 and 800 million tons of construction waste are generated per year in the United States and European Union [4,5]. The quantity of construction waste that is produced in China each year has been estimated at 1.8 billion tons [6]. The proper treatment of construction waste before disposal is vital. Besides occupying extensive land, untreated construction waste may produce harmful substances that pollute groundwater and air [7–10].

So far, two problems need to be addressed mainly relating to the rapid depletion of natural resources and extensive accumulation of construction waste each year. A common solution to these problems may be realized in the recycling of construction waste to produce new concrete. The present study focuses on the recycling of bricks to be used as a partial replacement for natural coarse aggregates. This is because a considerable ratio of the construction waste generated each year comprises bricks. It has been reported that the quantity of clay brick waste generated each year is increasing in a geometric manner [11]. Recycled brick aggregates are usually prepared by crushing bricks in jaw crushers having different sizes of openings. Further, the crushed bricks are then sieved into different sizes using mechanical sieves. A rough estimate indicates that 400 million tons of brick waste are generated each year in China, accounting for up to 45% of the total construction waste [12].

Early experimental investigations on the recycling of bricks as coarse aggregates date back to the late 1990s and early 20th century [13–15]. Several studies have highlighted the substandard properties of recycled brick aggregate concrete (RBAC). Desmyter [16] concluded that recycled aggregates absorb more water than natural aggregates. Therefore, the resulting concrete offers lower mechanical properties as compared to natural aggregate concrete (NAC). Further, the mortar adhering to the surface of recycled aggregates results in an increased porosity leading to a 5–10% higher water absorption. Debieb and Kenai [17] reported up to a 30% reduction of the compressive strength when 100% of the natural aggregates were replaced by recycled aggregates. Medina et al. [18] reported a 39% reduction of the compressive strength for a 40% replacement of natural aggregates. Yang et al. [19] found an 11% and 20% reduction of the compressive strength for 20% and 50% replacement of natural aggregates. Due to the low density of adhered mortar, recycled aggregate concrete exhibits 5% to 15% lower particle density [20]. Jiang et al. [21] concluded that the reduction in the mechanical properties of RBAC is minimal if the replacement ratio of natural aggregates is below 30%. The substandard mechanical properties of RBAC have so far limited its use to non-structural applications [22,23]. A prevalent solution to improve the substandard properties of concrete is external wrapping. Fiber-reinforced polymer (FRP) sheets are used for this purpose. Several studies have highlighted the improvement in the mechanical properties of concrete using external FRP wraps [24–31]. Gao et al. [32] investigated the role of carbon FRP and glass FRP (GFRP) sheets in improving the properties of RBAC. It was found that the compressive strength decreased as the replacement ratio of natural aggregates increased, whereas no effect on axial deformation was reported. The failure modes of carbon and glass FRP-confined RBAC specimens were similar to those of RAC. Tang et al. [33] confined geopolymer recycled aggregate concrete using CFRP jackets and tested it under static and cyclic compressive loads. Both the peak compressive strength and ductility were improved by the application of CFRP jackets. Han et al. [34] tested recycled aggregate concrete confined with recycled polyethylene naphthalate/terephthalate composites. The test results in terms of compressive strength and ultimate strain indicated that the confinement stiffness had a more substantial effect as compared to the replacement ratio of natural aggregates. Zeng et al. [35] strengthened recycled glass aggregate concrete using CFRP jackets. A similar behavior to CFRP-confined NAC for CFRP-confined recycled glass aggregate concrete was observed.

From the above discussion, it is recognized that synthetic FRPs are efficient in improving the substandard properties of RAC. However, the cost of synthetic FRPs has been recognized as a major hindrance in their applicability to small-scale projects [36–38]. Yoddumrong et al. [39] introduced locally available bi-directional low-cost glass-fiber-reinforced polymers (LC-GFRP) to strengthen low-strength reinforced concrete (RC) columns. A significant improvement in the hysteretic behavior of the strengthened RC column was observed. Rodsin et al. [40] strengthened extremely low-strength concrete cylinders (i.e., 5 MPa to 15 MPa) using LC-GFRP. A substantial improvement in the peak compressive stress and ductility was observed in the strengthened specimens. Rodsin [41] utilized LC-GFRP sheets to enhance the mechanical properties of circular specimens constructed with RBAC. The results revealed up to a 437% increase in the ultimate compressive stress and up to 1058% improvement in the ultimate strain of LC-GFRP-strengthened RBAC specimens. In a recent study, Rodsin et al. [42] strengthened square RBAC specimens by using LC-GFRP. A corner radius of 13 mm was provided to prevent stress concentrations near sharp corners. A considerable improvement in the compressive stress–strain curves of strengthened specimens was reported.

From the above discussion, it is clear that RBAC offers lower mechanical properties than NAC, which can be improved by providing lateral confining pressures. The present study investigates the role of low-cost glass fiber reinforced polymer (LC-GFRP) sheets in improving the substandard properties of square RBAC specimens. It has been suggested that the stress concentrations near sharp corners in rectilinear specimens can result in premature failure of external sheets [43,44]. The shape of the recycled brick aggregates used in this study is approximately round, which may cause an additional reduction in the properties of RBAC. Therefore, this study investigates the efficiency of GFRP sheets on specimens with and without the provision of a corner radius and incorporating the round shape of recycled brick aggregates.

2. Experimental Program

2.1. Test Matrix

The experimental program involved thirty-two square specimens tested mainly in two groups, as shown in Table 1. The two groups were separated depending upon the provision of the corner radius. The specimens in the first group were tested without any corner radius and identified with R0 in their notation, whereas a 26 mm corner radius was provided in the second group. Each group comprised eight different specimen types with two representative specimens for each type. The specimens in each group were differentiated by the concrete strength, i.e., low- or high-strength concrete specimens. Furthermore, in each group two specimens were considered as the control and tested without LC-GFRP wraps. Meanwhile, the remaining specimens were wrapped with two, four, and six layers of GFRP. The notation used to identify specimens involved four parts. The first part "SQ" was constant for all of the specimens identifying the cross-sectional shape as square, the second part was either "LS" or "HS" for low- or high-strength concrete strength, the third part identified the presence of the corner radius with R0 and R26 for no corner radius and a corner radius of 26 mm, respectively, and the last part identified the number of GFRP layers. For instance, the notation SQ-LS-R26-6GFRP referred to a square specimen constructed with low-strength concrete with its corner rounded to a 26 mm radius and strengthened with six layers of GFRP. Further details are provided in Table 1.

2.2. Material Properties

The solid clay bricks were crushed using a brick-crushing machine. The crushed brick aggregates were sieved to obtain coarse brick aggregates with sizes ranging from 5 mm to 20 mm (Figure 1a). The mechanical properties of the bricks were estimated by following the recommendations of ASTM C1314-21 and ASTM C140/C140M-22a [45,46]. The estimated mechanical properties of the bricks are reported in Table 2. However, the standard density of natural aggregates was approximately 2000–2900 kg/m^3. Actual density tests were not considered in this study for natural aggregates. The replacement ratio of natural coarse aggregates with brick aggregates was 50% (Figure 1b). The target concrete strengths were 15 MPa and 25 MPa for the low-strength and high-strength concrete, respectively. The mix proportions adopted for the two concrete strengths are shown in Table 3. The ultimate tensile strain and strength of LC-GFRP wraps were 2.04% and 377.64 MPa, respectively, which were determined by following ASTM D3039/D3039M-17 [45]. The thickness of the LC-GFRP sheet was 0.50 mm.

(a)

(b)

Figure 1. (**a**) Brick aggregates, and (**b**) natural aggregates.

Table 1. Summary of tested specimens.

Group	Name	Strength	Corner Radius	GFRP	Number
	SQ-LS-R0-CON	Low strength	R0	-	2
	SQ-LS-R0-2GFRP	Low strength	R0	2	2
	SQ-LS-R0-4GFRP	Low strength	R0	4	2
	SQ-LS-R0-6GFRP	Low strength	R0	6	2
1	SQ-HS-R0-CON	High strength	R0	-	2
	SQ-HS-R0-2GFRP	High strength	R0	2	2
	SQ-HS-R0-4GFRP	High strength	R0	4	2
	SQ-HS-R0-6GFRP	High strength	R0	6	2
	SQ-LS-R26-CON	Low strength	R26	-	2
	SQ-LS-R26-2GFRP	Low strength	R26	2	2
	SQ-LS-R26-4GFRP	Low strength	R26	4	2
2	SQ-LS-R26-6GFRP	Low strength	R26	6	2
	SQ-HS-R26-CON	High strength	R26	-	2
	SQ-HS-R26-2GFRP	High strength	R26	2	2
	SQ-HS-R26-4GFRP	High strength	R26	4	2
	SQ-HS-R26-6GFRP	High strength	R26	6	2

Table 2. Mechanical properties of solid clay bricks.

Property	Density (kg/m^3)	Compressive Strength (MPa)	Water Absorption (%)
Value	120	3.14	23.27

Table 3. Mix proportions for concrete.

Constituent (kg/m^3) / Strength	Cement	Sand	Natural Aggregates	Brick Aggregates	Water
Low	261	783	522	522	313
High	627	806	358	358	251

2.3. Details and Construction of Test Specimens

Each specimen measured 150 mm × 150 mm in cross-section and 300 mm in height to achieve the height-to-width ratio of 2.0. All of the specimens were cast in steel molds of the same dimensions. However, the steel molds were round to a corner radius of 26 mm for the Group 2 specimens, as shown in Figure 2a, whereas the rectilinear-shaped molds were used for the Group 1 specimens. The concrete was poured in three equal layers in each mold, whereas proper compaction (using vibrating poker) was applied to each layer. The specimens were taken out of the molds after one day of casting and cured for 28 days in laboratory environments (i.e., temperature was approximately 30–33 degree centigrade and humidity was 65–75%). The strengthening of specimens was performed after their curing by using a hand layout.

(a)

(b)

Figure 2. Steel molds for (**a**) zero corner radius and (**b**) 26 mm corner radius.

The surface of each specimen was properly cleaned and smoothed before applying GFRP. Epoxy resin was applied to the concrete surface by using a hand brush to impregnate the surface. This was followed by wrapping GFRP around the specimens. The first layer of GFRP was epoxy impregnated using a brush (see Figure 3a), before the application of the second layer. This process was repeated for the subsequent GFRP layers. Finally, the extended GFRP sheets above and below the top and bottom surfaces were ground, as shown in Figure 3b.

(a)

(b)

Figure 3. (**a**) Application of epoxy resin and (**b**) grinding of excessive GFRP portions and smoothening of the top and bottom surfaces.

2.4. Test Setup and Instrumentation

A monotonic compressive load was applied to each specimen by using a hydraulic Universal Testing Machine. The ultimate capacity of universal testing machine was 200 tons. A calibrated load cell was deployed to measure the intensity of the applied compressive load, whereas a logger recorded the measured load. A uniform application of the load was achieved by placing steel plates above and below the specimen, as shown in Figure 4. The axial shortening of the specimens under the applied compressive load was simultaneously measured by using two linear variable differential transducers (LVDTs). The recorded deflection was subsequently converted to the strain, whereas the recorded load was converted to compressive stress by utilizing the geometrical dimensions of specimens.

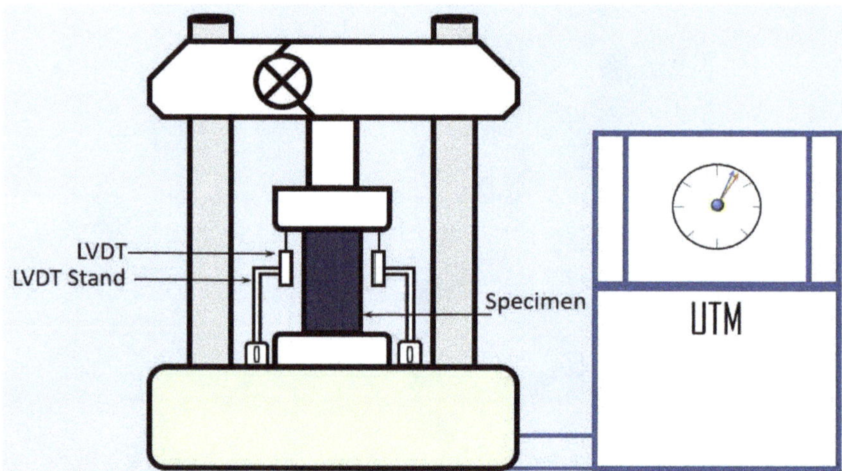

Figure 4. Typical test setup and instrumentation.

3. Experimental Results

3.1. Ultimate Failure Modes

The ultimate failure modes of Group 1 specimens are shown in Figure 5. The failure of the two control specimens was due to the sudden concrete crushing and splitting within the top half of their heights. The stress concentrations near the sharp corners of rectilinear concrete specimens are known to exist [43,44]. To avoid premature failure due to these stress concentrations, ACI 440.2R-02 [47] recommends a minimum corner radius of 13 mm. Since no corner radius was provided in the Group 1 specimens, the failure accompanied the rupture of GFRP sheets at the corners, as shown in Figure 5. The resulting failure was brittle, irrespective of the concrete strength and the number of GFRP layers. This suggests that premature failure due to stress concentrations could not be prevented even with the application of six GFRP layers. This could be associated with the relatively lower ultimate strain of GFRP, i.e., 2.04%. Future studies are required to further explore this phenomenon by using FRP composites with higher rupture strains, such as polyester fiber ropes [42].

The failure modes of the Group 2 specimens are shown in Figure 6. In the same way as the control specimens in Group 1, the failure of the Group 2 control specimens was brittle. However, the crushing and splitting propagated all along the full height. The low-strength-strengthened specimens in Group 2 failed due to the rupture of the GFRP sheets in the hoop direction. For the two and four GFRP layers, the tensile rupture of GFRP sheets was found to occur between the corners suggesting that the stress concentrations were mitigated successfully. Specimen SQ-LS-R26-6GFRP failed by the rupture of GFRP sheets at the corners. This can be attributed to the resulting high compressive strength of Specimen SQ-LS-R26-6GFRP due to six GFRP layers that may have eventually resulted in higher stress concentrations near the corners as compared to those in Specimens SQ-LS-R26-2GFRP and SQ-LS-R26-4GFRP. The failure of high-strength concrete specimens in Group 2 also exhibited rupture of GFRP sheets mainly near the corners, and a similar analogy of high-stress concentrations near the corners can be made due to the high concrete strength.

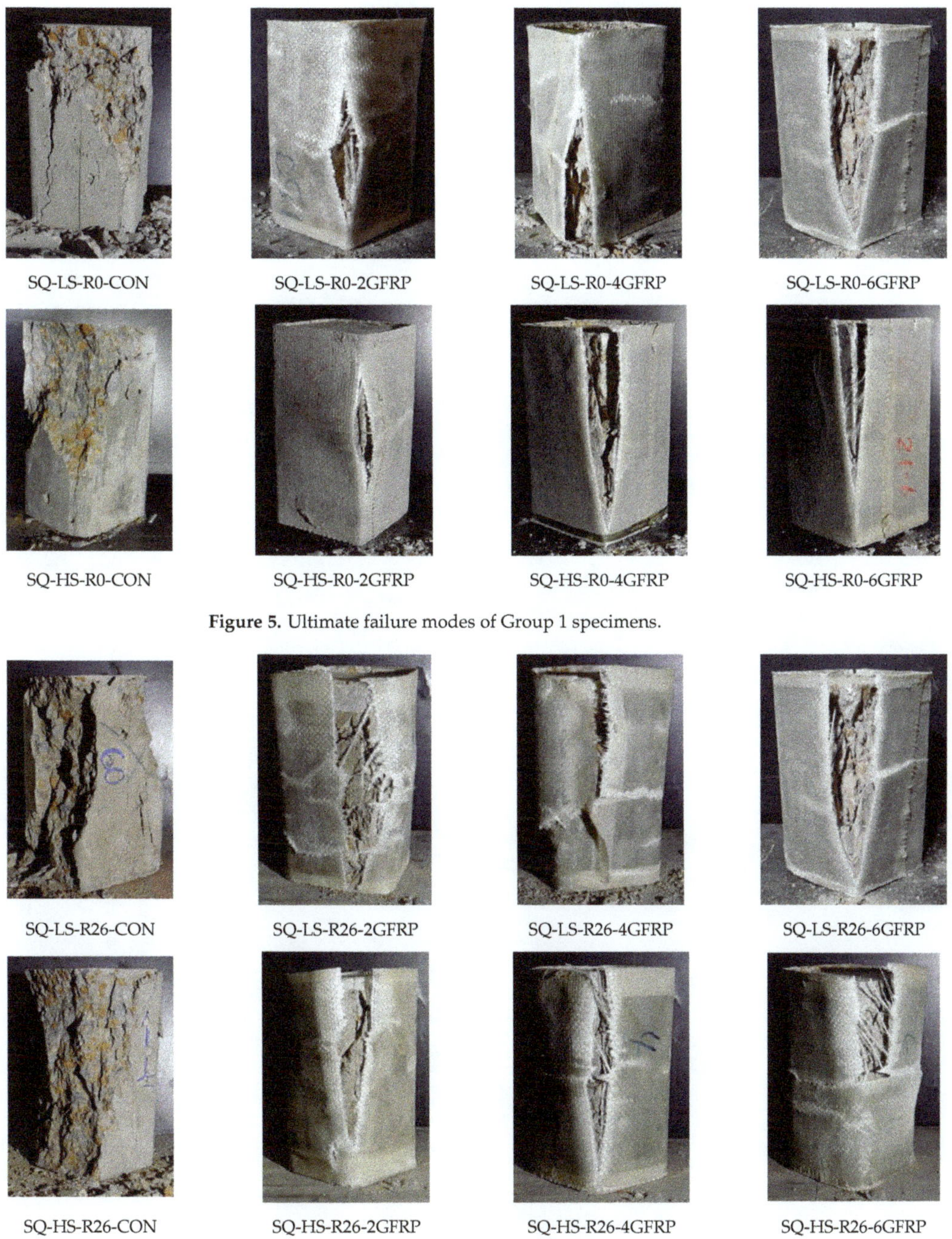

Figure 5. Ultimate failure modes of Group 1 specimens.

Figure 6. Ultimate failure modes of Group 2 specimens.

3.2. Peak Stress and Ultimate Strain

A summary of the peak compressive stress sustained and the ultimate strain for all of the specimens is presented in Table 4. The average peak stress for control specimen SQ-LS-R0-CON was slightly higher than the control specimen SQ-LS-R26-CON. This could be associated with the larger bearing area of the SQ-LS-R0-CON specimen as compared to the control specimen SQ-LS-R26-CON. The low-strength concrete specimens demonstrated an 83%, 103%, and 137% increase in the peak compressive stress due to two, four, and six GFRP wraps, respectively. The corresponding improvement in the ultimate strain was observed at 82%, 194%, and 658%, respectively. Similarly, a considerable improvement in the peak compressive stress of high-strength specimens in Group 1 was also observed as two, four, and six GFRP wraps enhanced the peak compressive stress by 50%, 71%, and 84%, respectively, whereas the enhancement in ultimate strain was 70%, 159%, and 134%, respectively. Group 2 specimens also demonstrated a substantial improvement in the peak compressive stress and strain. The peak compressive stress of the low-strength specimens was improved by 116%, 210%, and 278%, respectively. At the same time, the ultimate strain was enhanced by 222%, 495%, and 752%, respectively. The high-strength specimens exhibited an increase of 84%, 364%, and 563% for the peak compressive stress due to two, four, and six GFRP layers, respectively, whereas the ultimate strain was increased by 84%, 364%, and 563%, respectively. The above discussion concerning Table 4 suggests that the LC-GFRP resulted in a substantial improvement in the peak compressive stress and ultimate strain, which is crucial given the brittle nature of the concrete. Overall, the % increase in peak stresses of the GFRP-confined specimens with 0 mm corner radius was lower than the GFRP-confined specimens with a 26 mm corner radius due to the premature rupture of LC-GFRP at sharp corners.

Table 4. Summary of the peak stress and the corresponding strain.

Specimen ID	Peak Stress (MPa)	Standard Deviation	Increase in Peak Stress (%)	Ultimate Strain ϵ_{cc}	Increase in ϵ_{cc} (%)
SQ-LS-R0-CON	8.66	0.707	-	0.0053	-
SQ-LS-R0-2GFRP	15.87	4.243	83	0.0096	82
SQ-LS-R0-4GFRP	17.59	0.000	103	0.0155	194
SQ-LS-R0-6GFRP	20.51	1.414	137	0.0400	658
SQ-HS-R0-CON	14.41	4.950	-	0.0051	-
SQ-HS-R0-2GFRP	21.62	4.243	50	0.0086	70
SQ-HS-R0-4GFRP	24.67	0.707	71	0.0131	159
SQ-HS-R0-6GFRP	26.48	2.121	84	0.0119	134
SQ-LS-R26-CON	7.74	0.707	-	0.0084	-
SQ-LS-R26-2GFRP	16.71	1.414	116	0.0270	222
SQ-LS-R26-4GFRP	23.99	1.414	210	0.0500	495
SQ-LS-R26-6GFRP	29.29	8.485	278	0.0717	752
SQ-HS-R26-CON	13.39	2.121	-	0.0062	-
SQ-HS-R26-2GFRP	23.16	1.414	73	0.0115	84
SQ-HS-R26-4GFRP	32.07	1.414	140	0.0290	364
SQ-HS-R26-6GFRP	40.73	6.364	204	0.0414	563

3.3. Compressive Stress-Strain Curves

The recorded stress–strain behavior of Group 1 specimens is shown in Figure 7. Figure 7a shows the stress–strain curves for the low concrete strength specimens in Group 1. It is evident that a substantial improvement in the peak compressive stress was observed. The important parameter to be observed is the range of strain for which the peak sustained stress was maintained. This suggests that LC-GFRP imparted a considerable ductility to the concrete, which is crucial for strengthening against dynamic loads. The post-peak behavior of Specimens SQ-LS-R0-2GFRP and SQ-LS-R0-4GFRP was descending, whereas a stable second branch was observed for Specimen SQ-LS-R0-6GFRP. On the contrary, the second branch of the compressive stress–strain curves of the high-strength specimens in Group 1 was descending irrespective of the number of LC-GFRP layers. The second difference in

Figure 7a,b is the value of the ultimate strain. The high-strength specimens were able to sustain compressive stress to lower strain values as compared to low-strength specimens.

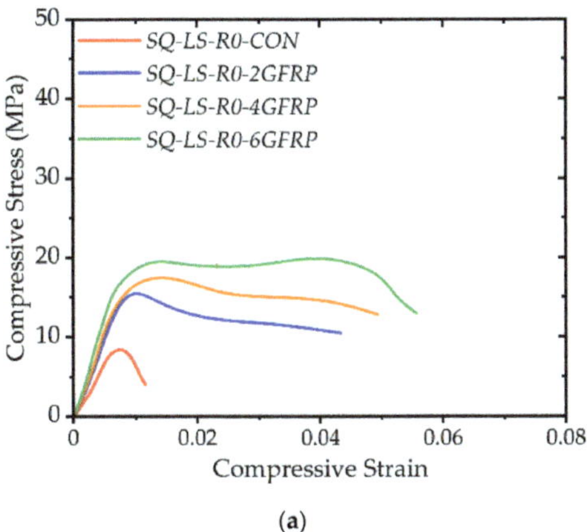
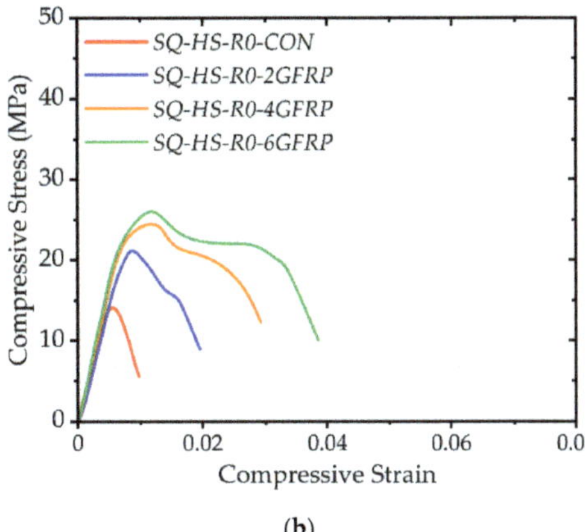

(**a**)

(**b**)

Figure 7. Compressive stress vs. strain response for Group 1 specimens with (**a**) low concrete strength and (**b**) high concrete strength.

Figure 8a,b shows stress–strain curves for Group 2 specimens with low and high concrete strength, respectively. A significant difference between Figures 7 and 8 is observed in the second branch. It is recalled that the sharp corners in the Group 2 specimens were rounded to a 26 mm corner radius. The corresponding result is depicted in Figure 8a, as the stiff initial branch was followed by an ascending branch highlighting the importance of the corner radius. The peak compressive stress and the ultimate strain were found to increase with the number of LC-GFRP sheets. However, the improvement in the axial ductility was limited by the concrete strength. It is evident in Figure 8 that the failure of the high concrete strength specimens for the same number of LC-GFRP layers occurred at lower strains than those of the low concrete strength specimens. This could be related with the lower lateral dilation of the high-strength concrete as compared to the low-strength concrete.

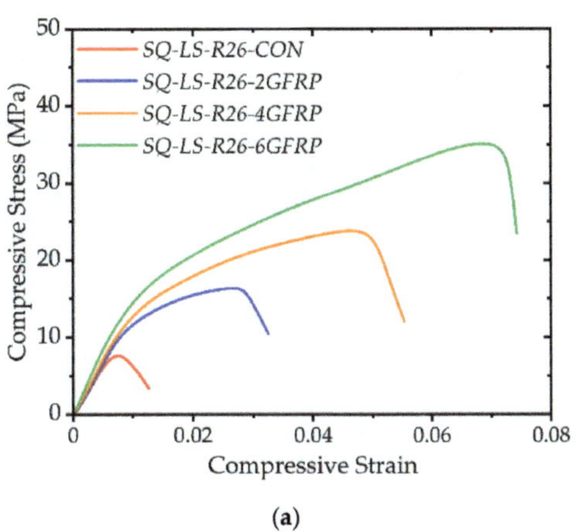
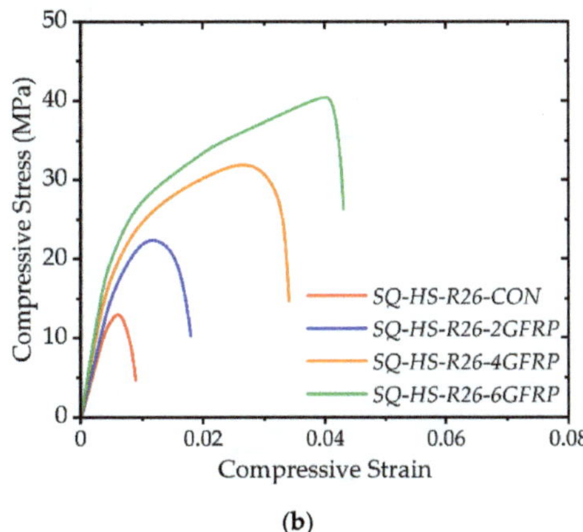

(**a**)

(**b**)

Figure 8. Compressive stress vs. strain response for Group 2 specimens with (**a**) low concrete strength and (**b**) high concrete strength.

3.4. Effect of Concrete Strength and Corner Radius

The effect of the concrete strength on the gain in peak compressive stress due to LC-GFRP confinement is shown in Figure 9. It is evident that the gain in peak compressive stress is dependent on the concrete strength. For both of the groups and for the same number of LC-GFRP layers, the specimens with low concrete strength demonstrated greater improvement in the peak compressive stress than high strength specimens. Another observation that can be made from Figure 9 is the effect of corner radius on the improvement in peak compressive stress. For the case of no corner radius (see Figure 9a), the maximum increase in the peak compressive stress observed for Specimen SQ-LS-R0-6GFRP was 137%, whereas the corresponding value for a 26 mm corner radius was 278%, observed for Specimen SQ-LS-R26-6GFRP.

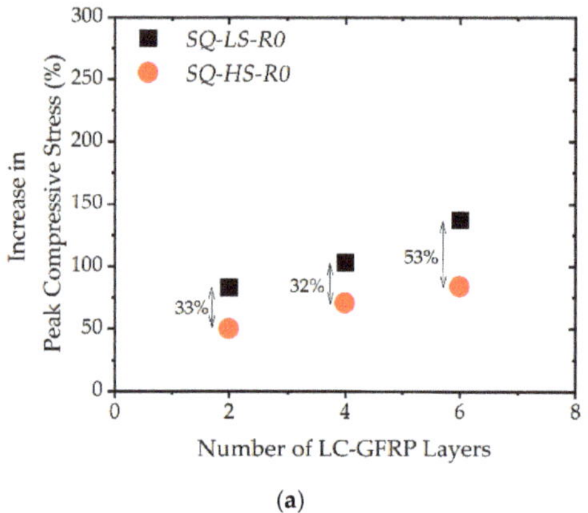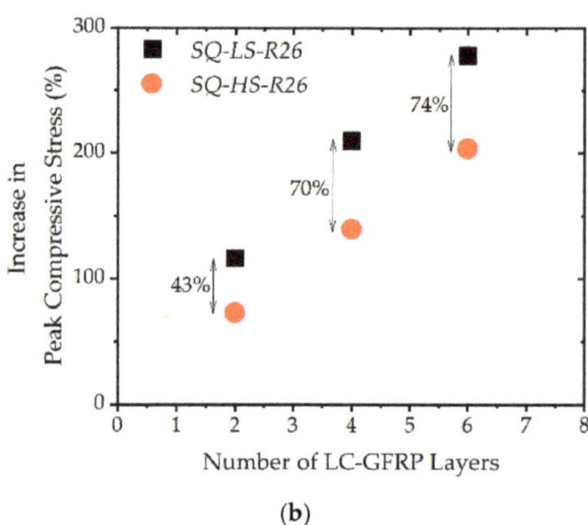

Figure 9. Increase in peak compressive stress in (**a**) Group 1 and (**b**) Group 2.

The effect of concrete strength on the improvement in ultimate strain is exhibited in Figure 10. A similar trend as that for the peak compressive stress is observed. This is evident as the low-strength specimens demonstrated a greater improvement in their ultimate strains than the high-strength specimens. The second branch of stress–strain curves of the Group 1 specimens was either descending or stable, whereas an ascending second branch was observed for the Group 2 specimens. As a result, the Group 2 specimens demonstrated a higher increase in the peak compressive stress.

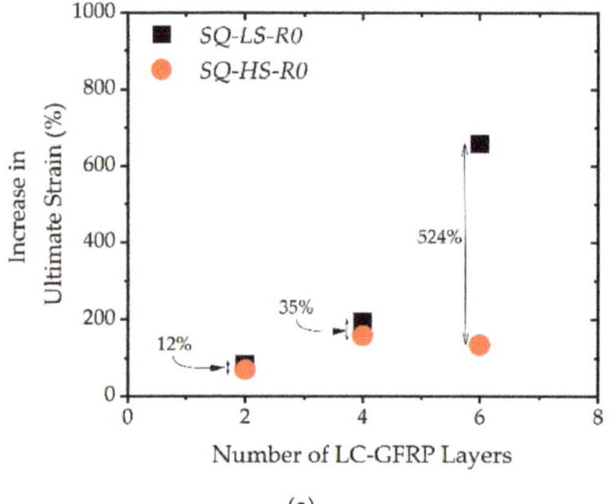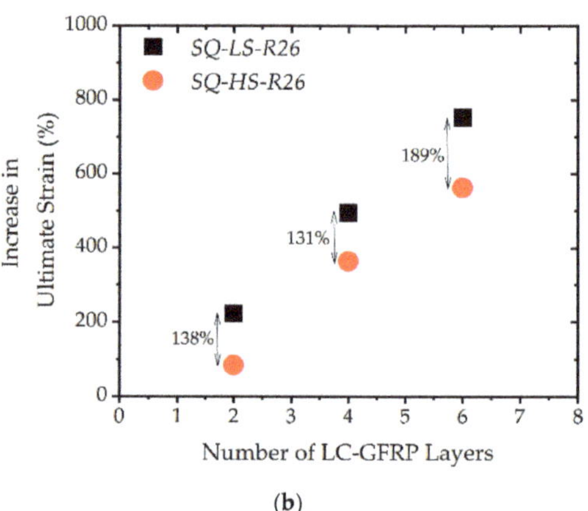

Figure 10. Increase in ultimate strain in (**a**) Group 1 and (**b**) Group 2.

4. Analytical Investigations

4.1. Existing Analytical Models

The accurate prediction of the peak compressive strength and the ultimate strain of strengthened concrete is important from both the design and analysis considerations. In existing studies, the confined concrete peak strength is often related to the lateral confinement pressure that is generated by the external confinement as:

$$\frac{f_{cc}}{f_{co}} = 1 + k_1 \left(\frac{f_l}{f'_{co}} \right) \tag{1}$$

where f'_{co} is the compressive strength of unconfined concrete; f_l is the lateral pressure generated by external confinement; and k_1 is the regression constant and varies for different existing models. The confining pressure f_l is computed by taking an equilibrium between the outward core pressure and the resulting forces generated within the confinement, as shown in Figure 11. The resulting equilibrium equation is defined as [44]:

$$f_l = \frac{2 f_{frp} t}{D} \times \rho \tag{2}$$

where f_{frp} is the tensile strength of external wrap; t is the thickness of external wrap; and D is the diagonal length of the rectilinear section, which is defined as [48]:

$$D = \frac{2bd}{b+d} \tag{3}$$

where b and d are the cross-sectional dimensions of the section. The parameter ρ in Equation (2) is defined as [48]:

$$\rho = 1 - \frac{(b-2r)^2 + (d-2r)^2}{3A} \tag{4}$$

where r is the corner radius and A is the cross-sectional area defined as:

$$A = bd - (4-\pi)r^2 \tag{5}$$

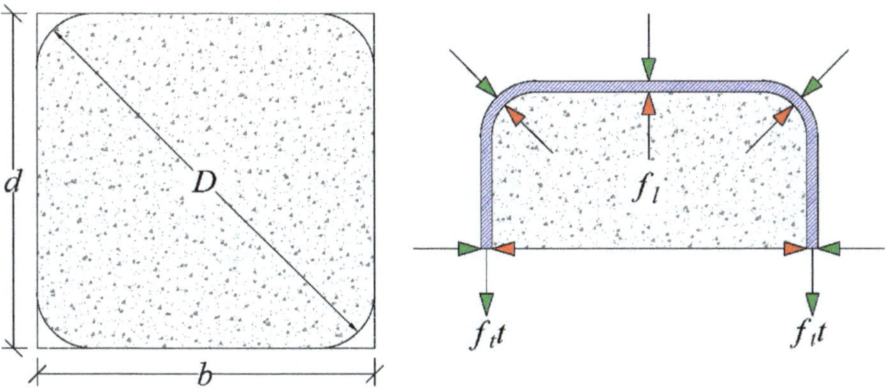

Figure 11. Equilibrium between core pressure and the resulting confining forces (b = *width*, d = *depth*, D = *diagonal diameter*, f_l = *lateral confining pressure*, t = *thickness of FRP*).

The ultimate strain of the confined concrete ϵ_{cc} can be expressed in a similar way as:

$$\frac{\epsilon_{cc}}{\epsilon_{co}} = 1 + k_2 \left(\frac{f_l}{f'_{co}} \right) \tag{6}$$

where ϵ_{co} is the ultimate strain of unconfined concrete and k_2 is the regression constant. Several numerical models are available in the literature to relate the peak compressive stress f_{cc} and the ultimate strain ϵ_{cc} to the confining pressure exerted by external FRPs. Several existing peak compressive stress and ultimate strain models are presented in Table 5. In a recent study by Rodsin et al. [42], it was found that the accuracy of the models in Table 5 varied with the concrete strength. Further, the model of Hussain et al. [44] closely approximated the peak compressive stress of LC-GFRP-confined concrete. It was further found that none of the models in Table 5 were able to provide good agreement with experimental ultimate strain results. Therefore, further studies were recommended to increase the database of LC-GFRP-confined concrete specimens to propose equations for the peak compressive stress and ultimate strain.

Table 5. Existing compressive stress–strain models.

ID	Model	Ultimate Stress	Ultimate Strain
1	Hussain et al. [44]	$\dfrac{f_{cc}}{f'_{co}} = 1 + 2.70\rho^{0.90}\left(\dfrac{f_l}{f'_{co}}\right)$	$\dfrac{\epsilon_{cc}}{\epsilon_{co}} = 2 + 10\rho^{1.10}\left(\dfrac{f_l}{f'_{co}}\right)$
2	ACI 2002 [48]	$\dfrac{f_{cc}}{f'_{co}} = -1.254 + 2.254\sqrt{1 + \dfrac{7.94f_l}{f'_{co}}} - 2\dfrac{f_l}{f'_{co}}$	$\dfrac{\epsilon_{cc}}{\epsilon_{co}} = 1.5 + 13\left(\dfrac{f_l}{f'_{co}}\right)\left(\dfrac{\epsilon_{fe}}{\epsilon_{co}}\right)^{0.45}$
3	Shehata et al. [49]	$\dfrac{f_{cc}}{f'_{co}} = 1 + 0.85\left(\dfrac{f_l}{f'_{co}}\right)$	$\dfrac{\epsilon_{cc}}{\epsilon_{co}} = 1 + 13.5\left(\dfrac{f_l}{f'_{co}}\right)$
4	Touhari and Mitiche [50]	$\dfrac{f_{cc}}{f'_{co}} = 1 + \left(1 - \dfrac{\left(\left(\frac{\pi}{2}\right)-1\right)(b-2r)^2}{b^2}\right)\dfrac{f_l}{f'_{co}}$	$\dfrac{\epsilon_{cc}}{\epsilon_{co}} = 2.3 + 7\left(1 - \dfrac{\left(\left(\frac{\pi}{2}\right)-1\right)(b-2r)^2}{b^2}\right)\dfrac{f_l}{f'_{co}}$
5	Mirmiran et al. [51]	$\dfrac{f_{cc}}{f'_{co}} = 1 + 6.0\left(\dfrac{2r}{D}\right)\left(\dfrac{f_l^{0.7}}{f'_{co}}\right)$	-
6	Lam and Teng [52]	$\dfrac{f_{cc}}{f'_{co}} = 1 + 3.30\left(\dfrac{f_l}{f'_{co}}\right)$	$\dfrac{\epsilon_{cc}}{\epsilon_{co}} = 1.75 + 12\left(\dfrac{f_l}{f'_{co}}\right)\left(\dfrac{\epsilon_{fe}}{\epsilon_{co}}\right)^{0.45}$

The accuracy of the models in Table 5 is shown in Figure 12 to predict the peak compressive stress of LC-GFRP-confined concrete. Unlike the findings of Rodsin et al. [42], the model of Hussain et al. [44] underestimated the peak compressive stress of LC-GFRP-confined specimens with zero corner radius (see Figure 12a,b), whereas the models of ACI 2002 [48] and Lam and Teng [52] seem to provide good agreement with experimental results. For specimens in Group 2, the model of Hussain et al. [44] seems to correlate well with the experimental results along with the model of ACI 2002 [48]. From the study by Rodsin et al. [42] and the present study, it can be seen that none of the considered models were able to provide good agreement with the experimental peak compressive stresses on a consistent basis.

The accuracy of the considered models to predict the ultimate strain of LC-GFRP-confined RBAC specimens is shown in Figure 13. In general, none of the models was able to predict the ultimate strains on a consistent basis. Therefore, it was desired to propose peak compressive stress and ultimate strain models for LC-GFRP-confined RBAC.

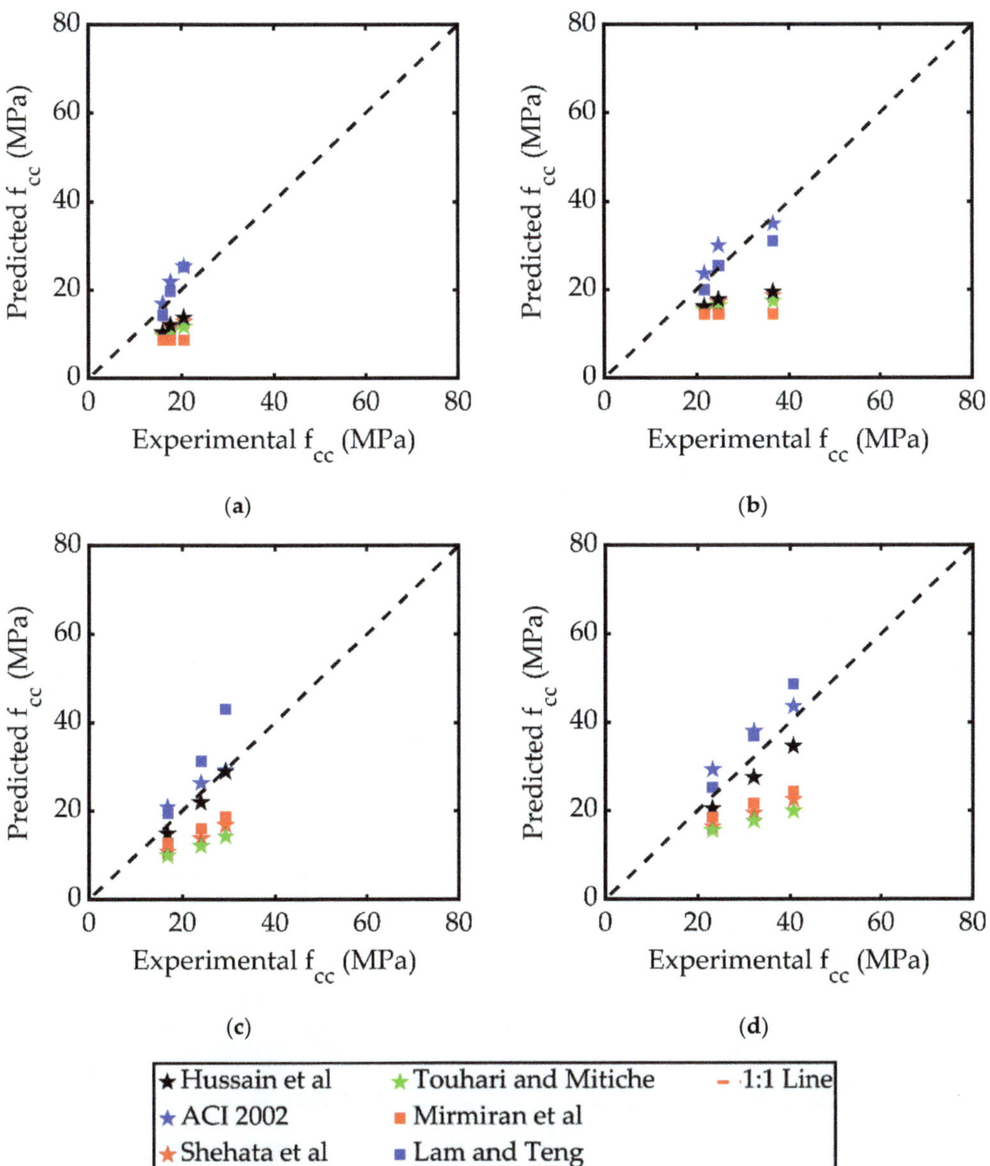

Figure 12. Comparison of predicted vs. experimental peak compressive stresses of subgroups (**a**) SQ-LS-R0, (**b**) SQ-HS-R0, (**c**) SQ-LS-R26, and (**d**) SQ-HS-R26 [42,46–50].

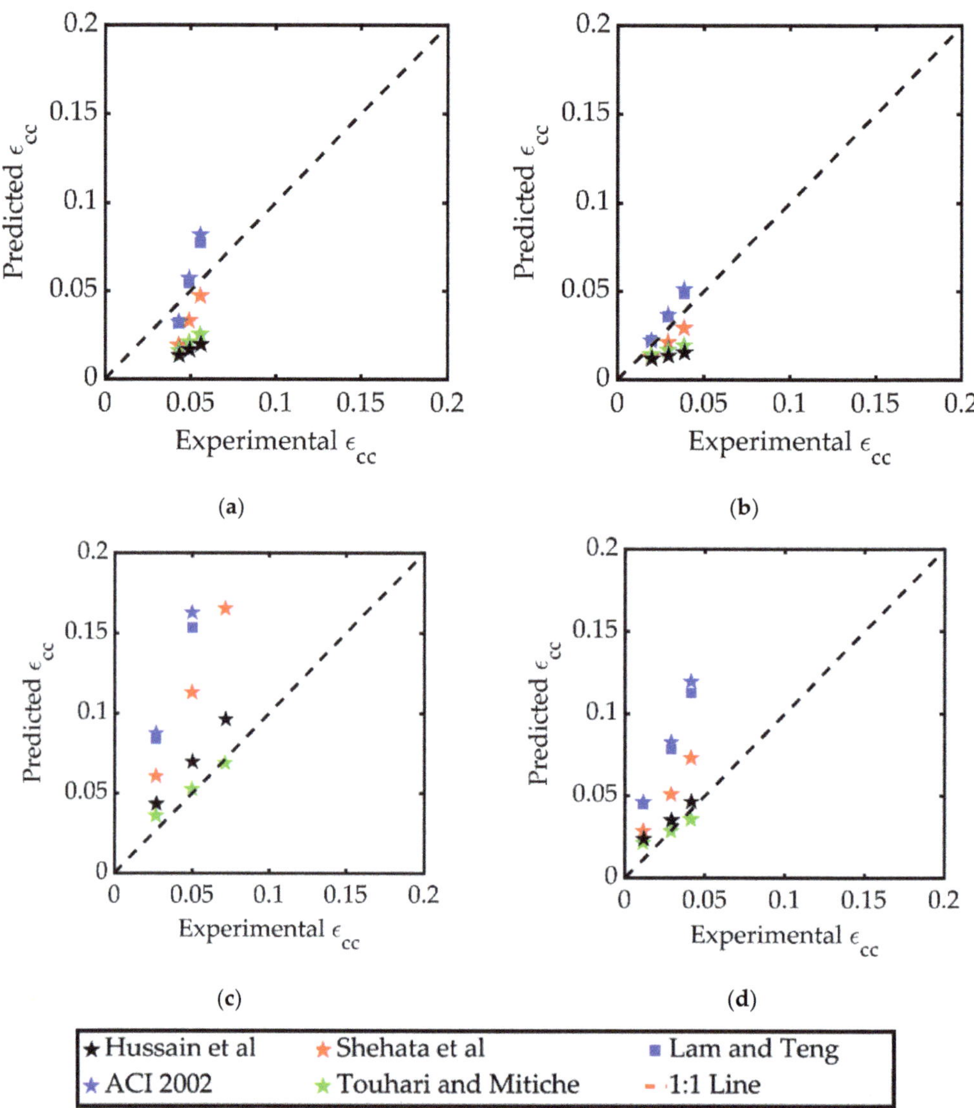

Figure 13. Comparison of predicted vs. experimental ultimate strains of subgroups (**a**) SQ-LS-R0, (**b**) SQ-HS-R0, (**c**) SQ-LS-R26, and (**d**) SQ-HS-R26 [42,46–48,50].

4.2. Proposed Model

Regression analysis was performed to propose equations for the peak compressive stress, and the ultimate strain of LC-GFRPP-confined RBAC. Six specimens tested by Rodsin et al. [42] were also included to increase the sample size. It should be mentioned that the samples tested by Rodsin et al. [42] incorporated a corner radius of 13 mm as opposed to the 26 mm corner radius in the present study. Figure 14 presents the effect of the corner radius on the increase in the peak compressive stress of RBAC due to LC-GFRP confinement. In general, it is observed that the increase in the peak compressive stress for the same layers of LC-GFRP and concrete strength is improved as the corner radius is increased. Therefore, the effect of the corner radius must be included in the proposed equation of peak compressive stress. Secondly, it is observed that the increase in the peak compressive stress is more in Figure 14a (low-strength concrete) than in Figure 14b (high-strength concrete). Therefore, the effect of unconfined concrete must also be included. Finally, the effect of the lateral confining pressure due to LC-GFRP confinement is also evident in Figure 14 as the compressive stress increases with the increase in the layers of LC-GFRP.

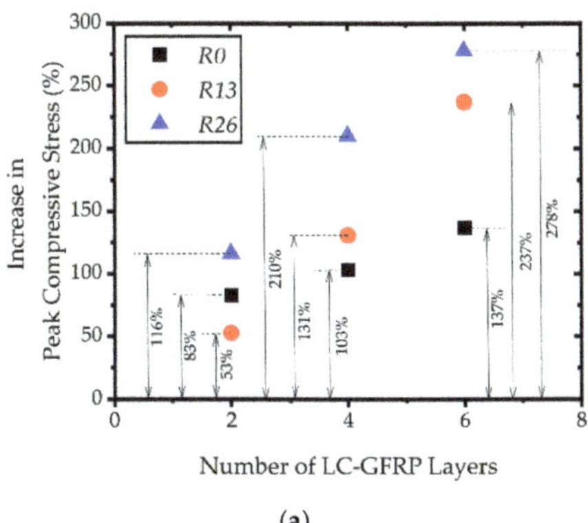

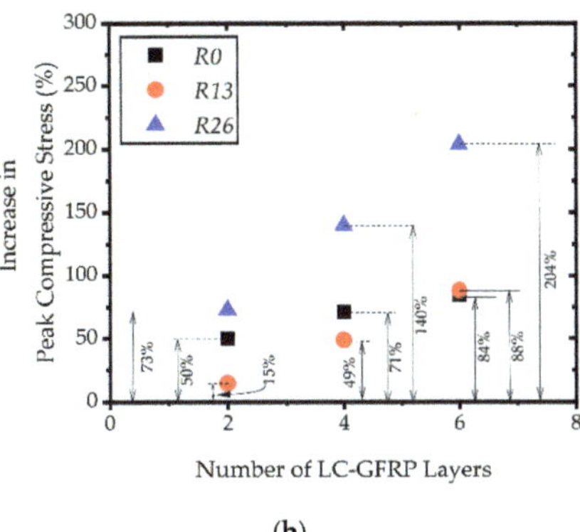

Figure 14. Effect of corner radius on the increase in compressive stress (**a**) low strength concrete and (**b**) high strength concrete.

The regression analysis was performed to predict equations for four quantities, mainly peak compressive stress f_1 and corresponding strain ϵ_1, ultimate stress f_2 and the corresponding strain ϵ_2 as shown in Figure 15. The nonlinear regression analysis was conducted using the classical Gauss–Newton method, and the analysis was performed using SPSS Statistics. It was discussed in Section 3 that both ascending and descending behavior was observed in the second branch of the stress–strain curves depending upon the concrete strength and corner radius. Equations (7) and (8) were found to correlate well with the experimental results of f_1 and f_2 by considering unconfined concrete strength f_{co}, the confining pressure f_l, and corner radius r:

$$\frac{f_1}{f_{co}} = 1 + 1.841 \left(\frac{b-r}{b} \right)^{-0.51} \left(\frac{f_l}{f_{co}} \right)^{0.55} \tag{7}$$

$$\frac{f_2}{f_{co}} = 1 + 0.101 \left(\frac{b-r}{b} \right)^{-7.63} \left(\frac{f_l}{f_{co}} \right)^{0.73} \left(\frac{f_1}{f_{co}} \right)^{2.95} \tag{8}$$

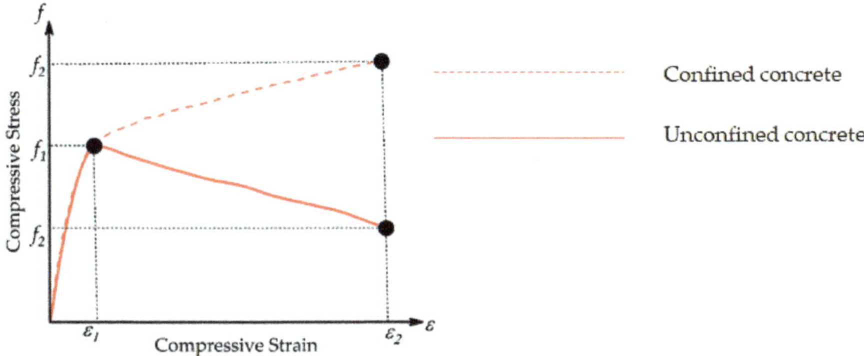

Figure 15. Proposed parameters on the stress–strain curve.

The accuracy of Equations (7) and (8) is shown in Figure 16a,b, respectively. A good correlation between the experimental and predicted values of f_1 and f_2.

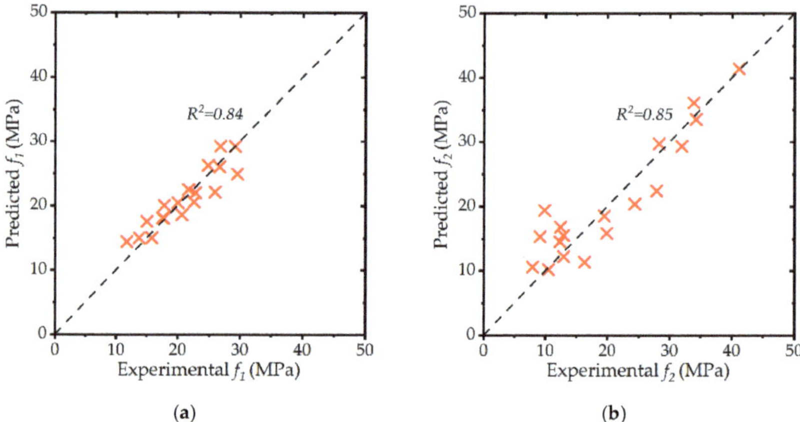

Figure 16. Comparison of experimental and predicted values of (**a**) f_1 and (**b**) f_2.

Equations (9) and (10) were found to correlate with experimental ϵ_1 and ϵ_2 values and their accuracy is shown in Figure 17a,b, respectively. It can be seen that a good agreement between experimental and predicted ϵ_1 and ϵ_2 values is obtained.

$$\epsilon_1 = \epsilon_{co}\left(1 + 0.295\left(\frac{f_l}{f_{co}}\right)^{-0.464}\left(\frac{b-r}{b}\right)^{-3.304}\right) \tag{9}$$

$$\epsilon_2 = \epsilon_{co}\left(1 + 7.853\left(\frac{f_l}{f_{co}}\right)^{1.30}\left(\frac{b-r}{b}\right)^{-2.579}\right) \tag{10}$$

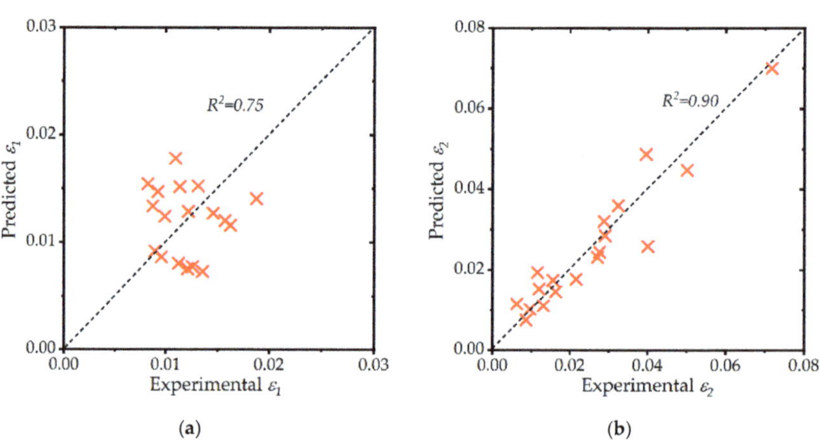

Figure 17. Comparison of experimental and predicted values of (**a**) ϵ_1 and (**b**) ϵ_2.

5. Conclusions

This study investigated the role of low-cost glass-fiber-reinforced polymer (LC-GFRP) sheets as external passive confinement to enhance the mechanical properties of recycled brick aggregate concrete (RBAC). Thirty-two square RBAC specimens were constructed and tested in two groups depending upon the corner radius. The following important conclusions are drawn:

1. The peak compressive stress and ultimate strain were found to increase with the number of LC-GFRP wraps. The maximum increase in the peak compressive stress and the ultimate strain was observed for six LC-GFRP wraps;
2. It was found that the provision of a 26 mm corner radius significantly improved the efficiency of LC-GFRP wraps. Premature failure due to stress concentrations near sharp corners in specimens with zero corner radius undermined the efficacy of LC-GFRP wraps. Therefore, it is suggested to provide a corner radius to prevent this premature failure;
3. The shape of the compressive stress–strain curves of LC-GFRP-confined RBAC was bilinear. The initial stiff branch was identical in all of the strengthened specimens. The second branch was ascending and descending for the zero- and 26-mm corner radius, respectively. As a result, the ultimate stress of 26 mm corner radius specimens was higher;
4. A comparison between the strengthened specimens based on the concrete strength revealed that low-strength specimens demonstrated a higher increase in the peak compressive stress and strain as compared to high-strength specimens;
5. Existing FRP stress–strain models were found inconsistent in reproducing experimental results. Nonlinear regression analysis was conducted to propose equations for the key parameters in the stress–strain curves of LC-GFRP-confined RBAC.

6. Future Research Directions

This study employed the use of fired clay solid bricks to prepare the crushed brick aggregates. The density of these bricks is very low, i.e., 120 kg/m^3. In actual practices, there are different types of bricks with varying densities. Future studies are recommended to consider crushed brick aggregates with higher densities to further investigate the role of LC-GFRP composites to enhance the strength of confined concrete.

Author Contributions: Conceptualization, R.P., K.R., K.C. and S.S.; Investigation, N.A. and Q.H.; Methodology, K.C. and K.K.; Validation, K.K.; Writing—original draft, R.P., K.R., K.C., N.A., S.S. and Q.H.; Writing—review & editing, R.P., K.R., N.A., S.S., Q.H. and K.K. All authors have read and agreed to the published version of the manuscript.

Funding: This research received no external funding.

Acknowledgments: This work was financially supported by King Mongkut's Institute of Technology Ladkrabang Research Fund. The authors are thankful to Asian Institute of Technology (AIT) for supporting test facilities.

Conflicts of Interest: The authors declare no conflict of interest.

References

1. Mehta, P.K.; Monteiro, P.J.M. *Concrete: Microstructure, Properties, and Materials*, 3rd ed.; McGraw-Hill: New York, NY, USA, 2006.
2. Monteiro, P.J.; Miller, S.A.; Horvath, A. Towards sustainable concrete. *Nat. Mater.* **2017**, *16*, 698–699. [CrossRef]
3. Noguchi, T.; Kitagaki, R.; Tsujion, M. Minimizing Environmental Impact and Maximizing Performance in Concrete Recycling. *Struct. Concr.* **2011**, *12*, 36–46. [CrossRef]
4. Wong, C.L.; Mo, K.H.; Yap, S.P.; Alengaram, U.J.; Ling, T.C. Potential Use of Brick Waste as Alternate Concrete-Making Materials: A Review. *J. Clean Prod.* **2018**, *195*, 226–239. [CrossRef]
5. Tang, Q.; Ma, Z.; Wu, H.; Wang, W. The Utilization of Eco-Friendly Recycled Powder from Concrete and Brick Waste in New Concrete: A Critical Review. *Cem. Concr. Compos.* **2020**, *114*, 103807. [CrossRef]
6. He, Z.; Shen, A.; Wu, H.; Wang, W.; Wang, L.; Yao, C.; Wu, J. Research Progress on Recycled Clay Brick Waste as an Alternative to Cement for Sustainable Construction Materials. *Constr. Build. Mater.* **2021**, *274*, 122113. [CrossRef]
7. Liu, Q.; Li, B.; Xiao, J.; Singh, A. Utilization Potential of Aerated Concrete Block Powder and Clay Brick Powder from C&D Waste. *Constr. Build. Mater.* **2020**, *238*, 117721. [CrossRef]
8. Zhang, J.; Ding, L.; Li, F.; Peng, J. Recycled Aggregates from Construction and Demolition Wastes as Alternative Filling Materials for Highway Subgrades in China. *J. Clean Prod.* **2020**, *255*, 120223. [CrossRef]

9. Andrew, G.H.T.; Tay, Y.W.I.; Annapareddy, A.; Li, M.; Tan, M.J. Effect of Recycled Glass Gradation in 3D Cementitious Material Printing. In Proceedings of the 3rd International Conference on Progress in Additive Manufacturing, Nanyang Technological University, Singapore, 14–17 May 2018; pp. 50–55.

10. Ting, G.H.A.; Quah, T.K.N.; Lim, J.H.; Tay, Y.W.D.; Tan, M.J. Extrudable Region Parametrical Study of 3D Printable Concrete Using Recycled Glass Concrete. *J. Build. Eng.* **2022**, *50*, 104091. [CrossRef]

11. Zhu, L.; Zhu, Z. Reuse of Clay Brick Waste in Mortar and Concrete. *Adv. Mater. Sci. Eng.* **2020**, *2020*, 6326178. [CrossRef]

12. Li, H.; Dong, L.; Jiang, Z.; Yang, X.; Yang, Z. Study on Utilization of Red Brick Waste Powder in the Production of Cement-Based Red Decorative Plaster for Walls. *J. Clean Prod.* **2016**, *133*, 1017–1026. [CrossRef]

13. Mansur, M.A.; Wee, T.H.; Cheran, L.S. Crushed Bricks as Coarse Aggregate for Concrete. *Mater. J.* **1999**, *96*, 478–484. [CrossRef]

14. Khalaf, F.M. Using Crushed Clay Brick as Coarse Aggregate in Concrete. *J. Mater. Civ. Eng.* **2006**, *18*, 518–526. [CrossRef]

15. Khalaf, F.M.; DeVenny, A.S. Recycling of Demolished Masonry Rubble as Coarse Aggregate in Concrete: Review. *J. Mater. Civ. Eng.* **2004**, *16*, 331–340. [CrossRef]

16. Vrijders, J.; Desmyter, J. *Een Hoogwaardig Gebruik van Puingranulaten Stimuleren*; OVAM: Mechelen, Belgium, 2008.

17. Debieb, F.; Kenai, S. The Use of Coarse and Fine Crushed Bricks as Aggregate in Concrete. *Constr Build. Mater.* **2008**, *22*, 886–893. [CrossRef]

18. Medina, C.; Zhu, W.; Howind, T.; de Rojas, M.I.S.; Frías, M. Influence of Mixed Recycled Aggregate on the Physical—Mechanical Properties of Recycled Concrete. *J. Clean. Prod.* **2014**, *68*, 216–225. [CrossRef]

19. Yang, J.; Du, Q.; Bao, Y. Concrete with Recycled Concrete Aggregate and Crushed Clay Bricks. *Constr Build. Mater.* **2011**, *25*, 1935–1945. [CrossRef]

20. Nováková, I.; Mikulica, K. Properties of Concrete with Partial Replacement of Natural Aggregate by Recycled Concrete Aggregates from Precast Production. *Procedia Eng.* **2016**, *151*, 360–367. [CrossRef]

21. Jiang, T.; Wang, X.M.; Zhang, W.P.; Chen, G.M.; Lin, Z.H. Behavior of FRP-Confined Recycled Brick Aggregate Concrete under Monotonic Compression. *J. Compos. Constr.* **2020**, *24*, 04020067. [CrossRef]

22. Kox, S.; Vanroelen, G.; van Herck, J.; de Krem, H.; Vandoren, B. Experimental Evaluation of the High-Grade Properties of Recycled Concrete Aggregates and Their Application in Concrete Road Pavement Construction. *Case Stud. Constr. Mater.* **2019**, *11*, e00282. [CrossRef]

23. Yang, Y.F.; Ma, G.L. Experimental Behaviour of Recycled Aggregate Concrete Filled Stainless Steel Tube Stub Columns and Beams. *Thin-Walled Struct.* **2013**, *66*, 62–75. [CrossRef]

24. Ameli, M.; Ronagh, H.R.; Dux, P.F. Behavior of FRP Strengthened Reinforced Concrete Beams under Torsion. *J. Compos. Constr.* **2007**, *11*, 192–200. [CrossRef]

25. Jiangfeng, D.; Shucheng, Y.; Qingyuan, W.; Wenyu, Z.; Jiangfeng, D.; Shucheng, Y.; Qingyuan, W.; Wenyu, Z. Flexural Behavior of RC Beams Made with Recycled Aggregate Concrete and Strengthened by CFRP Sheets. *J. Build. Struct.* **2019**, *40*, 71–78. [CrossRef]

26. Smith, S.T.; Teng, J.G. FRP-Strengthened RC Beams. I: Review of Debonding Strength Models. *Eng. Struct.* **2002**, *24*, 385–395. [CrossRef]

27. Harajli, M.H. Axial Stress–Strain Relationship for FRP Confined Circular and Rectangular Concrete Columns. *Cem. Concr. Compos.* **2006**, *28*, 938–948. [CrossRef]

28. Spoelstra, M.R.; Monti, G. FRP-Confined Concrete Model. *J. Compos. Constr.* **1999**, *3*, 143–150. [CrossRef]

29. Pimanmas, A.; Saleem, S. Dilation Characteristics of PET FRP–Confined Concrete. *J. Compos. Constr.* **2018**, *22*, 04018006. [CrossRef]

30. Al-Salloum, Y.A. Compressive Strength Models of FRP-Confined Concrete. In Proceedings of the 1st Asia-Pacific Conference on FRP in Structures, APFIS 2007, Hong Kong, China, 12–14 December 2007; Volume 1, pp. 175–180.

31. Eid, R.; Paultre, P. Compressive Behavior of FRP-Confined Reinforced Concrete Columns. *Eng. Struct.* **2017**, *132*, 518–530. [CrossRef]

32. Gao, C.; Huang, L.; Yan, L.; Kasal, B.; Li, W. Behavior of Glass and Carbon FRP Tube Encased Recycled Aggregate Concrete with Recycled Clay Brick Aggregate. *Compos. Struct.* **2016**, *155*, 245–254. [CrossRef]

33. Tang, Z.; Li, W.; Tam, V.W.Y.; Yan, L. Mechanical Behaviors of CFRP-Confined Sustainable Geopolymeric Recycled Aggregate Concrete under Both Static and Cyclic Compressions. *Compos. Struct.* **2020**, *252*, 112750. [CrossRef]

34. Han, Q.; Yuan, W.Y.; Ozbakkaloglu, T.; Bai, Y.L.; Du, X.L. Compressive Behavior for Recycled Aggregate Concrete Confined with Recycled Polyethylene Naphthalate/Terephthalate Composites. *Constr. Build. Mater.* **2020**, *261*, 120498. [CrossRef]

35. Zeng, J.J.; Zhang, X.W.; Chen, G.M.; Wang, X.M.; Jiang, T. FRP-Confined Recycled Glass Aggregate Concrete: Concept and Axial Compressive Behavior. *J. Build. Eng.* **2020**, *30*, 101288. [CrossRef]

36. Iskander, M.G.; Hassan, M. State of the Practice Review in FRP Composite Piling. *J. Compos. Constr.* **1998**, *2*, 116–120. [CrossRef]

37. Wang, X.; Wu, Z. Evaluation of FRP and Hybrid FRP Cables for Super Long-Span Cable-Stayed Bridges. *Compos. Struct.* **2010**, *92*, 2582–2590. [CrossRef]

38. Chaiyasarn, K.; Hussain, Q.; Joyklad, P.; Rodsin, K. New Hybrid Basalt/E-Glass FRP Jacketing for Enhanced Confinement of Recycled Aggregate Concrete with Clay Brick Aggregate. *Case Stud. Constr. Mater.* **2021**, *14*, e00507. [CrossRef]

39. Yoddumrong, P.; Rodsin, K.; Katawaethwarag, S. Seismic Strengthening of Low-Strength RC Concrete Columns Using Low-Cost Glass Fiber Reinforced Polymers (GFRPs). *Case Stud. Constr. Mater.* **2020**, *13*, e00383. [CrossRef]

40. Rodsin, K.; Hussain, Q.; Suparp, S.; Nawaz, A. Compressive Behavior of Extremely Low Strength Concrete Confined with Low-Cost Glass FRP Composites. *Case Stud. Constr. Mater.* **2020**, *13*, e00452. [CrossRef]

41. Rodsin, K. Confinement Effects of Glass FRP on Circular Concrete Columns Made with Crushed Fired Clay Bricks as Coarse Aggregates. *Case Stud. Constr. Mater.* **2021**, *15*, e00609. [CrossRef]

42. Rodsin, K.; Ali, N.; Joyklad, P.; Chaiyasarn, K.; Zand, A.W.A.; Hussain, Q. Improving Stress-Strain Behavior of Waste Aggregate Concrete Using Affordable Glass Fiber Reinforced Polymer (GFRP) Composites. *Sustainability* **2022**, *14*, 6611. [CrossRef]

43. Wang, L.M.; Wu, Y.F. Effect of Corner Radius on the Performance of CFRP-Confined Square Concrete Columns. *Eng. Struct.* **2008**, *30*, 493–505. [CrossRef]

44. Hussain, Q.; Ruangrassamee, A.; Tangtermsirikul, S.; Joyklad, P.; Wijeyewickrema, A.C. Low-Cost Fiber Rope Reinforced Polymer (FRRP) Confinement of Square Columns with Different Corner Radii. *Buildings* **2021**, *11*, 355. [CrossRef]

45. *ASTM C1314-21*; Standard Test Method for Compressive Strength of Masonry Prisms. ASTM: West Conshohocken, PA, USA, 2021.

46. *ASTM C140/C140M-22a*; Standard Test Methods for Sampling and Testing Concrete Masonry Units and Related Units. ASTM: West Conshohocken, PA, USA, 2022.

47. *ASTM D3039/D3039M-17*; Standard Test Method for Tensile Properties of Polymer Matrix Composite Materials. ASTM: West Conshohocken, PA, USA, 2017.

48. Soudki, K.; Alkhrdaji, T. Guide for the Design and Construction of Externally Bonded FRP Systems for Strengthening Concrete Structures (ACI 440.2R-02). In Proceedings of the Structures Congress and Exposition, New York, NY, USA, 20–24 April 2005; pp. 1–8. [CrossRef]

49. Shehata, I.A.E.M.; Carneiro, L.A.V.; Shehata, L.C.D. Strength of Short Concrete Columns Confined with CFRP Sheets. *Mater. Struct.* **2002**, *35*, 50–58. [CrossRef]

50. Touhari, M.; Mitiche, R.K. Strength Model of FRP Confined Concrete Columns Based on Analytical Analysis and Experimental Test. *Int. J. Struct. Integr.* **2020**, *11*, 82–106. [CrossRef]

51. Mirmiran, A.; Shahawy, M.; Samaan, M.; Echary, H.E.; Mastrapa, J.C.; Pico, O. Effect of Column Parameters on FRP-Confined Concrete. *J. Compos. Constr.* **1998**, *2*, 175–185. [CrossRef]

52. Lam, L.; Teng, J.G. Strength Models for Fiber-Reinforced Plastic-Confined Concrete. *J. Struct. Eng.* **2002**, *128*, 612–623. [CrossRef]

 *sustainability*

Article

Engineered Stone Produced with Glass Packaging Waste, Quartz Powder, and Epoxy Resin

Gabriela Nunes Sales Barreto [1,*], Elaine Aparecida Santos Carvalho [1], Vitor da Silva de Souza [1], Maria Luiza Pessanha Menezes Gomes [1], Afonso R. G. de Azevedo [2], Sérgio Neves Monteiro [3] and Carlos Maurício Fontes Vieira [1]

[1] Advanced Materials Laboratory (LAMAV), State University of Northern Fluminense–UENF, Av. Alberto Lamego, 2000, Campos dos Goytacazes 28013-602, RJ, Brazil; elainesanttos@yahoo.com.br (E.A.S.C.); vitor.silvasouza_@hotmail.com (V.d.S.d.S.); malu_pmg@hotmail.com (M.L.P.M.G.); vieira@uenf.br (C.M.F.V.)

[2] Civil Engineering Laboratory (LECIV), State University of the Northern Rio de Janeiro-UENF, Av. Alberto Lamego, 2000, Campos dos Goytacazes 28013-602, RJ, Brazil; afonso@uenf.br

[3] Department of Materials Science, Instituto Militar de Engenharia–IME, Praça General Tibúrcio, 80, Praia Vermelha, Urca, Rio de Janeiro 22290-270, RJ, Brazil; snevesmonteiro@gmail.com

* Correspondence: gabibarreto.93@gmail.com

Citation: Barreto, G.N.S.; Carvalho, E.A.S.; Souza, V.d.S.d.; Gomes, M.L.P.M.; de Azevedo, A.R.G.; Monteiro, S.N.; Vieira, C.M.F. Engineered Stone Produced with Glass Packaging Waste, Quartz Powder, and Epoxy Resin. *Sustainability* **2022**, *14*, 7227. https://doi.org/10.3390/su14127227

Academic Editor: Mariateresa Lettieri

Received: 31 March 2022
Accepted: 8 June 2022
Published: 13 June 2022

Publisher's Note: MDPI stays neutral with regard to jurisdictional claims in published maps and institutional affiliations.

Abstract: Engineered stone (ENS) is a type of artificial stone composed of stone wastes bonded together by a polymeric matrix. ENS presents a profitable alternative for solid waste management, since its production adds value to the waste by reusing it as raw material and reduces environmental waste disposal. The present work's main goal is to produce an ENS based on quartz powder waste, glass packaging waste, and epoxy resin. The wastes were size-distributed by the fine sieving method. Then, the closest-packed granulometric mixture, as well as the minimum amount of resin that would fill the voids of these mixtures, was calculated. ENS plates were prepared with 15%wt (ENS-15) and 20%wt (ENS-20) epoxy resin by vibration, compression (10 tons for 20 min at 90 °C), and vacuum of 600 mmHg. The plates were sanded and cut for physical, chemical, and mechanical tests. Scanning electron microscopy analysis of fractured specimens was performed. ENS-15 presented 2.26 g/cm^3 density, 0.1% water absorption, 0.21% apparent porosity, and 33.5 MPa bend strength and was resistant to several chemical and staining agents. The results classified ENS as a high-quality coating material, technically and economically viable, with properties similar to commercial artificial stones. Therefore, the development of ENS based on waste glass and quartz powder meets the concept of sustainable development, as this proposed novel material could be marketed as a building material and simultaneously minimize the amount of these wastes that are currently disposed of in landfills.

Keywords: engineered stone; wastes; waste glass; quartz; epoxy resin

1. Introduction

An enormous issue with glass is its one-time use, such as in beverage packaging, squandering a non-biodegradable material that occupies significant space in landfills, reinforcing the reliance on natural resources, and causing damage to the environment [1]. Moreover, dumping glass waste in landfills adds to the disposal issue that is disturbing the biological equilibrium all over the planet [2].

Waste glass, in theory, after sorting and cleaning, could be remelted to manufacture new glass products. However, its variety of colors and contaminations makes recycling laborious, reducing the glass recycling rate. Thereby, waste glass is frequently sent to landfills, contaminating nature, harming the environment, and wasting resources. Around the globe, it is estimated that a whopping amount of glass waste is landfilled each year, approximately 200 million tons, drawing international attention and efforts to develop novel products based on waste glass [3].

Compared to other wastes, such as wood and plastic, waste glass presents chemical stability, which makes its reuse in the manufacture of construction materials of great interest due to some advantages over recycling, such as low energy consumption (as it does not need melting) and simplicity in the treatment of waste [4].

Another material that takes a million years to decompose, generating environmental harm when wrongly disposed of, is quartz powder, composed of 99.0% SiO_2, 0.8% Al_2O_3, 0.1% Fe_2O_3, and 0.1% TiO_2 [5].

Wastes, in general, are nowadays a huge challenge for the industrialized world, since they are usually discarded as useless due to poor technological handling [6]. Improperly disposed of wastes generate a series of disorders, such as destruction of landscape, fauna and flora, air and soil contamination, and populational health risks [7]. For that reason, it is important to propose new sustainable approaches for waste management, representing consistent and effective ways of minimizing these impacts [8].

In that scenario, reusing waste materials in building materials has been attracting research interests, aiming to diminish natural resources wastage, manufacturing costs, and landfilling [9]. With regard to civil construction, waste recycling at any stage would provide sustainable development in this sector, which is considered to be a major waste generator among the macro sectors of the economy, since its consumption of natural resources is approximately 75% [10].

In this sense, the combination of the construction industry with the reuse of waste has become one of the main directions of research [5], with the replacement of raw materials from civil construction with waste products encouraged by the industry, aiming to promote sustainability [11].

Among the civil construction materials, a possible solution to reuse glass and quartz wastes is the development of a novel composite material, called engineered stone (ENS), artificial stones manufactured with a high amount of fine mineral agglomerated polymeric resins [6].

In this way, particulate natural materials of ENS can be substituted by wastes. The wastes are crushed in different granulometries and mixed with resins and additives. This mixture forms a mass that goes through a molding process by means of vibration, compression, and vacuum. The thermosetting resin contained in the paste undergoes a hardening process resulting in the ENS, with dimensions defined by the mold size [12].

Lee and Shin [13] reported on the influence of vacuum, mold temperature, and cooling rate on the production of fiberglass/PET composites. It was observed that specimens preheated without vacuum registered 1.9% porosity, while those made with vacuum registered only 0.3–0.4% porosity. Vacuum-processed samples also achieved better tensile strength values, increasing from 141 Mpa (without vacuum) to 258 Mpa (with vacuum). The authors concluded that the vacuum decreases the porosity, consequently improving the composite's mechanical properties.

Lee et al. [14] studied the effects of compaction pressure, vacuum level, and vibration frequency on the properties of artificial stones produced by vacuum vibrocompression. The authors observed that by increasing the compaction pressure (up to a certain limit), vacuum and vibration diminish porosity and water absorption, enhancing the mechanical properties.

Several studies have proposed the reuse of natural wastes such as marbles, granites, quartzites, and brick waste [14–19] agglomerated by polymeric matrixes, such as epoxy and polyester, to develop engineered stones, producing stones with better mechanical, physical, and chemical properties compared to artificial and natural stones on the market.

This research is focused on producing ENS with quartz powder combined with colorless glass waste from beverage packaging. The decision to use only colorless glass came from the concern with the aesthetic appeal of the final product, since artificial stones with the highest market value are white. The use of glass from beverage packaging had the purpose of finding a sustainable destination for this single-use glass product.

This research's main goal is to confirm the technical feasibility of a novel ENS produced with beverage packaging waste, quartz powder waste, and epoxy resin. The investment in the ENS development could reinsert waste into a manufacturing process along with moving the economy by entering an unexplored and growing industrial area in Brazil [19].

2. Materials and Methods

2.1. Materials

The colorless beverage glass packaging was collected by the authors in bars and supermarkets located in Campos dos Goytacazes, RJ, Brazil. The quartz dust was supplied by EcologicStone, an artificial stone company located in Cachoeiro de Itapemirim, ES, Brazil. The epoxy resin (MC130) of type diglycidyl ether bisphenol A (DGEBA) with 1.15 g/cm^3 density and the hardener (FD 139), composed of triethylenetetramine (TETA) were both supplied by Epoxyfiber.

The size reduction of wastes was achieved by the fine sieving method according to ABNT NBR 7181 [20], in order to obtain three granulometric ranges: coarse (2.000–0.420 mm), medium (0.420–0.075 mm), and fine particles (<0.075 mm). The glass bottles were crushed in a ball mill in order to obtain coarse and medium granulations, and the quartz dust was already fitting the fine granulometry.

2.2. Parameters of Plates Development

To determine the closest-packed mixture, a ternary diagram based on the experimental numeric-modeling grid simplex (simplex-lattice design) (Figure 1), based on the ABNT/MB-3388 Brazilian standard [21], was proposed. Ten (10) different mixtures of the three granulometric ranges were proposed. The proportions are indicated in Figure 1. Each mixture was placed in a 1013.24 cm^3 steel vessel coupled to a 10 kg weight, under 60 Hz of vibration for 10 min.

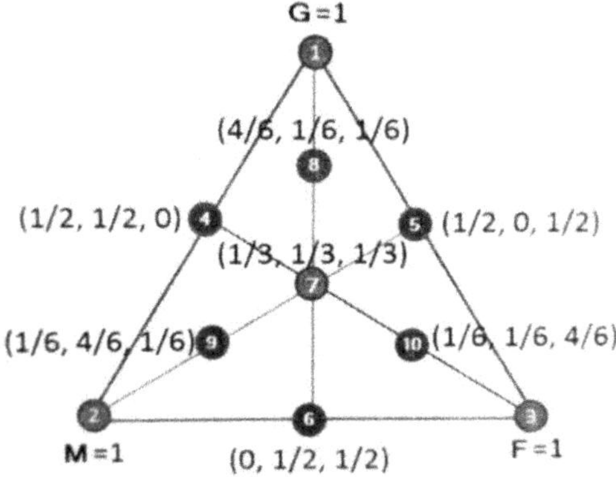

Figure 1. Ternary diagram with the simplex complete cubic model 10 mixtures composition. Proportions of coarse (G), medium (M), and fine (F) particles [7].

The mixture with the greatest vibrated density through packaging was calculated, representing the closest packing of particles, and was chosen for the development of the ENS plates.

The minimum amount of resin (MAR) needed to produce the ENS plates was calculated through the volume void (VV), which is found by Equation (1) [22]:

$$VV\% = \left(1 - \frac{particles\ apparent\ dry\ density}{glass\ density + quartz\ density}\right) \times 100 \qquad (1)$$

The minimum amount of resin (MAR), represented by the amount of resin enough to fill these volume void, were then calculated through Equation (2) [22]:

$$\text{MAR\%} = \frac{\text{VV\%} \times \rho_{resin}}{\text{VV\%} \times \rho_{resin} + (100 - \text{VV\%}) \times \rho_{(glass+quartz)}} \tag{2}$$

2.3. Engineered Stone Plate Development

The ENS plates, with $100 \times 100 \times 10$ mm size dimensions, were produced by the vacuum, vibration, and compression method. The raw materials were dried in an oven for 24 h at 100 °C, weighed in the appropriate proportions, and mixed with both the epoxy resin and hardener. The mixture was placed in a mold connected to a vacuum system (600 mmHg), which was on a vibrating table for 2 min. The mold, still under vacuum, was positioned in a hydraulic press, where it was compressed for 20 min under 90 °C heating. The ENS plates were cooled to room temperature and removed from the mold. The plates were sanded and then cut with a diamond disc in dimensions required by the standards to perform the physical, chemical, and mechanical tests. The epoxy contents in the ENS were 15%wt (ENS-15) and 20%wt (ENS-20).

2.4. Characterization of Engineered Stone Plates

To perform the physical indices test, 18 specimens of $30 \times 30 \times 30$ mm were cut from the ENS plates. The apparent density, water absorption, and apparent porosity were determined in accordance with Annex B of the ABNT/NBR 155845-2 [23], responsible for classifying stone materials for applications such as coatings in civil construction.

The dilatometry test aims to identify dimensional variations of the specimen under heating conditions. The test was performed on the Netzsch DIL402PC equipment under a heating rate of 10 °C/min and a temperature range of 30–1000 °C.

The 3-point bend test was performed according to the Annex F of ABNT/NBR 155845-6 [24] Brazilian standard and UNE-EN 14617-2 Spanish standard [25]. Six (6) specimens of $10 \times 25 \times 100$ mm were tested on the Instron branded universal mechanical testing machine, model 5582, under a 0.25 mm/min loading rate, 100 KN load cell, and 80 mm two-point distance, to evaluate the maximum bending stress, as well as the standard deviation.

Abrasive wear tests were performed according to the ABNT/NBR 12042 Brazilian standard [26], using 2 samples of $70 \times 70 \times 30$ mm, which had their thicknesses measured before and after abrasive wearing suffered in 500 and 1000 m track in a MAQTEST Amsler equipment.

The chemical attack resistance was determined according to an adaptation of Annex H of the standard ABNT/NBR ISO 10545-13 [27]. The attack was performed on 16 specimens with dimensions of $50 \times 50 \times 10$ mm, representing 4 specimens for each one of the 4 attack agents. The specimens were under chemical attack by ammonium chloride and citric acid for 24 h and by hydrochloric acid and potassium hydroxide for 96 h.

The stain resistance was performed on ENS-15 according to an adaptation of the guidelines described in Annex G of ABNT/NBR ISO 10545-14 [28], in addition to the staining agents already listed in the standard, namely penetrating agents (Cr_2O_3 green and Fe_2O_3 red), the oxidizing agent (iodine), and the film-forming agent (olive oil). Common household products were also used: grape juice, ketchup, mustard, and coffee. For each staining agent, 5 polished ENS specimens of approximately $40 \times 20 \times 10$ mm had half of their face covered with the staining agents for 24 h, and the material was then classified according to ease of stain removal.

The salt crystallization test was performed on ENS-15 according to an adaptation of the guidelines described in ABNT/NBR 8.094 [29]. Five 50×50 mm specimens were prepared, washed, and dried in an oven at 70 °C for 24 h. The specimens were identified, weighed, measured, and subsequently submerged in a sodium chloride solution for 6 h and oven-dried for 18 h, totaling a 24 h cycle. The process was repeated until 50 cycles were completed.

In the 10th, 30th, and 50th cycles, the samples were analyzed using the Olympus LEXT OLS400 confocal microscope to observe possible surface changes due to exposure to the sodium chloride solution.

The ENS pore characteristics before and after the salt crystallization test were obtained in an automatic mercury injection porosimeter, model Autopore IV 9500, from Micromeritics Instrument Company of America, in 6 × 6 × 6 mm samples. Mercury intrusion and extrusion were investigated under pressures between 0 and 33,000 Psi, which equates to approximately 228 MPa, with pore diameters reading between 0.005 µm and 360 µm.

The ENS fractured surface regions subjected to bending tests were observed through SHIMADZU's Super Scan SSX-550 scanning electron microscope (SEM) for microstructural analysis, at 20 kV of secondary electrons. The samples were prepared using an adhesive carbon tape enveloped by a gold surf. The microstructural analysis is important to determine the quality of the adhesion between the particles and the epoxy resin, as well as the presence of voids.

3. Results and Discussion

3.1. Development Parameters

The vibration densities of the 10 mixtures proposed in Figure 1 are displayed in Table 1.

Table 1. Compositions and vacuum and vibration density of the 10 mixtures proposed for ENS development.

Mixture	Density (g/cm^3)
1	1.253 ± 0.03
2	1.250 ± 0.02
3	1.232 ± 0.03
4	1.510 ± 0.01
5	1.418 ± 0.02
6	1.563 ± 0.01
7	1.676 ± 0.03
8	1.586 ± 0.03
9	1.439 ± 0.02
10	1.625 ± 0.04

As shown in Table 1, the mixture that presented the greatest vibration density was Mixture 7 (1/3 coarse, 1/3 medium, and 1/3 fine particles), representing the closest-packed mixture. Therefore, Mixture 7 was chosen as a proportion of particle sizes for the ENS development.

The MAR was calculated through Equation (2) as 15%wt for the epoxy resin. This amount of resin is considered to be the minimum amount of matrix necessary to fill both interstitial and surface voids of the closest-packed mixture.

The ENS plates were produced with 1/3 of coarse particles of glass waste, 1/3 of medium particles of glass waste, and 1/3 of fine particles of quartz waste. Plates were produced with an amount of 15%wt epoxy resin (ENS-15), corresponding to the MAR, and also with 20%wt (ENS-20) of epoxy resin, aiming to verify whether the addition of resin would result in a more effective void filling that would be determined by its mechanical properties.

3.2. Physical Indices

Table 2 presents the physical indices (apparent density, water absorption, and apparent porosity) results found after testing 18 specimens of both ENS-15 and ENS-20.

Table 2. Physical properties of apparent density, water absorption, and apparent porosity of ENS-15 and ENS-20.

Engineered Stone	Density (g/cm^3)	Water Absorption (%)	Apparent Porosity (%)
ENS-15	2.26 ± 0.01	0.10 ± 0.01	0.21 ± 0.03
ENS-20	2.14 ± 0.01	0.12 ± 0.01	0.25 ± 0.03

As can be seen in Table 2, by modifying the amount of resin, a difference in the physical indices was observed. ENS-20 had density values lower than ENS-15, which is due to the fact that the resin content increased, hence a decrease in the amount of waste. The rise in water absorption and porosity values observed in ENS-20 is explained by the resin binding action that brings the particles together. More resin means fewer waste particles in the mixture, forming more empty spaces, called voids, increasing both apparent porosity and water absorption.

Lee et al. [14] developed engineered stone based on glass and granite wastes under different conditions of pressure, temperature, and vacuum. In their work, the results ranged between 2.03 and 2.45 g/cm^3 for density and from 0.01 to 0.20% for water absorption. By observing Table 2, one can notice that the density and water absorption values found for both ENS-15 and ENS-20RAVQ 20% are within the range found by Lee et al.

A commercial stone named Stellar™, also produced by EcologicStone, the same company that supplied the quartz dust for this work, was characterized in the work of Carvalho et al. [15], presenting 2.38 g/m^3 density, 0.18% water absorption, and 0.44% apparent porosity. Both ENS-15 and ENS-20 are relatively lower than the commercial stone, with similar porosity and water absorption, indicating that the produced rock can be an alternative to replace the commercial Stellar™ stone, not only reusing the wastes but also with structure and logistic savings because of the lower density.

According to Chiodi and Rodrigues [30], lining or facing civil construction materials classified as high quality must have porosity below 0.5%. As can be seen in Table 2, the porosity values found for both ENS-15 and ENS-20 are below this value, evidencing the high quality of both engineered stones produced for coating applications.

As ENS-15 showed better physical indices, it was chosen to be tested for its mechanical and thermal properties.

3.3. Dilatometry Test

Numerous materials present dimensional variations after being exposed to a heating period, presenting dimensional variations under the effect of the contraction and expansion phenomena in their structures. These variations take place due to physical and chemical events that occur in the structures and chemical composition of their materials. In the development of engineered stone, thermal transformations occur in the curing stage because of the matrix/waste polymerization reaction [31].

The ENS, composed of glass and quartz in the epoxy matrix, had excellent test stability and resistance, probably due to the aggregate's properties and characteristics at high temperatures. Figure 2 presents the ENS-15 dilatometry analysis.

In Figure 2, one can observe the structural behavior of ENS-15 in a heated environment in a range of temperature from 0° to 1000 °C, identifying the structural phenomena presented such as the body temperature at the exact moment.

In the range of 0–300 °C, it is possible to see the adaptation of materials to the heated environment, favoring pore expansion and the structure's moisture loss, in addition to a molecular chemical activation, which causes movement expanding the entire structure. Owing to the adaptation, at 338 °C, the structure contracts, which is followed by an expansion at 362.6 °C and a significant decline at 382.4 °C, a temperature where the resin begins to decompose, with a weight loss demonstrated in previous works [17].

According to Silva et al. [17], the epoxy resin thermogravimetry curve registers the first 70% weight loss at about 380 °C due to the resin's decomposition. Subsequently, at 382.4 °C, there is a weight loss that continues until the end of the process, at 550 °C [17]. After 600 °C, the structure undergoes another contraction, and there is a total weight loss because of the resin and the melting temperature of the waste glass, which is around 600–1117 °C, and of the quartz dust, which, at around 870 °C, becomes tridymite, a high-temperature silica polymorph [17,31].

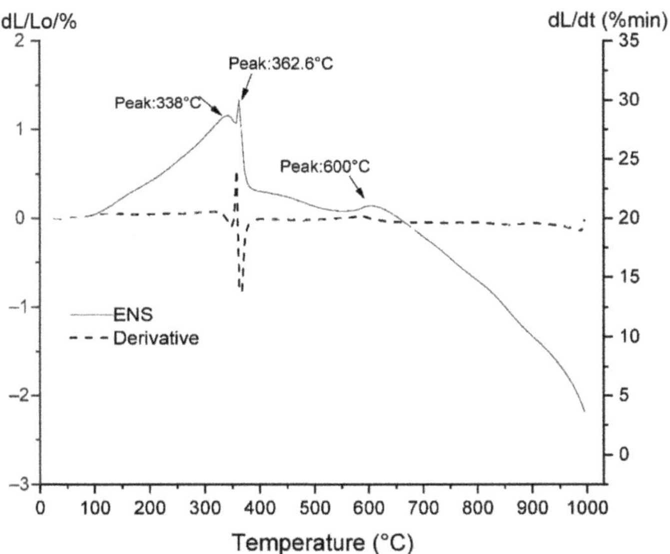

Figure 2. Dilatometry curve of ENS-15.

3.4. Three-Point Bend Strength

Table 3 presents the three-point bend strength values of ENS-15 and pure epoxy resin.

Table 3. Three-point bend strength of ENS-15 and epoxy resin.

Materials	3-Point Bend Strength (MPa)
ENS-15	33.54 ± 4.05
Epoxy	93.6 ± 4.70

Figure 3 shows the stress vs. strain curves obtained by the three-point bending test of ENS-15 developed and of the pure epoxy resin. Mechanical strength, or the stress at which the material breaks, is the most important property for structural materials.

The mechanical bending resistance of ENS-15 was 33.5 ± 4.0 MPa, with a strain ranging from 0.0 to 0.3.

When comparing the ENS bending strength results of those found by Carvalho et al. [15] for Stellar™ (36.61 ± 2.48 MPa) with those also found by Carvalho et al. [6], artificial stone developed with quarry waste and 15% epoxy resin (32 ± 2 MPa), and by Gomes et al. [16], AOS artificial stone developed with brick waste and quarry dust with 20% epoxy resin (30 ± 3 MPa), it is noticeable that the values for ENS-15 are in accordance with those found by other authors for similar materials.

According to Chiodi Filho and Rodriguez [30], ornamental stones used as coatings in civil construction, with bending strength above 20 MPa, are classified as highly resistant materials. As ENS-15 presented flexural strength above this value, it can be classified as a high-strength engineered stone, enabling it to be used on tops and countertops without restriction, as well as in internal and external walls.

The GSO ASTM C503/2015 [32] standard specifies that calcitic marble bend strength must be greater than 7 MPa. The bend strength values found by the ENS developed in this work represent about five times this value. Therefore, it is possible to affirm that the material developed represents a formidable alternative to replace calcitic marble, which could result in material savings, since the greater resistance would make it possible to produce the same materials with smaller dimensions.

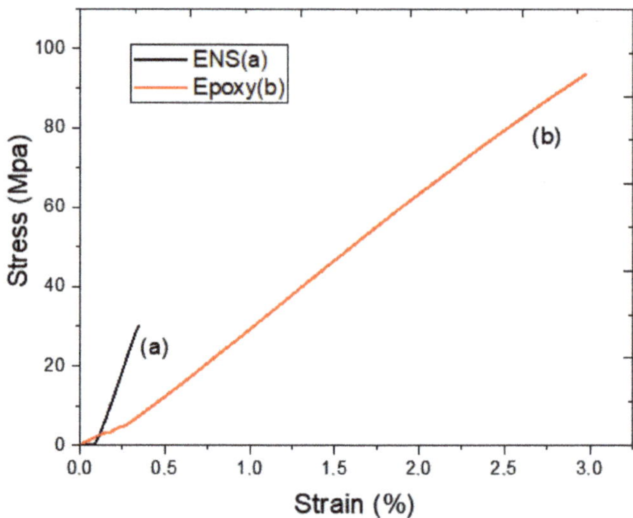

Figure 3. Mechanical behavior of ENS-15 and the pure epoxy resin in 3-point bend strength tests.

3.5. Abrasive Wear Test

Table 4 presents the thickness reduction of ENS-15 after two specimens were subjected to abrasive wear tests at two running distances, 500 and 1000 m.

Table 4. Amsler abrasive wear through with thickness reduction of ENS-15.

Material	Thickness Reduction (mm)	
	500 m	1000 m
ENS-15	1.41 ± 0.22	2.86 ± 0.25

For artificial stones to be applied for civil construction, there are no requirements, standards, or limits to thickness reduction in abrasive wear tests. Chiodi Filho and Rodrigues [30] proposed technological parameters for the use of artificial stones in pavements. The authors recommended that pavements with low traffic should display wear thickness reduction lower than 6 mm, medium traffic should be lower than 3 mm, and heavy traffic lower than 1.5 mm. Based on the described parameters, the novel ENS-15 (Table 4) can be used for medium traffic pavement, since it presented a thickness reduction below 3 mm on the 1000 m track.

3.6. Chemical Attack Resistance

Table 5 presents the relative weight loss (%) suffered by the specimens ENS-15 and ENS-20 after 24 h or 96 h under chemical attack. The reagents simulate the composition of some substances, such as cleaning products and food or atmospheric acids, that would possibly be in contact with the engineered stone if applied as a coating.

Table 5. Relative weight loss after 24 h/96 h of chemical attack of ENS-15 specimens.

Reagents	Relative Weight Loss (%)	Time Span
NH_4Cl	0.07 ± 0.01	24 h
$C_6H_8O_7$ (citric acid)	0.18 ± 0.04	24 h
HCl	0.41 ± 0.02	96 h
KOH	0.08 ± 0.01	96 h

After observing the results presented in Table 5, it can be seen that the ENS produced had a smaller weight loss when attacked by basic reagents than when attacked by acidic ones, probably due to the matrix, since epoxy resin has a high resistance to chemical agents, especially alkali [33].

ENS-15 had a small relative weight loss in the face of chemical attacks, which can be attributed to the fact that the glass, quartz, and epoxy resin are materials that present certain chemical stability. Possibly, the glass, as its original application was beverage packaging, had a more chemically inert composition. Quartz at room temperature presents a stable phase of silica and is also, by nature, a chemically stable material [30].

Gomes et al. [16] evaluated the behavior of their AOS artificial stone also agglomerated by epoxy resin under chemical attack with the same agents used in the present work. However, the authors evaluated chemical attack resistance for all reagents for only 24 h, while in this present work, some chemical attack reagents acted for 96 h according to the standard [32]. Still, the reagent that caused the greatest weight loss was HCl, corroborating the conclusions of Gomes et al. [16].

3.7. Stain Resistance

Figure 4 presents a graph illustrating the stain resistance behavior of ENS-15 according to a scale of the difficulty degree in removing the stains after the cleaning procedures described in the standard [28]. In this scale, Number 5 represents the easiest stain removal, and Number 1 represents the harder stain removal. In fact, a stain at Number 1 is one that remains even after all the cleaning procedures.

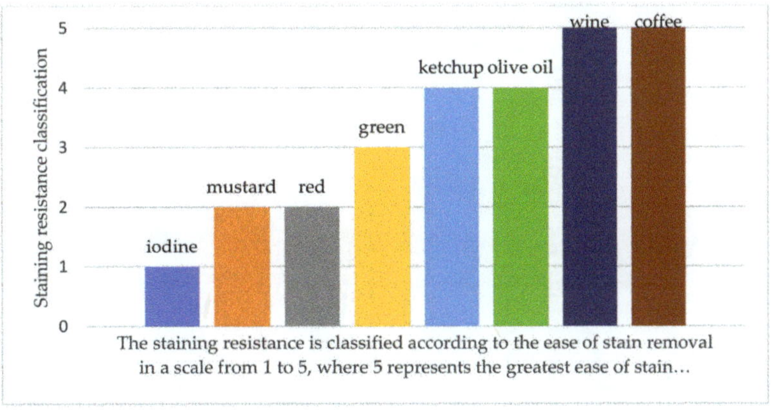

Figure 4. Stain resistance behavior of ENS-15 after 24 h covered with the staining agents.

ENS-15 showed high stain resistance for agents such as wine and coffee, which were cleaned only with hot water. Olive oil and ketchup stains were easily removed by detergent, and, after cleaning with an abrasive paste, the green agent (Cr_2O_3) stain was also removed. For cleaning the other stains left by red (Fe_2O_3), mustard, and iodine, the specimens were submerged in acidic solution for 24 h. ENS-15 did not show iodine stain resistance because this stain remained at Number 1, as shown in Figure 4.

Peixoto et al. [15,22] also developed an artificial stone with the incorporation of glass waste, but with another type of glass from the industrial lamination stage. They evaluated its stain resistance with the same method and same staining agents, but no stain remained after the test.

Borsellino et al. [34] also evaluated the stain resistance of their artificial marbles developed with marble and epoxy resin and also with polyester resin. However, they used different staining agents and another method. It was pointed out by the authors that plates with more epoxy resin content suffered fewer surface alterations after the stain resistance test, which was attributed to the higher epoxy resin stain resistance.

Stain resistance is kindred to surface pores because, after polishing the stone surface, the internal pores become surface pores, also known as open pores, which hinder the removal of stains [22].

The stain resistance test is important to evaluate the quality of engineered stones because of its application, which takes aesthetics into account, and because of its use in kitchens, bathrooms, and laboratory countertops, which constantly expose ENS to staining substances.

3.8. Salt Crystallization Test

Table 6 shows the ENS weight before the test and its weight after the salt crystallization cycles. In each cycle, the samples were submerged in saline water and taken to an oven for drying, respecting their respective times. It can be observed that there was a small weight variation during the cycles, which indicates little absorption of salinized water, evidencing the low porosity shown in the physical index.

Table 6. Weight variation after 10, 30, and 50 cycles of ENS salt crystallization test.

Test Conditions	Weight Variation (g)	Weight Variation (%)
10 cycles	0.02 ± 0.01	+0.04 ± 0.02
30 cycles	0.06 ± 0.03	+0.09 ± 0.05
50 cycles	0.50 ± 0.30	+0.87 ± 0.54

Figure 5 shows the engineered stone's pore distribution before and after the salt crystallization test.

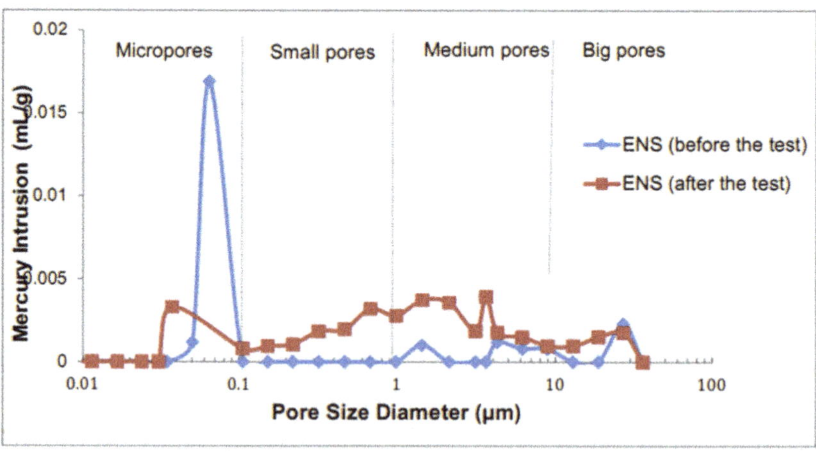

Figure 5. Pore diameter distribution of ENS before and after the salt crystallization test.

The ENS had four pore diameter ranges after the test, large, medium, small, and micropores, whereas the intact ENS had no range of small pores. Mercury intrusion and volume increased for ENS in the micropore range, probably due to sample failures. The intact ENS obtained a 0.005 mL/g mercury intrusion volume and 0.89% porosity, and the ENS after the test had 0.0064 mL/g mercury intrusion and 1.29% porosity, indicating the appearance of new pores.

The diameter of pores and the mercury intrusion volume increased in all diameter ranges. The elevation in mercury intrusion and pore diameter can be attributed to temperature variations and the crystallization of salts in the pores that expand, producing tensions that enlarge their diameter.

The studies by Benavente et al. [35] and Yu and Oguchi [36] reported that saline solutions could come from external sources, such as natural salts or anthropogenic discharges, such as marine spray, pesticides, and wastes from fossil fuel combustion. The exposure of artificial stones to these external saline solutions could induce a capillarity and porosity increase, consequently impairing the material's mechanical properties.

The literature also reports an enormous discrepancy between the limited damage observed in the laboratory and the severe deterioration that happens when artificial stones are contaminated with NaCl from external sources. When considering the pore damage due to salt crystallization, the lower destructive potential of NaCl in the laboratory could be attributed to its tendency to supersaturate [37], which can be observed in Table 6, as, despite repetitive cycles, there was no considerable weight variation in the tested specimens.

Flat and three-dimensional micrographs of the ENS before and after the test were obtained with confocal microscopy during cycles of 10, 30, and 50, as shown in Figures 6 and 7.

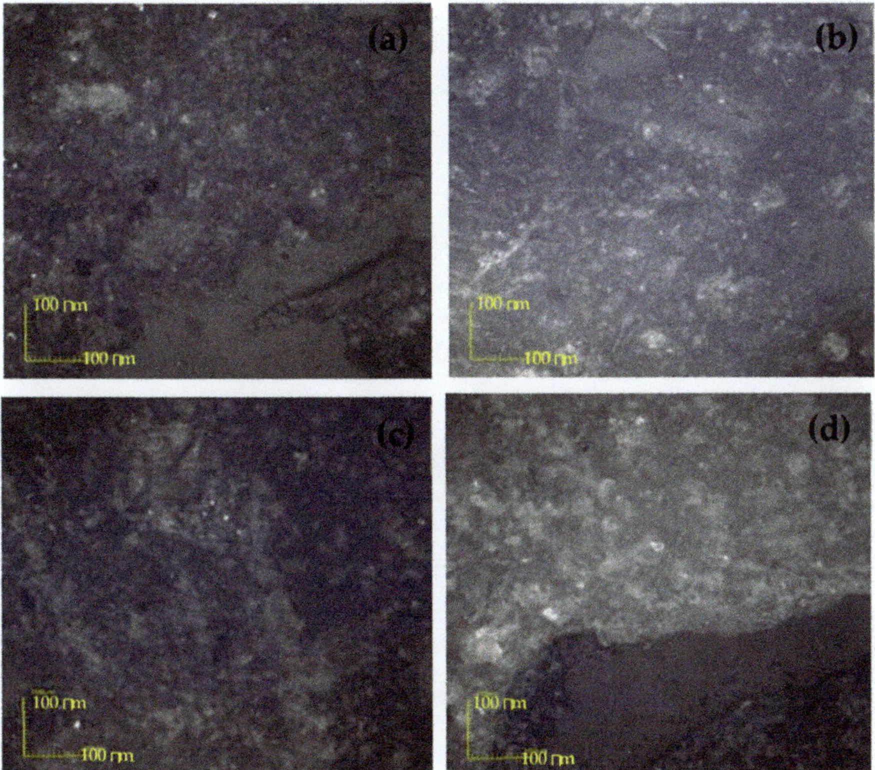

Figure 6. Photomicrograph of ENS under the salt crystallization test: ENS before test (**a**), after 10 cycles (**b**), after 30 cycles (**c**), and after 50 cycles (**d**), with 430× magnification.

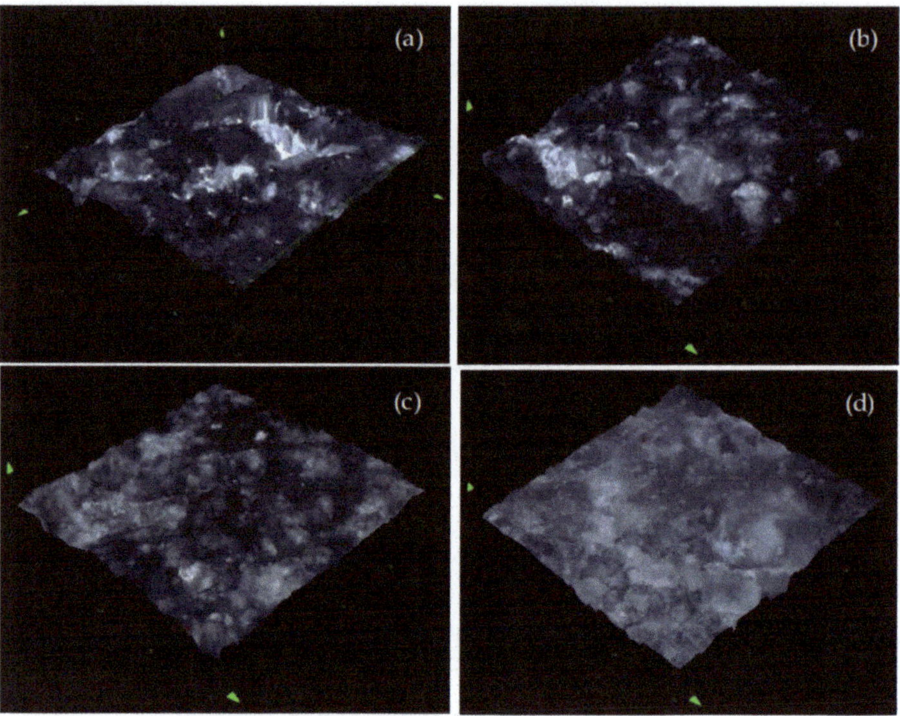

Figure 7. Three-dimensional photomicrograph of ENS under the salt crystallization test: ENS before test (**a**), after 10 cycles (**b**), after 30 cycles (**c**), and after 50 cycles (**d**), with 430× magnification.

As displayed in Figure 6, ENS after 50 cycles (Figure 6d) underwent a greater change on the surface, evidenced by the more whitish and yellowish color. The accelerated degradation in the laboratory brings the brightness loss, yellowing, porosity, microcracks, and the mineral alteration state caused by the joint action of factors such as temperature and sodium presence (saline water). This leads to salt crystallization in the exposed pores, generating tensions responsible for increasing the diameter of pores and physical origin surface changes [38]. At 10 and 30 cycles, there was no significant surface change.

The three-dimensional photomicrographs (Figure 7) show that the ENS after 30 cycles (Figure 7c) had a greater pore or cavity number, being more pronounced in the ENS after 50 cycles (Figure 7d), indicating the loss of surface cohesion as the number of cycles rise, leading to the detachment of the material in the form of dust and/or small particles [38].

3.9. SEM Observations

Figure 8a,b displays the ENS micrographs of surface fracture obtained by SEM.

In Figure 8, it is possible to notice that the presence of pores or cavities in the material (white arrows) is quite reduced, which was already expected due to the low porosity and water absorption and to the manufacturing process that uses vibration, compaction, and vacuum.

The occurrence of displaced particles (Figure 8a) of the matrix can also be observed, as indicated by the arrow. The voids in the interface region point to this displacement. There was little displacement of waste grains (Figure 8a,b), confirming the efficient wetting [17].

The compaction enhances the settlement and adhesion of particles, and the vacuum assists the resin to better penetrate the interstices of the waste particles, filling the void volume [3], which is clearly seen in the micrograph, by a good interfacial adhesion between the waste particles and the resin. The good interfacial interaction is evidence of effective interfacial wetting of the resin, which is directly related to the improvement of the material's mechanical properties, as reported by Miller et al. [39] and Debnath et al. [40]. Figure 8a,b both show intergranular fracture surfaces with evidence of mechanical stress failure, since the fracture cracks pass through the grains indicated by the arrow.

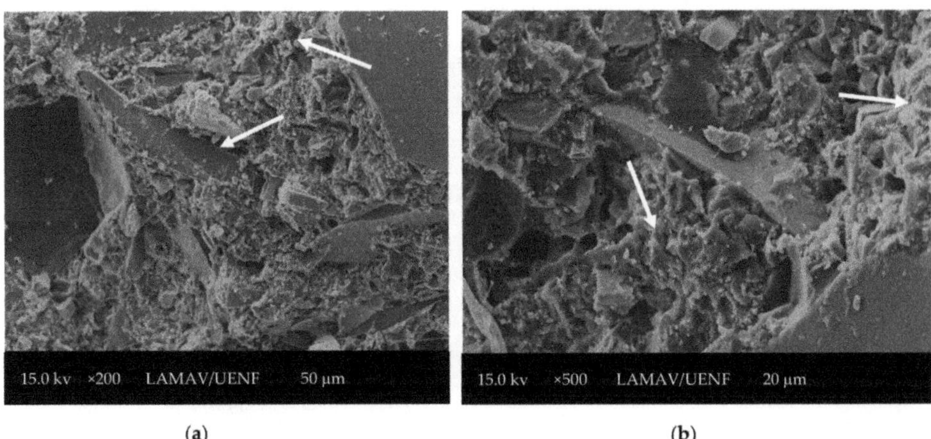

(a)   (b)

Figure 8. ENS-15 scanning electron fractured surface micrographs with different magnifications: (**a**) 200× and (**b**) 500×.

4. Conclusions

This work aimed to develop an ENS based on glass packaging waste, quartz dust, and epoxy resin (ENS-15 and ENS-20) through vacuum vibration and compression, as well as to determine its physical and mechanical properties.

From the experimental results obtained in this study, the following conclusions can be drawn:

- ENS-15 and ENS-20 are materials of low density, with 2.26 and 2.24 g/cm^3, respectively. The water absorption was 0.10 and 0.12%, considering both as low absorption ENS materials;
- Both ENS-15 and 20 are high-quality coating materials for lining civil construction due to their porosity below 0.5%;
- ENS-15 is a highly resistant material, owing to its 33.5 MPa bend strength, which exceeds the recommended value of at least 20 MPa for construction coating materials;
- ENS with a 2.86 mm thickness reduction on the 1000 m track can be used for pavement on floors with medium traffic, since it presented thickness reduction below the recommended value for this type of material;
- The ENS displayed good chemical stability, proved by its low relative weight loss after the chemical attack test.
- The ENS is likely to be used as a coating in kitchens and bathrooms, since it is highly resistant to some staining agents except iodine;
- SEM micrographs showed the low presence of pores and an optimal waste/resin interfacial adhesion, corroborating the physical indices and bend strength results.
- Therefore, the development of ENS based on glass, quartz dust, and epoxy resin has the potential to be used as an alternative material, competing with both natural and commercial artificial stones, which are currently marketed in civil construction, leading to environmental and economical improvements.

Author Contributions: Conceptualization, C.M.F.V. and E.A.S.C.; methodology, G.N.S.B., C.M.F.V., and E.A.S.C.; Validation, G.N.S.B., V.d.S.d.S., M.L.P.M.G., and E.A.S.C.; investigation, G.N.S.B. and V.d.S.d.S.; data curation, G.N.S.B.; writing—original draft preparation, G.N.S.B.; writing—review and editing, G.N.S.B., V.d.S.d.S., and E.A.S.C.; supervision, E.A.S.C. and A.R.G.d.A.; project administration, S.N.M. and C.M.F.V. funding acquisition, S.N.M., A.R.G.d.A., and C.M.F.V. All authors have read and agreed to the published version of the manuscript.

Funding: This research was funded by FAPERJ (Grant Number: E-26/200.847/2021) and CNPQ (Grant Number: 301634/2018-1). The participation of A.R.G.A. was sponsored by FAPERJ through the research fellowships Proc. Nos.: E-26/210.150/2019, E-26/211.194/2021, E-26/211.293/2021, and an-dE-26/201.310/2021, and by CNPq through the research fellowship PQ2 307592/2021-9.

Institutional Review Board Statement: Not applicable.

Informed Consent Statement: Not applicable.

Data Availability Statement: Not applicable.

Conflicts of Interest: The authors declare no conflict of interest.

References

1. Harrison, E.; Berenjian, A.; Seifan, M. Recycling of waste glass as aggregate in cement-based materials. *Environ. Sci. Ecotechnol.* **2020**, *4*, 100064. [CrossRef]
2. Bisht, K.; Syed Ahmed Kabeer, K.I.; Ramana, P.V. Gainful utilization of waste glass for production of sulphuric acid resistance concrete. *Constr. Build. Mater.* **2020**, *235*, 117486. [CrossRef]
3. Dong, W.; Li, W.; Tao, Z. A comprehensive review on performance of cementitious and geopolymeric concretes with recycled waste glass as powder, sand or cullet. *Resour. Conserv. Recycl.* **2021**, *172*, 105664. [CrossRef]
4. Guo, P.; Meng, W.; Nassif, H.; Gou, H.; Bao, Y. New Perspectives on Recycling Waste Glass in Manufacturing Concrete for Sustainable Civil Infrastructure. *Constr. Build. Mater.* **2020**, *257*, 119579. [CrossRef]
5. Sadowski, L.; Piechówka-Mielnik, M.; Widziszowski, T.; Gardynik, A.; Mackiewicz, S. Hybrid ultrasonic-neural prediction of the compressive strength of environmentally friendly concrete screeds with high volume of waste quartz mineral dust. *J. Clean. Prod.* **2019**, *212*, 727–740. [CrossRef]
6. Santos Carvalho, E.A.; de Figueiredo Vilela, N.; Monteiro, S.N.; Fontes Vieira, C.M.; da Silva, L.C. Novel Artificial Ornamental Stone Developed with Quarry Waste in Epoxy Composite. *Mater. Res.* **2018**, *21* (Suppl. S1), e20171104. [CrossRef]
7. Barreto, G.N.S.; Babisk, M.P.; Delaqua, G.C.G.; Gadioli, M.C.B.; Vieira, C.M.F. Evaluation on the Effect of the Incorporation of Blends of Fuel and Fluxing Wastes. *Mater. Res.* **2019**, *22* (Suppl. S1), e20190199. [CrossRef]
8. Adediran, A.; Lemougna, P.N.; Yliniemi, J.; Tanskanen, P.; Kinnunen, P.; Roning, J.; Illikainen, M. Recycling glass wool as a fluxing agent in the production of clay- and waste-based ceramics. *J. Clean. Prod.* **2021**, *289*, 125673. [CrossRef]
9. Gao, X.; Yu, Q.; Li, X.S.; Yuan, Y. Assessing the modification efficiency of waste glass powder in hydraulic construction materials. *Constr. Build. Mater.* **2020**, *263*, 120111. [CrossRef]
10. Cordon, H.C.F.; Cagnoni, F.C.; Ferreira, F.F. Comparison of physical and mechanical properties of civil construction plaster and recycled waste gypsum from São Paulo, Brazil. *J. Build. Eng.* **2019**, *22*, 504–512. [CrossRef]
11. Kumar, S.; Gupta, R.C.; Shrivastava, S. Long term studies on the utilization of quartz sandstone wastes in cement concrete. *J. Clean. Prod.* **2017**, *143*, 634–642. [CrossRef]
12. Lameiras, F.S.; da Silva, G.D.L. Methodology to control the influence of processing factors during composite stone fabrication. *Rem Rev. Esc. Minas* **2015**, *68*, 69–75. [CrossRef]
13. Lee, D.J.; Shin, I.J. Effects of vacuum, mold temperature and cooling rate on mechanical properties of press consolidated glass fiber/PET composite. *Compos. Part A Appl. Sci. Manuf.* **2002**, *33*, 1107–1114. [CrossRef]
14. Lee, M.-Y.; Ko, C.-H.; Chang, F.-C.; Lo, S.-L.; Lin, J.-D.; Shan, M.-Y.; Lee, J.-C. Artificial stone slab production using waste glass, stone fragments and vacuum vibratory compaction. *Cem. Concr. Compos.* **2008**, *30*, 583–587. [CrossRef]
15. Carvalho, E.A.S.; Marques, V.R.; Rodrigues, R.J.S.; Ribeiro, C.E.G.; Monteiro, S.N.; Vieira, C.M.F. Development of Epoxy Matrix Artificial Stone Incorporated with Sintering Residue from Steelmaking Industry. *Mater. Res.* **2015**, *18* (Suppl. S2), 235–239. [CrossRef]
16. Gomes, M.L.P.M.; Carvalho, E.A.S.; Sobrinho, L.N.; Monteiro, S.N.; Rodriguez, R.J.S.; Vieira, C.M.F. Production and characterization of a novel artificial stone using brick residue and quarry dust in epoxy matrix. *J. Mater. Res. Technol.* **2018**, *7*, 492–498. [CrossRef]
17. Silva, F.S.; Ribeiro, C.E.G.; Rodriguez, R.J.S. Physical and Mechanical Characterization of Artificial Stone with Marble Calcite Waste and Epoxy Resin. *Mater. Res.* **2017**, *21*, e20160377. [CrossRef]
18. Da Cunha Demartini, T.J.; Rodríguez, R.J.S.; Silva, F.S. Physical and mechanical evaluation of artificial marble produced with dolomitic marble residue processed by diamond-plated bladed gang-saws. *J. Mater. Res. Technol.* **2018**, *7*, 308–313. [CrossRef]

19. Gomes, M.L.P.; Carvalho, E.A.; Demartini, T.J.; de Carvalho, E.A.; Colorado, H.A.; Vieira, C.M.F. Mechanical and physical investigation of an artificial stone produced with granite residue and epoxy resin. *J. Compos. Mater.* **2021**, *55*, 1247–1254. [CrossRef]
20. Brazilian Association of Technical Norms—ABNT. Soil—Grain Size Analysis ABNT NBR 7181. September 2016. Available online: https://www.abntcatalogo.com.br/ (accessed on 20 March 2022).
21. Brazilian Association of Technical Norms—ABNT. NBR MB 3388: Soil—Determination of Minimum Index Void Ratio of Cohesionless Soils—Method of Test. Rio de Janeiro. 1991. Available online: https://www.abntcatalogo.com.br/ (accessed on 20 March 2022).
22. Peixoto, J.; Santos Carvalho, E.A.; Menezes Gomes, M.L.P.; da Silva Guimarães, R.; Monteiro, S.N.; de Azevedo, A.R.G.; Fontes Vieira, C.M. Incorporation of Industrial Glass Waste into Polymeric Resin to Develop Artificial Stones for Civil Construction. *Arab. J. Sci. Eng.* **2022**, *47*, 4313–4322. [CrossRef]
23. Brazilian Association of Technical Norms—ABNT. ABNT NBR15845-2: Rocks for Cladding. Part 2: Determination of Bulk Density, Apparent Porosity and Water Absorption. August 2015. Available online: https://www.abntcatalogo.com.br/ (accessed on 20 March 2022).
24. Brazilian Association of Technical Norms—ABNT. ABNT NBR15845-6: Rocks for Cladding. Part 6: Determination of Modulus of Rupture (Three Point Bending). July 2015. Available online: https://www.abntcatalogo.com.br/ (accessed on 20 March 2022).
25. Spanish Association of Standards and Certification. *UNE-EN14617-2-08*; Test Methods. Part 2: Determination of the Flexural Strength; Spanish Association of Standards and Certification: Madrid, Spain, 2008.
26. Brazilian Association of Technical Norms—ABNT. ABNT NBR12042: Inorganic Materials—Determination of the Resistance to Abrasion. December 2012. Available online: https://www.abntcatalogo.com.br/ (accessed on 20 March 2022).
27. Brazilian Association of Technical Norms—ABNT. ABNT NBRISO10545-13: Ceramic Tiles—Part 13: Determination of Chemical Resistance. July 2020. Available online: https://www.abntcatalogo.com.br/ (accessed on 20 March 2022).
28. Brazilian Association of Technical Norms—ABNT. ABNT NBRISO10545-14: Ceramic Tiles—Part 14: Determination of Resistance to Stains. November 2017. Available online: https://www.abntcatalogo.com.br/ (accessed on 20 March 2022).
29. Brazilian Association of Technical Norms—ABNT. ABNT NBR 8.094: Coated and Uncoated Metallic Material—Corrosion from Salt Fog Exposure—Test Method. 1983. Available online: https://www.abntcatalogo.com.br/ (accessed on 20 March 2022).
30. Chiodi Filho, C.; de Paula Rodrigues, E. *Guide Application of Stone Coverings*; ABIROCHAS: São Paulo, Brazil, 2009.
31. Carvalho, E.A.S.; da Silva de Souza, V.; Barreto, G.N.S.; Monteiro, S.N.; Rodriguez, R.J.S.; Vieira, C.M.F. Incorporation of Porcelain Powder and Mineral Wastes in Epoxy Matrix for Artificial Stone Purchase. In *Characterization of Minerals, Metals, and Materials*; Springer: Cham, Switzerland, 2021; pp. 435–443. [CrossRef]
32. American Society for Testing and Materials—ASTM. *ASTM C503-03*; Standard Specification for Marble Dimension Stone (Exterior); ASTM: West Conshohocken, PA, USA, 2003.
33. Tesser, E.; Lazzarini, L.; Bracci, S. Investigation on the chemical structure and ageing transformations of the cycloaliphatic epoxy resin EP2101 used as stone consolidant. *J. Cult. Herit.* **2018**, *31*, 72–82. [CrossRef]
34. Borsellino, C.; Calabrese, L.; di Bella, G. Effects of powder concentration and type of resin on the performance of marble composite structures. *Constr. Build. Mater.* **2009**, *23*, 1915–1921. [CrossRef]
35. Benavente, D.; García del Cura, M.A.; Fort, R.; Ordóñez, S. Durability estimation of porous building stones from pore structure and strength. *Eng. Geol.* **2004**, *74*, 113–127. [CrossRef]
36. Yu, S.; Oguchi, C.T. Role of pore size distribution in salt uptake, damage, and predicting salt susceptibility of eight types of Japanese building stones. *Eng. Geol.* **2010**, *115*, 226–236. [CrossRef]
37. Lubelli, B.; Cnudde, V.; Diaz-Goncalves, T.; Franzoni, E.; van Hees, R.P.J.; Ioannou, I.; Menendez, B.; Nunes, C.; Siedel, H.; Stefanidou, M.; et al. Towards a more effective and reliable salt crystallization test for porous building materials: State of the art. *Mater. Struct.* **2018**, *51*, 55. [CrossRef]
38. Costa, K.O.B. *Microstructural, Physical and Mechanical Characterization and Durability of Fluminense Ornamental Stones*; Campos dos Goytacazes: Lisbon, Portugal, 2021. (In Portuguese)
39. Miller, J.D.; Ishida, H.; Maurer, F.H.J. Dynamic-mechanical properties of interfacially modified glass sphere filled polyethylene. *Rheol. Acta* **1988**, *27*, 397–404. [CrossRef]
40. Debnath, S.; Ranade, R.; Wunder, S.; McCool, J.; Boberick, K.; Baran, G. Interface effects on mechanical properties of particle-reinforced composites. *Dent. Mater.* **2004**, *20*, 677–686. [CrossRef]

Article

Development of Sustainable Artificial Stone Using Granite Waste and Biodegradable Polyurethane from Castor Oil

Maria Luiza Pessanha Menezes Gomes [1],*, Elaine Aparecida Santos Carvalho [1], Gabriela Nunes Sales Barreto [1], Rubén Jesus Sánchez Rodriguez [1], Sérgio Neves Monteiro [2] and Carlos Maurício Fontes Vieira [1]

[1] Advanced Materials Laboratory—LAMAV, State University of the Northern Rio de Janeiro—UENF, Av. Alberto Lamego 2000, Campos dos Goytacazes 28013-602, Brazil; elainesanttos@yahoo.com.br (E.A.S.C.); gabibarreto93@gmail.com (G.N.S.B.); sanchez@uenf.br (R.J.S.R.); vieira@uenf.br (C.M.F.V.)

[2] Department of Materials Science, Military Engineering Intitute—IME, Praça General Tibúrcio 80, Praia Vermelha, Urca, Rio de Janeiro 22290-270, Brazil; snevesmonteiro@gmail.com

* Correspondence: malu_pmg@hotmail.com

Abstract: Brazil is one of the world's major ornamental stone producers. As a consequence, ornamental stone wastes are generated on a large scale and are usually open air disposed. Thus, it is important to develop novel material reusing these accumulated wastes, aiming to minimize environmental impact. The development of artificial stones made with ornamental stone wastes agglomerated by a synthetic polymer represents an excellent alternative and, therefore, is currently the subject of several works. This work seeks to develop an innovative artificial stone containing 85%wt of granite waste and 15%wt of vegetable polyurethane from castor oil, a biodegradable resin, from a renewable source. The purpose is creating a sustainable material, technically viable to be applied as a civil construction coating. To manufacture the artificial stone plates, granite and polyurethane were mixed and transferred to a metallic mold subjected to vibration, vacuum and, later, hot compression. The artificial stone presented low water absorption (0.13%) and apparent porosity (0.31%) as well as a favorable 17.31 MPa bend strength. These results were confirmed through the excellent particles/matrix adhesion displayed in the micrographs, in addition to great chemical resistance.

Keywords: artificial stone; vegetable polyurethane; granite waste

Citation: Gomes, M.L.P.M.; Carvalho, E.A.S.; Barreto, G.N.S.; Rodriguez, R.J.S.; Monteiro, S.N.; Vieira, C.M.F. Development of Sustainable Artificial Stone Using Granite Waste and Biodegradable Polyurethane from Castor Oil. *Sustainability* **2022**, *14*, 6380. https://doi.org/10.3390/su14116380

Academic Editor: Miguel Bravo

Received: 1 April 2022
Accepted: 9 May 2022
Published: 24 May 2022

Publisher's Note: MDPI stays neutral with regard to jurisdictional claims in published maps and institutional affiliations.

1. Introduction

The building materials demand is rapidly increasing, especially in developing countries, on account of the rapid urbanization and the improving life standart. Consequently, decorative stone mining has grown to be an remarkably prosperous activity [1]. Brazil is one of the world's largest ornamental stone producers, having produced nine million tons of stone for coating and decoration in 2020, despite the pandemic that devastated the world, only 2.2% lower than the previous year [2]. The large amount of waste generated by the ornamental stone industry produces several environmental impacts such as water, landscape and flora pollution, since most of this waste is improperly disposed of [3,4]. Furthermore, according to Cordon et al. about 75% of natural resources are consumed for the development of this activity [5].

In the current scenario, environmental preservation is crucial, bringing prominence to the development of sustainable building materials that contribute to its preservation while simultaneously fulfilling their functions. Several researches around the world have been carried out in order to reuse granite wastes, thus reducing their environmental impacts [1,3]. It is worth mentioning that, under regular working conditions, granite waste is not environmentally deleterious, being classified as a non-inert and non-hazardous waste, i.e., it does not react and does not dissolve in water [6].

Some specific granite waste characteristics favor its industrial use as raw material for novel products, such as: fine granulometry, predefined chemical composition and the absence of mixed grains between the basic components [1]. For instance, in mortar manufacturing, it is possible to use granite wastes to replace natural aggregates, saving natural resources and providing a good mechanical performance without harming the mortar's appearance, once granite contains in its composition [1]. For concrete, it is possible to reuse granite waste to partially replace the fine aggregates, which even increases its compressive and flexural strength [7]. For ceramic products manufacturing, fine granite waste can be mixed in the clay mass, producing ceramic bodies with promising properties, depending on the firing temperature, and without altering its color, due to oxides content in granite's chemical composition [8]. Thereby, granite wastes are able to partially replace the fine aggregates in concrete, increasing its bending and compression strength [1,7,8].

Another alternative for recycling granite waste is its reuse as a raw material in the development of other construction materials, such as artificial stone. Indeed, artificial stones are made of a high content of particulate aggregates bonded by polymeric resins. The aesthetics and properties of artificial stones are distinct since their manufacturing occurs under different conditions, from a wide range of compositions, different types of aggregates and resins binders as well as different types of processing [9,10].

Artificial stones are, in general, technically superior compared to natural ornamental stones. As the particles are bonded together by a polymeric resin, which results in a lighter material. In fact, the artificial stones' density is lower. To improve even more, the resin fulfills most of the voids and pores, shaping a material to be further resistant to liquid infiltration, granting its use in humid environments for a long term without its performance being impaired. Likewise, the mechanical strength is superior, once voids can work as stress concentrators and facilitate crack expansion [11].

Several researches have been carried out in order to develop the most differentiated artificial stones, varying the types of aggregates and polymeric resins [12–15]. Shishegaran et al. [12] evaluated the mechanical properties of three artificial stones containing: traveline powder and sludge, traveline powder and sand and traveline powder, sand and sludge, using two distinct epoxy resins. Sarami and Mahdavian [13] evaluated the properties of artificial stones using travertine, marmarit stone with cement, water and unsaturated polyester resin. Borsellino et al. [14] produced artificial marbles using polyester or epoxy resin varying the power concentration of marble particles. Carvalho et al. [15] developed an artificial stone with an epoxy matrix incorporated with sintering steel industry waste. Notwithstanding, the researches carried out until then have used synthetic resins, derived from petroleum, in their polymer matrices, mainly epoxy and polyester resins.

Petroleum-derived polymers consumption has been growing over time, also giving rise to a large amount of waste that is often discarded inappropriately and, in aggravation, takes years to degrade. To minimize this problem, it is possible to use biodegradable polymers as their technical and economic feasibility has great potential for expansion [16]. Polyurethane (PU) castor oil is a widely found low-cost renewable oil that is attracting attention from researchers due to its utilization in coatings, adhesives, paints, sealants and encapsulating compounds. Castor oil-based PUs ordinarily have low mechanical strength and limited ductility due to their flexible, highly crosslinked and permanent network structure. In a sustainability concepts, it is interesting to replace polymers by this type of PU from a renewable source [17,18].

This work's main goal is to develop and characterize a novel sustainable artificial stone that, together with the granite waste recycling, also proposes application of a biodegradable PU from castor oil, a renewable and non-toxic source, as matrix. The scope of this work is also the evaluation of whether it can be used as a coating in civil construction. The artificial stone waste/polyurethane resin (ASPU) developed composite allows the creation of sustainable material with suitable properties to be used as coating and reducing the harmful environmental impacts that would occur with the disposal of granite waste and the synthetic polymers utilization.

2. Materials and Methods

2.1. Materials

White granite waste from the slab finishing step that would be discarded was supplied by the Brumagran company located in Cachoeiro de Itapemirim, Brazil. The waste was hammered and subjected to a jaw crusher in order to be sieved according to the technical standard ABNT NBR 7181 [19] and thus classified into three granulometric ranges: coarse (2 to 0.42 mm), medium (between 0.42 to 0.075 mm), and fine (grains with particle sizes less than 0.075 mm).

Based on the three granulometric ranges, ten mixtures with different percentages of coarse, medium and fine particles were considered. A ternary diagram according to the Simplex Centroid Model was used in order to determine the mixture with the highest dry bulk density and consequently the one with the greatest packing. Therefore, the mixture selected for the development of ASPU was the one containing 4/6 of a coarse particle, 1/6 of a medium particle and 1/6 of a fine particle.

The resin used was vegetable polyurethane (PU) from castor oil. This resin is formulated by a prepolymer (component A) and a polyol (component B) cold mixing and a polyol (component B). The resin was supplied by company IMPERVEG, Brazil.

2.2. Methods

2.2.1. Development of Artificial Stone Plates

Initially, the waste was dried in an oven at 100 °C for 24 h to remove moisture. According to the resin manufacturer's guidelines, 1wt.% of component A for each 1.2 wt.% of component B, for 15 wt.% production. The ASPU plates were manufactured in dimensions $1000 \times 1000 \times 12 \, mm^3$.

The mixture was made in an automatic cylindrical mixer and then the mass was placed into a mold that was previously connected to a vacuum system and over a vibrating table, Produtest, Brazil, to better spread the mass in the cavity and facilitate the removal of air bubbles.

After 2 min of vibration time in the mold, still under vacuum, the mass was compressed in a hydraulic press at a 10 tons compaction pressure, at 80 °C for 20 min. After pressing, the mold was disconnected from the vacuum system and cooled to room temperature [11] to remove the artificial stone plate.

The artificial stone plates were then subjected to post curing as shown in Table 1 to promote better cross-linking of polymer chains.

Table 1. ASPU post cure steps.

	Temperature (°C)	Time (h)
1° Stage	60	72
2° Stage	80	24

The 60 °C was adopted as the minimum temperature based on the resin glass transition temperature (Tg). The maximum temperature was 80 °C because allophanate and biuret cross-links occur secondarily at high temperatures and are important in the polyurethane post-cure, increasing hardness, tensile strength and elastic modulus. The biuret cross-links are even more important because they occur at 80 °C range with the remaining isocyanate groups [20,21].

2.2.2. Thermogravimetric Analysis (TGA)

TGA was performed for natural granite, vegetable polyurethane and ASPU in a TGA-Q5000 TA Instruments, New Castle, DE, USA. The analysis was done with approximately 10 mg of each over a 30 to 935 °C temperature range with a heating rate of 10 °C/min, using an air flow of 60 mL/min during the test.

2.2.3. Physical Properties

Granite and ASPU density, apparent porosity and water absorption were determined by the method described in annex B of the NBR 15845 [22] standard. For the test, 12 specimens of each material with dimensions $30 \times 30 \times 10$ mm^3 were cut and sanded.

2.2.4. Mechanical Properties

Granite, polyurethane and ASPU 3-point bend rupture stress were determined. The test was performed in a Instron model 5582 universal machine, following the EN 14617-2 [22] and annex F of the NBR 15845 [23] standards guidelines. For each material, six specimens were used in the dimensions $10 \times 25 \times 100$ mm^3. The test was performed under 0.25 mm/min loading rate, 100 kN load cell and 80 mm distance between the two points.

The bend stress was calculated using Equation (1):

$$\sigma_F = \frac{(3F \times L)}{2b \times e^2} \tag{1}$$

where: σ_F = bend rupture stress (MPa); F = load (N); L = distance between support bars (mm); b = width of specimen after test (mm); e = minimum thickness of specimen (mm).

2.2.5. Microstructural Analysis

The granite and ASPU fractured surface regions of specimens previously subjected to bending tests were observed through SHIMADZU's Super Scan SSX-550 scanning electron microscope (SEM) for microstructural analysis, at 20 kV of secondary electrons. The microstructural analysis is important to determine the quality of the adhesion between the particles and the resin and the presence of voids.

2.2.6. Abrasive Wear

Abrasive wear was determined for granite and for ASPU, according to the NBR 12042 [24] standard, using MAQTEST Amsler equipment. Two specimens with $70 \times 70 \times 40$ mm had their thicknesses measured before and after undergoing 500 and 1000 m track.

2.2.7. Chemical Attack Resistance

The granite and ASPU resistance to chemical attacks determination was performed according to a Brazilian standard NBR 13818 Annex H adaptation [25]. Sixteen specimens of each stone were used, four specimens for each reagent, namely: ammonium chloride, hydrochloric acid, citric acid and potassium hydroxide. Table 2 shows the reagents concentration and exposure time used according to the standard [25]. The specimens were weighed before and after the chemical attack in order to measure the weight loss.

Table 2. Concentration and time of exposure of chemical attack agents.

Reagents	Concentration	Time (h)
Ammonium choride	100 g/L	24
Ciric acid	100 g/L	24
Hydrochloric acid	3% v	96
Potassium hudroxide	30 g/L	96

3. Results and Analisys

3.1. Final Dimensions and Overall Aspects of ASPU

The $100 \times 100 \times 12$ mm^3 artificial stone plates were produced within the standards. Later, they were cut to the dimensions specified for carrying out the tests. Figure 1 shows an ASPU plate.

Figure 1. ASPU plate.

3.2. Thermogravimetric Analysis

Figure 2 shows the TG and DTG curves for granite, polyurethane and ASPU.

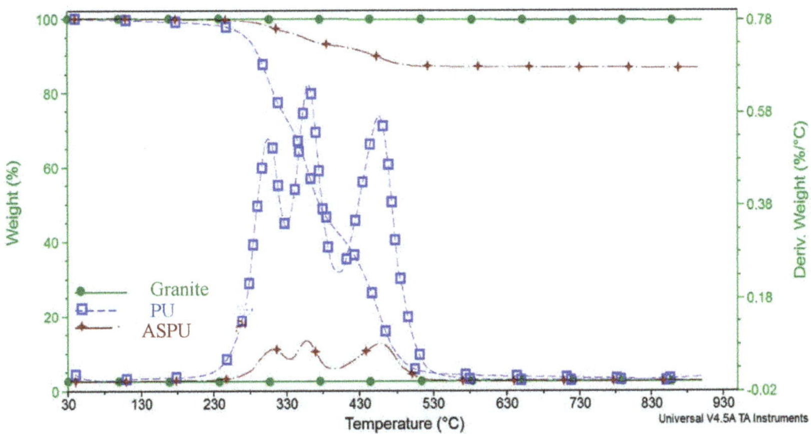

Figure 2. Thermogravimetric Analysis for granite, PU and ASPU.

In Figure 2, it was possible to verify that the PU has a three the decomposition stages, which is justified by releasing several gaseous products.

The first stage was around 230 °C, with a 23% mass loss. As urethane bonds break and degrade around 200 °C, this stage is likely associated with this event. The second stage takes place around 330 °C, with a 20% weight loss, probably due to CO degradation. The third and last stage, around 380 °C with a 57% mass loss, was attributed to the compounds produced after the second stage degradation [26,27].

The granite remained stable, with negligible mass loss, approximately 0.78%. Furthermore, it was possible to confirm the composition of ASPU, considering that after the analysis, 86.58% of waste remained, as the resin was the only material that suffered a degradation process.

3.3. Physical Properties

Table 3 presents the density, water absorption and apparent porosity values of granite and ASPU.

Table 3. Physical Properties of granite and ASPU.

Physical Properties	Granite	ASPU
Density (g/cm^3)	2.62 ± 0.01	2.24 ± 0.01
Water absorption (%)	0.38 ± 0.02	0.13 ± 0.02
Apparent Porosity (%)	0.99 ± 0.06	0.31 ± 0.05

The granite and ASPU density were 2.62 ± 0.01 g/cm^3 and 2.24 ± 0.01 g/cm^3, respectively. ASPU presented a density 15% lower than granite, probably due to the polymer in its composition, which is, in general, a low density material among the others, including minerals.

The lower density results in a material with lower weight per square meter, i.e., lighter. The importance of density relies on the fact that the higher its value, the better the adhesion between the particles and the polymeric matrix, which reduces the occurrence of voids [28]. In their work, Lee et al. [29] used glass and PET (polyethylene terephthalate) crystals for the development of artificial stone varying the levels of compression, vacuum and vibration and the density values varied between 2.03 and 2.45 g/cm^3. Granite and ASPU density, when compared to Lee et al., are within the expected range.

Regarding granite and ASPU water absorption, the values found were respectively 0.38 ± 0.02% and 0.13 ± 0.02%. The artificial stone developed had a result approximately three times lower than the ornamental stone. Lower water absorption levels and an impermeable surface are crucial for artificial stone once this material will be constantly in contact with water. Coating with low water absorption, following the guidelines of Chiodi Filho and Rodriguez [30], must be between 0.1–0.4% of water absorption. Ribeiro et al. [11] in its artificial marble developed with polyester resin obtained 0.19%, while Borsellino et al. [14] developed a 0.25% water absorption artificial marble, but with epoxy resin. With these comparisons, it is verified the good performance of ASPU, confirming the possibility of its use in humid environments.

The apparent porosity value was 0.99 ± 0.06% for granite and 0.31 ± 0.02% for ASPU. The ASPU water absorption was three times lower than that of granite. Chiodi Filho and Rodriguez [30] classify that high quality coating materials must have porosity below 0.5%, denoting the ASPU good performance. ASPU's low porosity can be attributed to the homogeneity of the material, with granite particles well adhered to the polyurethane matrix, ensuring the filling of the voids.

3.4. Three-Points Bend Stress

Figure 3 shows the bending stress vs. strain curves for ASPU, white granite and vegetable polyurethane.

The tensile stress of granite, ASPU and vegetable polyurethane were respectively 13.50 ± 1.10, 17.31 ± 0.82 and 23.01 ± 5.30 MPa.

According to the ABNT NBR 15844 standard [31], which determines the necessary requirements for the use of granite as a coating, the 3-point bend strength must not be less than 10 MPa. Even though granite presented an elevated strength (13.5 MPa), its resistance was inferior to ASPU due to its high porosity shown by physical tests results.

Comparing the polyurethane behavior with its derived artificial stone, it is possible to observe that the granite inclusion as a resin filler contributed to increase the material stiffness stiffness. This is an expected behavior, considering that the rigid particles incorporation in a polymer matrix commonly enhances the material's elasticity modulus [32].

According to Chiodi Filho and Rodriguez [30], artificial stones applied as coating are classified as high-strength materials when their ultimate bending strength is between

16–20 MPa, and are considered very high strength materials when this value exceeds 20 MPa. Therefore, ASPU was considered to be a high-strength material for coating, since its bending strength was 17.31 MPa, associated with its little porosity and proving the artificial stones quality performance as a coating.

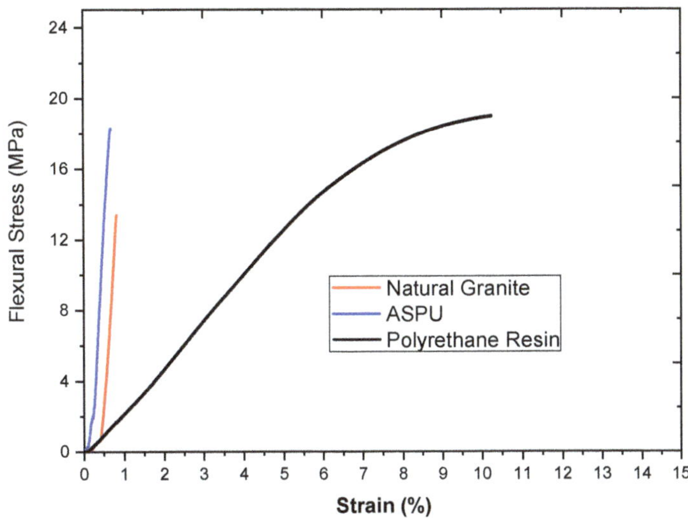

Figure 3. Stress vs. Strain curves for ASPU, granite and polyurethane (PU).

3.5. Microstructural Analysis

Figure 4 presents SEM micrographs of ASPU (a,b) and granite (c,d) fracture surfaces after thr three-point bending test.

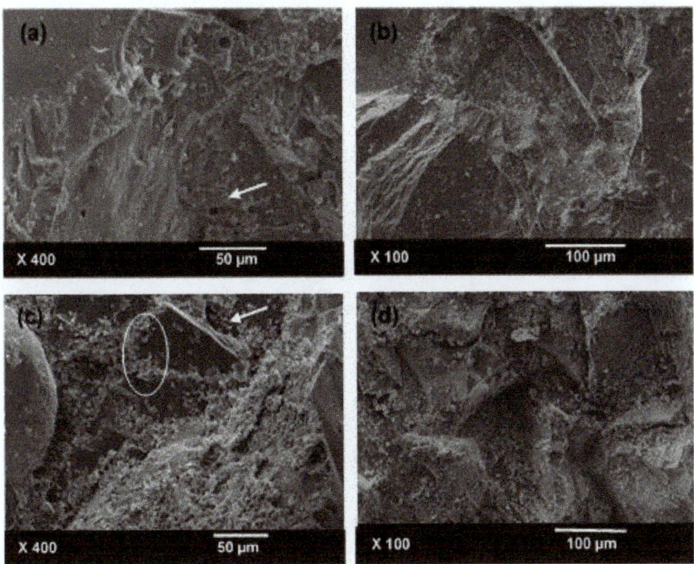

Figure 4. (**a–d**): Microscopic images of the fracture region obtained through scanning electron microscope (**a**) of the artificial stone ASPU with zoom 400× (**b**) ASPU with zoom 100×; (**c**) of the granite with zoom 400× and (**d**) granite with zoom 100×.

Figure 4a,b shows the ASPU's optimal load/matrix interaction. According to Debnath et al. [33], the quality of interfacial interaction is directly related to the improvement of a composite's mechanical strength. This is due to the fact that good interfacial wettability means higher adhesive strength.

Figure 4c,d shows the natural granite micrographs. Comparing granite to ASPU, one can notice that granite stone presented a higher incidence of voids and a greater particle detachment after the bending stress. The SEM micrographs confirmed the results found in porosity tests, with the ASPU presenting three times less porosity than the granite.

3.6. Abrasive Wear

Table 4 presents the results for ASPU and granite abrasive wear tests.

Table 4. Abrasive wear test results.

	Wear Thickness Reduction (mm)	
Running distance	500 m	1000 m
Granite	0.53	1.13
ASPU	0.97	1.75

According to a study carried out by Chiodi Filho and Rodriguez [30] that classified materials to be applied as floors according to the volume of traffic and the wear thickness reduction: high traffic floors must be less than 1.5 mm, medium traffic floors must be less than 3mm and low traffic floors must be less than 6mm.

Therefore, according to this study, granite can be used for high-traffic flooring since it presented a 0.53 mm thickness reduction after a 500 m run and 1.13 mm after 1000 m. However, ASPU can be used for medium traffic flooring considering a 0.97 mm and 1.75 mm wear after a 500 and 1000 m run, respectively.

Abrasive wear tests evaluate the material pullout when it undergoes wear friction. The inferior performance of ASPU can be explained by the greater ease of removal of granite particles from the polyurethane matrix [34].

However, according to the NBR 15844 standard [31], which classifies granites by to their wear properties, to guarante the granite's quality, the thickness reduction must not exceed 1.0 mm after 1000 m run. Thereby, considering this specific standart, the both granite and ASPU presented wear higher than recommended, limiting its application as a pavement.

3.7. Chemical Attack Resistance

Table 5 displays the percentage weight loss suffered by the granite and ASPU under chemical attack.

Table 5. Weight loss after chemical attack of granite and ASPU samples.

	Weight Loss (%)	
Reagents	**Granite**	**ASPU**
NH_4Cl	0.007	0.010
$C_6H_8O_7$ (citric acid)	0.019	0.011
HCl	0.043	0.039
KOH	0.025	0.022

The norm NBR 13818 [25] proposed reagents are ammonia chloride and potassium hydroxide, used to make cleaning products, hydrochloric acid, which is found in floor, stone and tiles heavy cleaner products and citric acid, contained in several fruits, must be in constant contact with the specimens.

Through the obtained results it is possible to verify that the exposure to all the reagents provoked a weight loss in the stones. Although these losses could be considered insignificant, in the long term the constant use of these substances can compromise their performance and aesthetics.

Hydrochloric acid (HCl) was the most aggressive reagent for both granite and ASPU, with a 0.043 g weight loss, followed by potassium hydroxide (KOH), citric acid ($C_6H_8O_7$) and ammonium chloride (NH4Cl). Hydrochloric acid is a strong and high degree of ionization acid manifesting the greatest weight loss. Therefore, cleaning materials containing this substance and potassium hydroxide should be especially avoided and it is recommended to clean with care using only a damp cloth with diluted neutral soap.

4. Conclusions

Through this study, it was possible to develop a sustainable polyurethane resin artificial stone (ASPU), using as matrix, for the first time, a polymer resin from renewable and biodegradable source, and as raw material, granite waste. The ASPU represents an innovative novel material, considering that so far, the artificial stones developed used only petroleum-derived resins as matrix.

Furthermore, the manufacturing process was an important factor to produce a technically viable material, with properties enabling its use as a coating in civil construction.

The values of density (2.24 g/cm^3), water absorption (0.13%) and apparent porosity (0.31%) presented the possibility of using ASPU in humid environments and denoted the material homogeneity and good load/matrix adhesion.

The bending strength value of 17.31 MPa ranked ASPU as a higher-strength coating material. The SEM micrographs displayed an adequate interaction between granite particles and PU and the little presence of voids, confirming the physical indices and mechanical strength results.

Through the test results, it can be concluded that, owing to ASPU's mechanical and physical properties, the novel artificial stone can be used as a wall covering, on bathroom's and kitchen's countertops and its use in floors must be avoided, specially for high traffic places such as airports and supermarkets.

For ASPU's maintenance, the most aggressive reagents were HCl and KOH, causing the greatest weight loss, therefore, it should be avoided cleaning the stone with cleaning materials that contain these reagents in the components.

From this study, further research can be carried out aiming to evaluate the behavior of ASPU in cold weather countries through a bending test after freezing and thawing cycles. Besides, the changeability of the material can be studied by performing other tests like dilatometry, salt spray and UV rays exposure.

In addition, the PU from castor oil feasibility to be used as matrix resin in artificial stone allows further studies aimed at the development of other types of artificial stones varying the aggregates, wastes, matriz/load proportions and processing parameters.

Author Contributions: Conceptualization, C.M.F.V.; methodology, M.L.P.M.G. and E.A.S.C.; formal analysis, M.L.P.M.G.; investigation, R.J.S.R.; resources, C.M.F.V. and M.L.P.M.G.; data curation, G.N.S.B.; writing—original draft preparation, M.L.P.M.G.; writing—review and editing, G.N.S.B.; visualization, E.A.S.C.; supervision, R.J.S.R. and S.N.M.; project administration, E.A.S.C.; funding acquisition, M.L.P.M.G. All authors have read and agreed to the published version of the manuscript.

Funding: This research was funded by Rio de Janeiro State Research Foundation—FAPERJ (E-26/200.139/2022).

Informed Consent Statement: Not applicable.

Data Availability Statement: Not applicable.

Acknowledgments: The authors thank the support of the Brazilian agencies: FAPERJ (E-26/200.847/2021), CNPQ (301634/2018-1) and UENF.

Conflicts of Interest: The authors declare no conflict of interest.

References

1. Nayak, S.K.; Satapathy, A.; Mantry, S. Use of waste marble and granite dust in structural applications: A review. *J. Build. Eng.* **2022**, *46*, 103742. [CrossRef]
2. ABIROCHAS—Brazilian Association of the Ornamental Stone Industry. Available online: https://abirochas.com.br/wp-content/uploads/2021/02/Informe-01_2021-Balanco-2020.pdf (accessed on 8 May 2022).
3. Gautan, L.; Jain, J.K.; Kalla, P.; Danish, M. Sustainable utilization of granite waste in the production of green construction products: A review. *Mater. Today Proc.* **2020**, *44*, 4196–4203. [CrossRef]
4. Demartini, T.J.C.; Rodriguez, R.J.S.; Silva, F.S. Phisycal and mechanical evalution of artificial marble produced with dolomitic marble residue processed by Diamond-plated bladed gang-saws. *J. Mater. Res. Technol.* **2018**, *7*, 308–313. [CrossRef]
5. Cordon, H.C.F.; Cagnoni, F.C.; Ferreira, F.F. Comparison of physical and mechanical properties of civil construction plaster and recycled waste gypsum from São Paulo, Brazil. *J. Build. Eng.* **2019**, *22*, 504–512. [CrossRef]
6. Braga, F.S.; Buzzi, D.C.; Couto, M.C.L. Environmental characterization of ornamental stone processing sludge. *Sanit. Environ. Eng.* **2010**, *15*, 237–244. (In Portuguese)
7. Ahmadi, S.F.; Reisi, M.; Amiri, M.C. Reusing granite waste in eco-friendly foamed concrete as aggregate. *J. Build. Eng.* **2022**, *46*, 103566. [CrossRef]
8. Singh, S.; Nagar, R.; Agrawal, V.; Rana, A.; Tiwari, A. Sustainable Utilization of Granite Cutting Waste in High Strength Concrete. *J. Clean. Prod.* **2016**, *126*, 223–235. [CrossRef]
9. Hamoush, S.; Abu-Lebdeh, T.; Picornell, M.; Amer, S. Development of sustainable engineered stone cladding for toughness, durability, and energy conservation. *Constr. Build. Mat.* **2011**, *25*, 4006–4016. [CrossRef]
10. Silva, S.F.; Ribeiro, C.E.G.; Rodriguez, R.J.S. Physical and Mechanical Characterization of Artificial Stone with Marble Calcite Waste and Epoxy Resin. *Mater. Reser.* **2018**, *21*, 6. [CrossRef]
11. Ribeiro, C.E.G.; Rodriguez, R.J.S.; Vieira, C.M.F.; de Carvalho, E.A.; Candido, V.S.; Monteiro, S.N. Production of synthetic ornamental marble as a marble waste added polyestercomposite. In *Materials Science Forum*; Hotza, D., Novaes de Oliveira, A.P., Eds.; Trans Tech Publications: Zürich, Switzerland, 2014; pp. 341–345.
12. Shishegaran, A.; Saeedi, M.; Mirvalad, S.; Korayem, A.H. The mechanical strength of the artificial stones, containing the travertine wastes and sand. *J. Mater. Res. Technol.* **2021**, *11*, 1688–1709. [CrossRef]
13. Sarami, N.; Mahdavian, L. Effect of inorganic compound on artificial stones' properties. *Int. J. Ind. Chem.* **2015**, *6*, 213–219. [CrossRef]
14. Borselino, C.; Calbrese, L.; di Bella, G. Effects of powder concentration and type of resin on the performance of marble composite structures. *Constr. Build. Mater.* **2009**, *23*, 1915–1921. [CrossRef]
15. Carvalho, E.A.S.; Marques, V.R.; Rodriguez, R.J.S.; Ribeiro, C.E.G.; Monteiro, S.N.; Vieira, C.M.F. Development of epoxy matrix artificial stone incorporated with sintering residue from steelmaking industry. *Mater. Res.* **2015**, *18*, 235–239. [CrossRef]
16. Hablot, E.; Zheng, D.; Bouquey, M.; Avérous, L. Polyurethanes based on castor oil: Kinetics, Chemical, mechanical and termal properties. *Macromol. Mater. Eng.* **2008**, *293*, 922–929. [CrossRef]
17. Clarinval, A.-M.; Halleux, J. Classification of biodegradable polymers. In *Biodegradable Polymers for Industrial Applications*; Woodhead Publishing: Sawston, UK, 2005; pp. 3–31. [CrossRef]
18. Rajalakshmi, P.; Marie, J.M.; Maria Xavier, A.J. Castor oil-derived monomer ricinoleic acid based biodegradable unsaturated polyesters. *Polym. Degrad. Stab.* **2019**, *170*, 109016. [CrossRef]
19. Brazilian Association of Technical Norms—ABNT. *ABNT NBR 7181: Soil—Grain Size Analysis*; ABNT: Rio de Janeiro, Brazil, 2016. (In Portuguese)
20. Hepburn, C. *Polyurethane Elastomers*, 2nd ed.; Elsevier: New York, NY, USA, 1992.
21. Reis, J.M.; Motta, E.P. Mechanical properties of castor oil polymer mortars. *Mat. Res.* **2014**, *17*, 1162–1166. [CrossRef]
22. Brazilian Association of Technical Norms—ABNT. *ABNT NBR 15845-2: Rocks for Cladding. Part 2: Determination of Bulk Density, Apparent Porosity and Water Absorption*; ABNT: Rio de Janeiro, Brazil, 2015. (In Portuguese)
23. Spanish Association of Standards and Certification. *UNE-EN 14617-2-08: Test Methods. Part 2: Determination of the Flexural Strength*; UNE-EN: Madrid, Spain, 2008. (In Spanish)
24. Brazilian Association of Technical Norms—ABNT. *ABNT NBR 12042: Inorganic Materials—Determination of the Resistance to Abrasion*; ABNT: Rio de Janeiro, Brazil, 2012. (In Portuguese)
25. Brazilian Association of Technical Norms—ABNT. *ABNT NBR 13818: Ceramic Tiles—Specification and Methods of Test*; ABNT: Rio de Janeiro, Brazil, 1997. (In Portuguese)
26. Ristic, I.S.; Bjelovic, Z.D.; Holló, B.; Szécsényi, K.M.; Budinski-Simendić, J.; Lazic, N.; Kicanovic, M. Thermal stability of polyurethanane materials based on castor oil as polyol componente. *J. Therm. Anal. Calorin.* **2013**, *111*, 1083–1091. [CrossRef]
27. Moura Neto, F.N.; Fialho, A.C.V.; Moura, W.L.; Rosa, A.G.F.; Matos, J.M.E.; Reis, F.S.; Mendes, M.T.A.; Sales, E.S.D. Castor polyurethane used as osteosynthesis plates: Microstructural and thermal analysis. *Polímeros* **2019**, *29*. [CrossRef]
28. Gomes, M.L.P.M.; Carvalho, E.A.S.; Sobrinho, L.N.; Monteiro, R.J.S.R.; Vieira, C.M.F. Production and characterization of a novel artificial stone using brick residue and quarry dust in epoxy matrix. *J. Mater. Res. Technol.* **2018**, *7*, 492–498. [CrossRef]
29. Lee, D.J.; Shin, I.J. Effects of vacuum, mold temperature and cooling rate on mechanical properties of press consolidated glass fiber/PET composites. *Compos. Part A* **2002**, *33*, 1107–1114. [CrossRef]
30. Chiodi Filho, C.; Rodriguez, H. *Guide Application of Stone Coverings*, 3rd ed.; ABIROCHAS: São Paulo, Brazil, 2009; p. 108.

31. Brasilian Association of Technical Norms—ABNT. *NBR 15844: Rocks for Cladding—Requiriments for Granites*; ABNT: Rio de Janeiro, Brazil, 2015.

32. Sujin Jose, A.; Athijayamani, A.; Jani, S.P. A review on the mechanical properties of bio waste particulate reinforced polymer composites. *Mater. Today Proc.* **2020**, *37*, 1757–1760. [CrossRef]

33. Debnath, S.; Ranade, R.; Wunder, S.L.; Mccool, J.; Boberick, K.; Baran, G. Interface effects on mechanical properties of particle-reinforced composites. *Dent. Mater.* **2004**, *20*, 677–686. [CrossRef] [PubMed]

34. Karaca, Z.; Yılmaz, N.G.; Goktan, R.M. Abrasion wear characterization of some selected stone flooring materials with respect to contact load. *Constr. Build. Mater.* **2012**, *36*, 520–526. [CrossRef]

 *sustainability*

Article

Comparison between Synthetic and Biodegradable Polymer Matrices on the Development of Quartzite Waste-Based Artificial Stone

Carlos Paulino Agrizzi [1,2], Elaine Aparecida Santos Carvalho [2], Mônica Castoldi Borlini Gadioli [3], Gabriela Nunes Sales Barreto [2], Afonso R. G. de Azevedo [4,*], Sérgio Neves Monteiro [5] and Carlos Maurício Fontes Vieira [2]

1. Mechanical Engineering Department, Federal Institute of Espirito Santo (IFES), Cachoeiro de Itapemirim 29075-910, ES, Brazil; carlosagrizzi@gmail.com
2. Advanced Materials Laboratory (LAMAV), State University of Northern Fluminense—UENF, Av. Alberto Lamego 2000, Campos dos Goytacazes 28013-602, RJ, Brazil; elainesanttos@yahoo.com.br (E.A.S.C.); gabibarreto93@gmail.com (G.N.S.B.); vieira@uenf.br (C.M.F.V.)
3. Mineral Technology Center—CETEM, Ministry of Science, Technology and Innovations, Rod. Cachoeiro x Alegre, Km 05, Cachoeiro de Itapemirim 29311-970, ES, Brazil; mborlini@cetem.gov.br
4. Civil Engineering Laboratory (LECIV), State University of the Northern Rio de Janeiro—UENF, Av. Alberto Lamego 2000, Campos dos Goytacazes 28013-602, RJ, Brazil
5. Department of Materials Science, Instituto Militar de Engenharia—IME, Praça General Tibúrcio 80, Praia Vermelha, Urca, Rio de Janeiro 22290-270, RJ, Brazil; snevesmonteiro@gmail.com
* Correspondence: afonso@uenf.br

Abstract: The development of artificial stone from the agglutination of polymeric resin using industrial wastes can be a viable alternative from a technical, economic, and sustainable point of view. The main objective of the present work was to evaluate the physical, mechanical, and structural properties of artificial stones based on quartzite waste added into a synthetic, epoxy, or biodegradable polyurethane polymer matrix. Artificial stone plates were produced through the vacuum vibration and compression method, using 85 wt% of quartzite waste. The material was manufactured under the following conditions: 3 MPa compaction pressure and 90 and 80 °C curing temperature. The samples were characterized to evaluate physical and mechanical parameters and microstructure properties. As a result, the artificial stone plates developed obtained $\leq$0.16% water absorption, $\leq$0.38% porosity, and 26.96 and 10.7 MPa flexural strength (epoxy and polyurethane resin, respectively). A wear test established both artificial quartzite stone with epoxy resin (AS-EP) and vegetable polyurethane resin (AS-PU) high traffic materials. Hard body impact resistance classified AS-EP as a low height material and AS-PU as a very high height material. The petrographic slides analysis revealed that AS-EP has the best load distribution. We concluded the feasibility of manufacturing artificial stone, which would minimize the environmental impacts that would be caused by this waste disposal. We concluded that the production of artificial rock shows the potential and that it also helps to reduce environmental impacts.

Keywords: artificial stone; quartzite; resin; coating

Citation: Agrizzi, C.P.; Carvalho, E.A.S.; Borlini Gadioli, M.C.; Barreto, G.N.S.; de Azevedo, A.R.G.; Monteiro, S.N.; Vieira, C.M.F. Comparison between Synthetic and Biodegradable Polymer Matrices on the Development of Quartzite Waste-Based Artificial Stone. *Sustainability* **2022**, *14*, 6388. https://doi.org/10.3390/su14116388

Academic Editor: Antonio Caggiano

Received: 30 March 2022
Accepted: 11 May 2022
Published: 24 May 2022

Publisher's Note: MDPI stays neutral with regard to jurisdictional claims in published maps and institutional affiliations.

1. Introduction

Developing countries that have accelerated industrialization and urbanization processes are producing billions of tons of waste per year. As a major global ornamental stone producer, Brazil is also a major waste generator, due to the daily extraction of thousands of tons of blocks in quarries. However, much of this type of waste is discarded in an uncontrolled manner, causing environmental problems [1,2].

The waste generated from the extraction and processing of ornamental stones represents around 83% of raw material losses. To produce 330 m^2 of slabs (average from a 10 m^3 block), approximately 30 m^3 of stone is extracted from the massif, of which 20 m^3 remains in the quarry as waste. Most of them are thick blocks, normally non-standard, irregular, and faulty. In addition, stone chips and hulls from the remaining stocks block rigging, as they can have direct use in the smaller plates, less value tile production, or other structural or decorative pieces [3].

Some of these losses are inevitable, as they are linked to the natural stone quality, the deposit characteristics, or the cutting process. However, they can be reduced by improving deposit knowledge, by mine planning, and, mainly, by transforming the waste generated into by-products [3]. Even though the artificial stone's unit price (m^2) and the volume used are relatively high compared to the natural materials, this waste destination, as a by-product, is still the application with the best cost benefit [4].

Artificial stones have aroused the interest of consumers due to certain advantages compared to natural stones. According to data from the Brazilian Association of Dimension Stones Industry—ABIROCHAS, in 2019, Brazilian exports of artificial stones totaled 12.75 thousand tons. Regarding imports, 70,000 tons present average prices of US\$ 530/t and US\$ 1470/t, respectively, while natural stones presented an average price of US\$ 523/t in the same year. Even with a slightly higher price than natural stones, the artificial stone import is much greater than the export [5]. Amongst the advantages of this material are the mechanical and aesthetic properties as well as the use of wastes that reduces irregular disposal, valorizing a traditionally unwanted product, creating jobs, and contributing to economic development [6].

1.1. Artificial Stone

Artificial stones are produced by a high percentage of particles agglomerated by a small percentage of polymeric material. Marble, granite, glass crystals, and quartz sand are examples of stones that, when particulate, are considered natural aggregates. Artificial stone's mechanical properties are considered to be higher than those of natural stone because of the artificial stone's lower water absorption and porosity. As such, artificial stone is a more suitable option for demanding work conditions, such as floors and walls [7–10].

These materials are molded by an aggregate mixture, compacted and agglomerated, which is placed into a mold through mechanical pressure and/or vibration, using vacuum-assisted or open molding methods. The process currently used by industries for the compact artificial stone production is called "vacuum vibro-compression". In this process, the mineral fillers are mixed with a polymeric resin and, afterwards, the mass is compacted under vibration and vacuum. The materials are generally subjected to a heat treatment at temperatures in the range of 70 to 110 °C, in order to promote the final curing of the resin, that is, the interlocking between the polymeric chains [11,12].

Suta et al. [13] used the method of vibration and compaction in a vacuum environment to produce a compact projected stone slab whose main solid components were glass waste and fine granite aggregate. The final product showed <0.02% water absorption and 51 MPa flexural strength, which were higher than those of natural construction slabs. The research results revealed that vibration was the most important processing step to adjust the aggregate orientation to become more compacted.

Lee and Shin [14] reported fiberglass/PET composites produced under the influence of vacuum, mold temperature, and cooling rate. The authors observed that specimens produced without vacuum during the preheating stage registered values of approximately 1.9% porosity, whereas specimens produced with the aid of vacuum recorded much lower values of 0.3–0.4% porosity.

The use of vacuum in manufacturing artificial stone plates is extremely important, since it facilitates the removal of air that, in the process of molding the composite, could be attached to the load and the mass, lowering the plate porosity degree. Once porosity harms the performance of artificial materials, the use of vacuum helps to produce better quality artificial materials [15].

1.2. Development of New Materials Based on Artificial Stones

The creation of novel materials using waste from the ornamental stone industry is economically interesting, in addition to meeting a sustainable ideology in the industrial environment. Today, many researchers use petroleum-based resins, wastes from natural stones cutting process, fine gravel particles, steel, rubber, PET, and iron ore mixed up with epoxy or polyester resin in artificial stone development, obtaining stones with superior properties when compared to natural stones [4,8–10,16].

Erfan [17] developed artificial stone with the incorporation of calcite marble residue and epoxy resin. The results showed 0.01% water absorption and a good filler/resin adhesion.

The use of a polymer matrix from a renewable source, which does not emanate toxic substances, is also another attractive alternative in view of the environmental impact attenuation. Gomes et al. [18] developed artificial stone with waste from natural granite cutting and polyurethane resin, and obtained an artificial stone with properties superior to natural granite.

1.3. Objective and Originality of the Research

The ornamental stone sector, which through its production and extraction generates about 40 to 60% of waste, was investigated [19]. It is estimated that there is a loss of 84% from the beginning to the final production of commercial dimension stone slabs. As Brazil is one of the largest ornamental stones producers, there is still a lack of concern to avoid waste and a neglect with the waste destination, even though companies in the sector are responsible for developing projects for their recovery and reuse [15].

Quartzite waste was obtained as raw stone, with a massive structure and coarse grain, usually cracked and milk white in color. The material is petrographically called "Quartzolite or quartzites" due to the presence of more than 90% silicon, and they are classified as "Class II B—Inert" wastes.

Vegetable oil-derived polyurethanes have been extensively studied in recent years, as a non-polluting alternative to replace petroleum-derived polyurethanes [20]. The development of new polyester polyols and prepolymers derived from fatty acids cause this polyurethanes class to present themselves with different chemical, physical, and mechanical properties, in addition to being from a renewable source [20]. It is important for the development of a new material such as artificial stone to use biodegradable natural resin and quartzite residue.

Epoxy resins are thermo-rigid materials widely used in adhesives, matrices for composites, electrical materials, and materials for coatings, among other applications, as well as in the production of artificial stones, because of their excellent adhesive, mechanical, thermal, and electrical properties [21]. For these reasons, it is important to investigate the influence of both types of synthetic and biodegradable resins on the development of artificial stone.

This new alternative, in addition to possibly reducing raw materials costs since the waste is costless (excluding its transportation costs), would also transform the waste back into raw material. The polyurethane resin used as a binder originates from castor oil, a renewable vegetable source, and together with the inert waste of natural quartzite stone, it allows the manufacture of a totally ecological and sustainable artificial stone [20].

Epoxy resin is an excellent thermoset polymer to be used as a binder in artificial stone development. The mechanical properties can be accentuated, providing an optimal adhesion to the quartz waste particles. As the epoxy resin is characterized by good chemical resistance, it will generate an environmentally harmless artificial stone with adding value to wastes that would be discarded while reducing the production costs [9].

This work's main goal is to compare the development of artificial stones from quartzite waste, using epoxy and plant polyurethane resins as binders, with good physical and mechanical properties. This waste is also compared with the performance of our artificial stone and with other artificial stones in order to confirm the technical feasibility of producing these novel sustainable materials with economic potential.

2. Materials and Methods

2.1. Materials

The "quartzite" natural stone (NS) waste, commercially called "Cristallo", was supplied by Pettrys company, located in Cachoeiro de Itapemirim, ES, Brazil. The raw stone material was collected directly from the company.

As a comparison material, a commercial stone (CS) called "Branco Aldan" was used. It was supplied by Empresa Guidoni, located in São Domingos do Norte, ES, Brazil. The commercial artificial stone is composed of quartz powder, additives, and polyester resin, information provided by the company.

The quartzite waste was classified by the sieves 8 to 200 mesh and divided into three different particle sizes. The larger particles (coarse) were classified in the range of 2.38 to 2.00 mm, the medium particles were between 2.00 and 0.630 mm, and fine particles were less than 0.630 mm [7–9,21].

The waste particles were agglutinated by two types of resins. The natural resin, vegetable polyurethane derived from castor oil, is bicomponent, that is, it results from the mixture between a component A, prepolymer (methylene diphenyl diisocyanate), with 1.22 g/cm^3 relative density, and a component B, a polyol (castor oil-base) with a 0.96 g/cm^3 typical density. The resin in 1:1.2 ratio of component B was supplied by the company IMPERVEG (Aguaí, Brazil). The other resin, epoxy, was a bisphenol A diglycidyl ether-MC130 and the hardener, FD129, triethylenetetramine (TETA), used in proportion of 13 wt% of resin, were both supplied by the company EPOXYFIBER (Rio de Janeiro, Brazi).

2.2. Determination of the Highest Packaging Granulometric Composition

Based on three ranges of grains obtained, 10 different mixtures with different percentages of rough, medium, and fine particles were proposed (Table 1).

Table 1. Experiment matrix to determine apparent dry best-packed.

Composition	Large (Coarse)	Medium	Fine
1	0	0	1
2	0	1	0
3	1	0	0
4	0	1/2	1/2
5	1/2	1/2	0
6	1/2	0	1/2
7	1/3	1/3	1/3
8	2/3	1/6	1/6
9	1/6	1/6	2/3
10	1/6	2/3	1/6

Figure 1 shows a complete ternary diagram developed in the experimental numeric-modeling grid Simplex (Simplex-Lattice Design) [22] to obtain greater packaging in ternary mixtures. To determine the proportion of the greatest packaging of the 10 mixtures, all were tested based on the standard ABNT/MB-3388 Brazilian standard (1991)—Determination of the minimum index void ratio of non-cohesive [23].

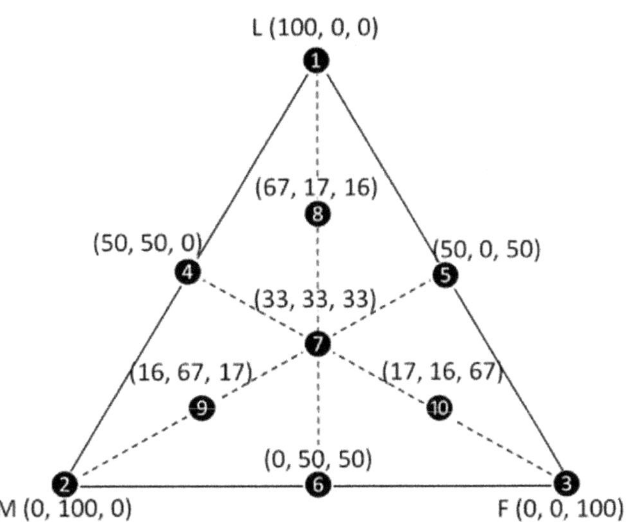

Figure 1. Ternary diagram of the Simplex complete cubic model [9].

The mixtures were placed in a device with a container of 1013.24 cm^3 and a 10 kg piston was used on the material, which was subjected to vibration through a vibrating table. This procedure was repeated three times for each of the 10 compositions. As it can be seen in Table 1, mixture 8 (4/6 rough, 1/6 medium, and 1/6 fine particles), was the one with the greatest apparent dry density representing the most close-packed one. Consequently, it was chosen for the production of artificial stone plates.

As described by Ribeiro [11], it was necessary to calculate the minimum amount of resin (MAR) necessary for the artificial stone production, using Equations (1) and (2):

$$VV\% = \left(1 - \frac{\rho_{PA}}{\rho_Q}\right) * 100 \tag{1}$$

where:
$VV\%$ = Void volume present in the mixture of particles;
ρ_{PA} = Apparent density of particles, calculated by the packaging method;
ρ_Q = Quartzite density, calculated by pycnometry.

From obtaining the void volume ($VV\%$) value, it was possible to calculate the minimum amount of resin (MAR), through Equation (2) below:

$$MAR\% = \frac{VV\% * \rho_{resin}}{VV\% * \rho_{resin} + (100 - VV\%) * \rho_Q} \tag{2}$$

where:
$MAR\%$ = Minimum amount of resin to fill the void volume;
$VV\%$ = Void volume present in the mixture of particles;
ρ_{resin} = Epoxy resin and polyurethane resin density;
ρ_Q = Quartzite density, calculated by pycnometry.

2.3. Production of Artificial Stone Plates

Artificial stone plates with a dimension of 100 × 100 × 10 mm were developed, with epoxy and natural vegetable polyurethane polymer resins, using the vacuum, vibration, and compression method. Initially, the quartzite particles were dried in an oven for 24 h at 100 °C to release moisture, then weighed and mixed with the resin in a vibration system under vacuum. Two plates were made of each resin in the dimensions 200 × 200 × 10 mm, for the hard body impact test required by the standard.

The mixture was taken to a Marcone MA 098-A hydraulic press for the plates production, with 3 MPa compression pressure and temperature of 90 °C for plates produced with epoxy resin and 80 °C for plates produced with polyurethane resin [7,24]. After pressing, the mold was disconnected from the vacuum system and cooled to room temperature to remove the plate. The plates with 85 wt% of artificial quartzite stone were developed with epoxy resin (AS-EP) and vegetable polyurethane resin (AS-PU), were subjected to a finishing step by polishing with sandpaper, and then cut with diamond disc to prepare samples for the tests according to the standards.

2.4. Characterization of Artificial Stone Plates

The apparent density and porosity as well as the water absorption were evaluated based on Annex B of the ABNT/NBR 15845 Brazilian standard, which establishes the applications of stone materials for the coating of building constructions [25].

The three-point flexural strength test was performed on our INSTRON universal testing machine, model 5582, based on F Annex of ABNT/NBR 15845 Brazilian standard [25]. The specimen's dimensions were 100 × 25 × 10 mm and the test was performed under 0.25 mm/min loading rate, 100 kN load cell, and 80 mm distance between the two points (Figure 2).

Figure 2. Specimen during the three-point bending test.

A stone bend strength is directly related to the stone's material porosity, structure, and texture. The test was carried out in dry conditions. The first stage consists of putting the specimens in a ventilated oven for 48 h at 70 °C and waiting 1 h for them to cool down to room temperature. Figure 3 shows the specimen before and after the bend test.

Figure 3. Image of the specimens, before and after the bending test. (**a**) AS-PU; (**b**) fractured AS-PU; (**c**) AS-EP; and (**d**) Fractured AS-EP.

The second stage is performed by applying a slow and steady load with increment speed defined by the standards. The bend strength is calculated by Equation (3), and then the arithmetic average is calculated.

$$R = \frac{3PL}{2bd^2} \tag{3}$$

where:
R = bend rupture stress (MPa);
P = rupture load (N);
L = distance between the action cleavers (mm);
b = width of the specimen after test (mm);
d = minimum thickness of the specimen (mm).

Abrasive wear tests were performed using a MAQTEST Amsler equipment on three samples with $70 \times 70 \times 40$ mm, according to ABNT/NBR 12042 Brazilian standard [26]. The samples' thicknesses were measured before the wear test and then measured again after abrasive wearing suffered in 500 and 1000 m track.

The hard body impact test was performed using three $10 \times 200 \times 200$ mm samples, according to Annex H of ABNT/NBR 15845 Brazilian standard [25]. It consists of releasing a 1 kg steel ball, at increasing heights, from 20 cm on, over the sample. The stone plate, seated on a sand mattress, receives the impacts until it cracks and rupture occurs. The height at which the material breaks occur is used to calculate the breaking energy using Equation (4).

$$W = m * g * h \tag{4}$$

W = breaking energy (J);
m = steel ball weight (Kg);
g = gravity acceleration (9.806 m/s^2);
h = breaking height (m).

The microstructure of the fracture surface of bend-ruptured specimens was analyzed by a scanning electron microscope (SEM) in a model Super Scan SSX-550 from Shimadzu. The samples were prepared using an adhesive carbon tape enveloped by a gold surface.

Petrography, the main diagnostic technique for pathologies of stone materials, is the analysis of thin sections of stone by transmitted light microscopy, in order to identify, among other things, changes of minerals and microcracks. Using Buehler Petrothin Section System sectioning equipment, the petrographic slides of AS-EP and AS-PU were prepared. The analysis was performed by a geologist in a ZEISS polarized light petrographic microscope.

3. Results

3.1. Determination of the Highest Packaging Mixture and the Minimum Amounf of Resin

To determine the materials' vibration density, 10 different mixtures from the three granulometric ranges were proposed, as shown in Table 2.

Table 2. Results of vibration density.

Mixture	Large (Coarse)	Medium	Fine	Vibration Density (g/cm^3)
1	0	0	1	1.37 ± 0.01
2	0	1	0	1.53 ± 0.01
3	1	0	0	1.13 ± 0.01
4	0	1/2	1/2	1.82 ± 0.02
5	1/2	1/2	0	1.57 ± 0.04
6	1/2	0	1/2	1.64 ± 0.06
7	1/3	1/3	1/3	1.68 ± 0.01
8	2/3	1/6	1/6	1.93 ± 0.01
9	1/6	1/6	2/3	1.87 ± 0.01
10	1/6	2/3	1/6	1.41 ± 0.01

The mixture with the highest vibration density determined the proportion of the waste's granulometric ranges used to manufacture the artificial stone plates. Higher vibration density means better packing and, therefore, fewer voids, which contributes to improving the mechanical properties of the final product. Therefore, mixture number 8, with 1.93 g/cm^3, which, according to the mathematical model presented in Figure 1, corresponds to 67% coarse, 17% medium, and 16% of fine particles, was chosen to manufacture the AS plates. With the optimized apparent dry density (composition number 8), the volume of voids for the load studied was calculated, and with this data, the minimum resin content for the resins was used, according to Table 3.

Table 3. Resin contents of prepared stones.

Artificial Quartzite Stone		Quartzite Charge	
volume of voids (%)		25.95%	
Density (g/cm^3)		2.609	
	Epoxy Resin		Vegetable Polyurethane Resin
Density (g/cm^3)	1.134		1.078
Minimum amount of resin to fill the void volume (%)	13.22		12.65
Resin content used (safety) (%)	15		15

3.2. Water Absorption, Density, and Apparent Porosity

Table 4 shows the apparent density, water absorption, and apparent porosity values of the artificial stones developed with 85 wt% of quartzite waste and both epoxy resin (EP), 85 wt% of quartzite waste and polyurethane resin (AS-PU), natural stone "Cristallo" (NS), and the commercial stone "Branco Aldan" (CS).

Table 4. Physical properties of apparent density, water absorption, and apparent porosity of the artificial stones developed (AS-EP and AS-PU), the commercial stone and the natural stone.

	Physical Properties			
	AS-EP	AS-PU	"Branco Aldan" Commercial Stone (CS)	Cristallo "Natural Stone" (NS)
Water absorption (%)	0.16 ± 0.06	0.14 ± 0.06	0.05 ± 0.01	0.26 ± 0.53
Apparent porosity (%)	0.38 ± 0.15	0.31 ± 0.13	0.13 ± 0.03	0.68 ± 1.38
Density (g/cm^3)	2.35 ± 0.03	2.22 ± 0.04	2.41 ± 0.01	2.63 ± 0.01

According to the results in Table 4, it can be observed that, as expected, both the artificial stones and the commercial stone had lower density than the natural stone. This is attributed to the artificial stone's composition, consisting of low-density polymers, producing a lighter material and consequently reducing logistical costs [21]. Manufacturers of artificial stones report density varying in the range of 2.4 to 2.5 g/cm^3 [9]. AS-EP and AS-PU densities were below this range probably because of the low densities of the polymer resins, epoxy 1.20 g/cm^3 and polyurethane 1.08 g/cm^3.

The commercial stone "Branco Alda" has 2.41 g/cm^3 density that is within the range of the artificial stone's manufacturers and superior to that of developed stones. The commercial stone uses a polyester resin with 1.18 g/cm^3 density as a binder, which can be explained by changes in the values in the production process variables [21]. It is worth mentioning that these results were within the range found by Lee et al. [4], in which by varying compression pressure, vacuum, and vibration frequency, obtained an artificial stone with density values ranging from 2.03 to 2.45 g/cm^3.

The porosity values of the natural stone were above the Chiodi Filho and Rodriguez [27] recommendation, which can be explained by the natural occurrence of flaws and cracks in the material. Consequently, these defects generate a mechanical strength decrease. The commercial stone obtained better values, which is directly related not only to the waste particle variables in the size distribution process but also to the manufacturing processing and to the additive added to the mixture, which promoted better adhesion and modified the resin properties [28,29].

Gomes et al. [20] developed an artificial stone with granite waste and polyurethane resin, using the same technique as the present work, and produced a stone with 0.42% apparent porosity and 0.19% water absorption, relatively higher than those of AS-PU and AS-EP. The artificial stones' low porosity content may have contributed to an excellent adhesion of the granite particles/polymer matrix, with the voids being filled by the resin, forming a material as homogeneous as possible.

Water absorption includes liquids percolation through these voids. Therefore, once the pores are not 100% interconnected through the cracks, water absorption values will always be lower than the apparent porosity ones [30]. For the artificial stones developed, the water absorption values of 0.16% in AS-EP and 0.14% in AS-PU indicate superior properties when compared to natural stone, which justifies the use of the proposed method.

Chiodi Filho and Rodrigues [27] include the water absorption index as one of the three most important technological parameters (flexion, wear, and absorption). All stones obtained water absorption $\leq$0.4%, a requirement suggested for stone selection, which are considered as stone class A1, for indoor and outdoor environments, with frequent wetting and low to high pedestrian traffic for floors [31]. For applications as coating, the ASTM C615 [32] standard indicates that, for granites and marble, the value must be $\leq$0.4% while the NBR 15845 [25] standard indicates it must be $\leq$0.2%.

Borsellino et al. [31], also used epoxy resin in their research and obtained 0.16% water absorption values. Among some works reported, water absorption between 0.04–0.67% is considered an ideal range for building materials for wall and floor coverings [4,10,16,29,33–35].

3.3. Three-Point Flexural Strength

Table 5 presents the three-point flexural strength values of the artificial stones (AS-EP and AS-PU), commercial artificial stone, and the pure resins, epoxy, and natural vegetable polyurethane resin.

Table 5. Results of three-point flexural strength of the artificial stones developed (AS-EP and AS-PU), the commercial stone, the natural stone, and the pure resins.

	Three-Point Flexural Strength (MPa)
AS-EP	27.96 ± 1.86
AS-PU	10.77 ± 0.64
"Branco Aldan" Commercial Stone (CS)	56.25 ± 2.62
Epoxy Resin	93.6 ± 4.7
Vegetable Polyurethane Resin	23 ± 5.3
"Cristallo" Natural Stone (NS)	14.32 ± 1.79

The natural PU resin has lower flexural strength (23 MPa) compared to epoxy resin (93 MPa), which may be related to the polyurethane resin degree of polymerization. However, the study of the stoichiometric quantity between the NCO/OH ratio, responsible for the degree of polymerization, which involves the main reaction sites present in the prepolymer and polyol, was not carried out. Reactions with an excess isocyanate may occur along with parallel reactions that harden the material. On the other hand, reactions with an excess of hydroxyls cause the polymer to soften through the reduction in intercrossed bonds [36], which explains why AS-PU obtained a low flexural strength compared to the other tested stones.

In their study, Mileo et al. [37] classified polyurethane as a ductile material on account of the low flexural strength and large deformation without breakage at the maximum applied load. It may presuppose an inadequate cure or the bubbles excess caused by both the presence of moisture and a plasticizing effect. This validates the natural polyurethane non-rupture, the AS-PU at the maximum applied load, and the appearance of bubbles.

Figure 4 shows typical flexural stress x strain curves, obtained from three-point flexural tests. Comparing the behavior of the two polymeric resins and of the developed artificial stones, it is possible to observe that the addition of load contributed to the material hardening. This is an expected behavior, considering that the incorporation of rigid particles in a polymeric matrix generally increases the material's flexural modulus [4]. Maximum performance can be achieved if the polymer adhesion to the reinforcement is perfect. The stronger the matrix/particle interface, the better the mechanical properties of the developed artificial stone [38].

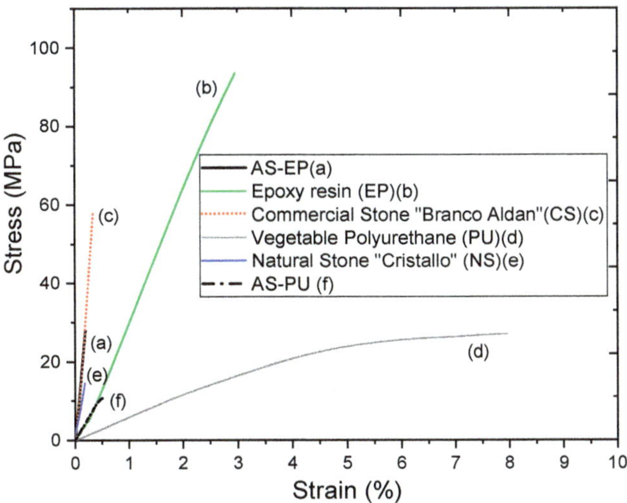

Figure 4. Mechanical behavior with 85% by weight of quartz residue, epoxy resin and natural polyurethane resin (AS-EP and AS-PU), natural stone commercially called "Cristallo" (NS), "Branco Aldan" commercial stone (CS), epoxy resin (EP), and polyurethane resin (PU), in three-point flexural strength tests.

Figure 5 shows the confidence interval (average ± standard error) for the artificial stone flexural strength test of artificial, natural, and commercial stones.

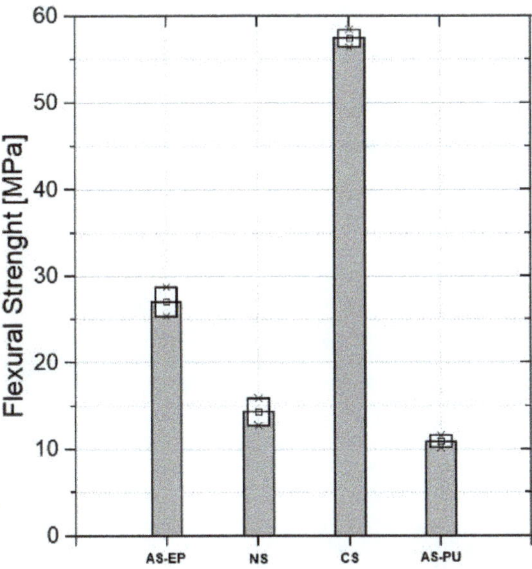

Figure 5. Confidence interval (average ± standard error) of the three-point stone flexural strength test.

The ABNT/NBR-15844 Brazilian standard [28] stipulates that granite stones for coating applications must have at least 10 MPa from the three-point flexural strength test. For coating applications, the ASTM C880 [39] standard indicates that, for granites and marble, the flexural strength must be ≥8.7 MPa, and the NBR 15845 standard [25] demands $\geq$10 MPa. Chiodi and Rodriguez [27] reported that ornamental stones with flexural strength above 20 MPa are classified as high quality for coating applications.

The AS-PU, despite the 10.7 MPa flexural strength, according to the referred standards and its specifications, can be used for civil construction. The "Branco Aldan" artificial commercial stone obtained better flexural strength, which can be attributed to the industrial and continuous manufacturing process as well as the type of additive, the granulometry, and the material used as aggregate, quartz powder, which is one of the hardest minerals (Mohs = 7).

Carvalho et al. [8] and Gomes et al. [9] developed artificial stone based in (quarry and red ceramic) wastes and epoxy resin with flexural strength values of 30 and 32 MPa, very close to those obtained in the AS-EP, of the same matrix, evidencing the influence of the polymeric binder in the final composite strength.

Gomes et al. [20] manufactured artificial stones using 85% granite particles and 15% polyurethane resin derived from castor oil, using the same methodology. The result was an 18 MPa maximum flexural strength, which was higher than AS-PU, possibly because this artificial stone may not have achieved a satisfactory cure.

3.4. Abrasive Wear

In terms of abrasive wear, for the application of ornamental stones on floor coverings, Chiodi Filho and Rodrigues [27] classify the material quality as: high traffic floor (<1.5 mm), medium traffic floor (<3 mm), and low traffic floor (<6 mm). Following technological parameters described above, AS-EP and AS-PU low wear values shown in Table 6 classify them for use in high traffic floors because of an efficient rearrangement in the minerals structure and texture during their agglutination by resins. A determining factor for this excellent result, which influenced the characteristics mentioned above, was the use of the highest close-package mixture to prepare the plates.

Table 6. Amsler wear associated with the thickness reduction of the artificial stones developed (AS-EP and AS-PU), CS, and NS.

Material	Wear Thickness Reduction (mm)	
	500 m	1000 m
AS-EP	0.58	1.20
AS-PU	0.60	1.21
"Branco Aldan" Commercial Stone (CS)	0.50	1.13
"Cristallo" Natural Stone (NS)	0.15	0.35

However, the natural quartzite stone value of 0.35 mm wear was the smallest found on the same route, possibly due to the presence of large mineral quartz crystals in the stone. Among the ornamental stones, quartz is one of the hardest minerals found (Mohs = 7) [38].

We compared the results of this work with some other works that cited that they developed artificial stones using the same method of vacuum, compression, and vibration. Silva et al. [10] produced an artificial stone based on calcite marble waste and 20% w epoxy resin. The artificial stone showed less than 1.5 mm wear on the 1000 m runway, suitable for high traffic floors. Carvalho et al. [29], when evaluating an artificial stone based on 20% w epoxy resin and 80% steel residue from the electrostatic precipitator from the sintering step, found 1.04 and 2.16 mm thickness reduction and 2.16 mm for 500 and 1000 m. The artificial stone was considered suitable for medium traffic floors.

Ribeiro et al. [40] developed an artificial stone using polyester resin and dolomitic marble waste. The developed artificial stone underwent a 5.23 and 8.83 mm thickness reduction for 500 and 1000 m. The wear presented by this material was related to its great porosity because the interfacial adhesion between the particles and the matrix were proven not to be good. It is observed that wear results differ according to types of wastes and resins as well as variables of the manufacturing method.

3.5. Hard Body Impact Resistance

Hard body impact resistance was performed in order to assess the AS-EP and AS-PU level of cohesion and toughness as well as how much energy is needed to dissipate until the material breaks, depending on the maximum drop height of objects supported in its surface. With this result, it is possible to scale the plates in the appropriate size to the usage requirements.

The average rupture height and dissipated energy of the artificial stone AS-PU is 4.4 times greater than that presented by AS-PE, with a 0.39 m height and 3.86 J energy (Table 7). Compared to the epoxy resin, the polyurethane resin has a less rigid behavior that can affect some properties such as flexural strength. On the other hand, it increases AS-PU toughness, precisely because it dissipates more energy. The porosity of AS-EP may also facilitate crack formation, since the pores are naturally stress concentrators [41].

Table 7. Resistance to hard body impact values.

Materials	Rupture Height (m)	Rupture Energy (J)
AS-EP	0.39	3.86
AS-PU	1.72	16.95
"Branco Aldan" Commercial Stone (CS)	>2.0	>19.62
"Cristallo" Natural Stone (NS)	0.2	1.96

Chiodi Filho and Rodrigues [27] proposed a stones classification according to their impact resistance in rupture height, ranging as: very low (<0.30 m), low (0.30–0.50 m), medium (0.50–0.70 m), high (0.70–0.95 m), and very high (>0.95 m) rupture height, and ABNT NBR 15844 [28] establishes that the stones must support a minimum height of 0.3 m. The AS-PU supported height greater than that specified by the standard and by Costa et al. [41] (0.95 m), proving its excellent quality and its feasibility to be applied in areas where there is a higher falling load incidence, such as supermarkets, airports, and industries in general.

The natural quartzite resisted a rupture height of 0.20 m, resulting in 1.96 J of energy, so the artificial stone impact resistance was higher than that of the natural stone. Silva et al. [10] and Gomes et al. [15] developed artificial marble and granite with epoxy resin, which presented cracks at heights of 0.43 m and 0.4 m with 4 J of activation rupture energy, similar to that of AS-EP.

3.6. Microstructure

Figures 6 and 7 present SEM micrographs of the fracture surface sections of AS-EP and AS-PU, respectively, after the three-point flexural test.

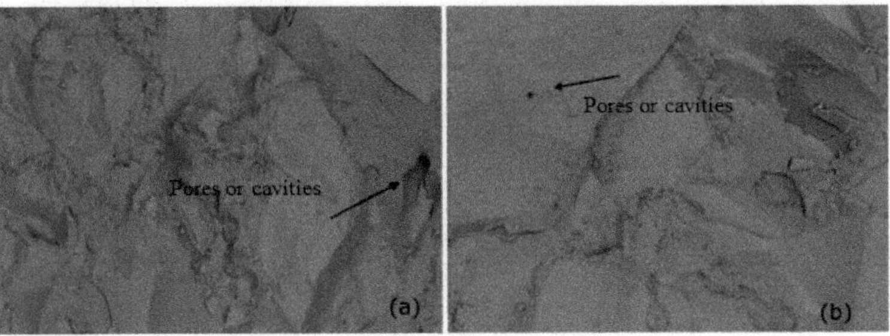

Figure 6. SEM micrographs of the fracture surface sections of AS-EP (**a**) 50×; (**b**)100×.

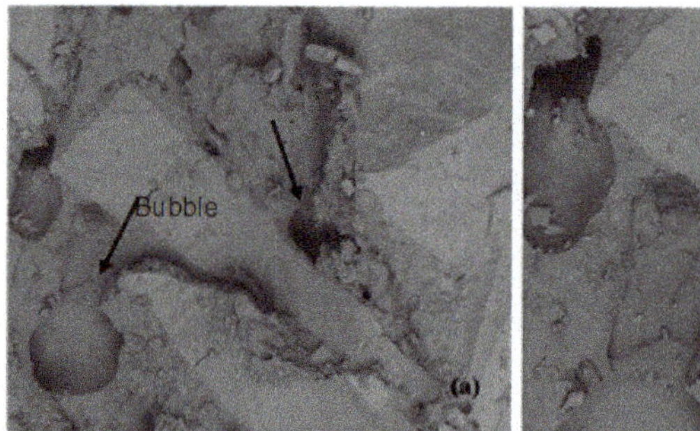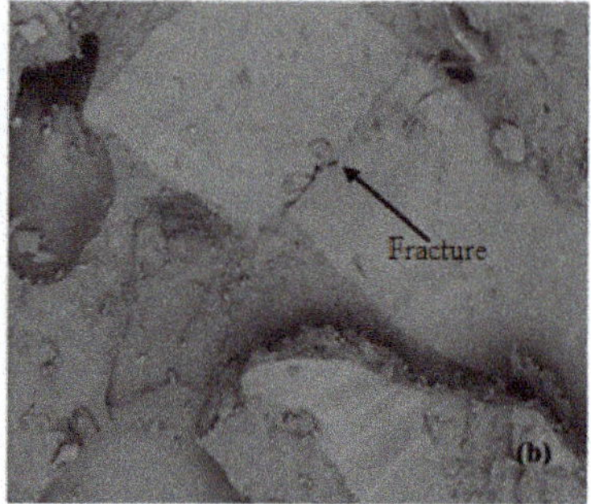

Figure 7. SEM micrographs of the fracture surface sections of AS-PU (**a**) 50×; (**b**)100×.

Figure 6a,b show AS-EP's smoother surface, evidencing a more defined fracture plane that is characteristic of a highly cohesive material. The arrow indicates few evident pores, proving the low water absorption both for the artificial stone with polyurethane and epoxy. According to Debnath et al. [38], the maximum performance of the system occurs through an optimal mineral load wetting by epoxy resin. Therefore, the greater the interface between the matrix and the load, the better the material mechanical properties, this interaction being directly linked to the adhesive bond strength provided by effective wetting in the interfacial regions.

Figure 7a,b show AS-PU's fracture surface SEM micrograph, illustrating the material filling all interstices between the larger grains. Bubbles not eliminated by the vacuum process appear within the matrix, which were already confirmed by the porosity test. It is possible that the bubbles were also generated during the polymerization process between the isocyanate (NCO) and hydroxyl (OH) groups, in which occur parallel reactions mainly involving $NCO + H_2O$ forming high molecular weight compounds such as poly (urethanes/ureas) with excellent mechanical properties. The $NCO + H_2O$ expansion reaction results in urea formation releasing CO_2 and forming bubbles. The release of curing process CO_2, added to the water vapor release at the time of pressing, are evidence of defects formation [42]. Figure 6b points out a quartz grain that fractured after the artificial stone was produced.

3.7. Petrography

Petrographic slides were analyzed for possible physical features that could decrease the mechanical properties of the material, such as microcracks and pores. Quartz crystals, when observed in natural light (polarized plane), are low relief, colorless, and have no pleochroism or cleavage. In polarized light, quartz grains are anhydrous with a prismatic habit, exhibiting wavy extinction and low interference colors. Therefore, as recognizable in Figures 8 and 9, all minerals present in the artificial stone of quartzite are quartz.

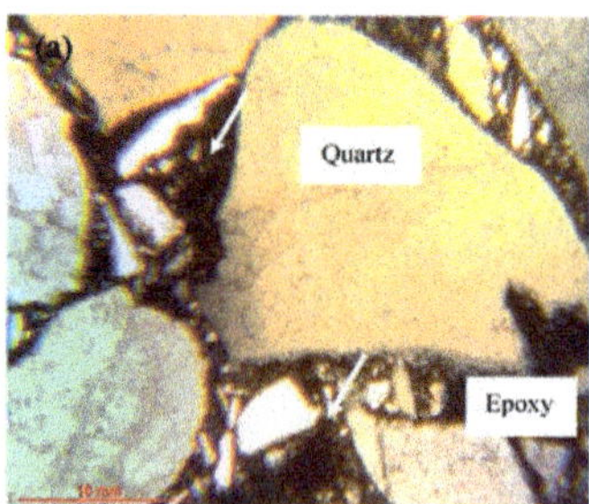

Figure 8. Petrographic slides of artificial stone with quartzite with epoxy AS-EP (**a**) and quartzite with polyurethane AS-PU (**b**).

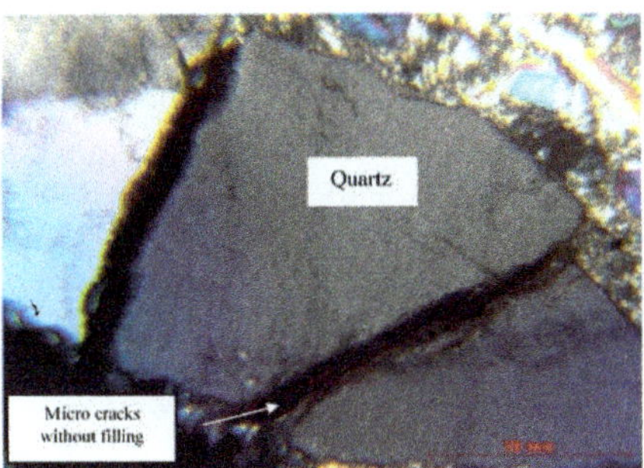

Figure 9. Petrographic slide showing failure in load allocation failure of artificial stone with quartz and polyurethane (AS-PU).

Figure 8a shows the AS-EP slide image, evidencing the optimal load distribution in the matrix, with coarse grains forming a network filled by the matrix and the medium grains and mixed with fine grains. Moreover, in the matrix dispersed region, the presence of small voids (pores) is noticeable. Figure 8b shows the AS-PU slide image, with arrows indicating the presence of small voids. The horizontal arrow indicates a space between the grains, pointing out that the load distribution could have been more efficient.

Figure 9 shows the image of the AS-PU petrographic slide, with the occurrence of microcracks. The microcracks can be attributed to a high compression pressure that, instead of packing the material, as it should, compressed it until it broke [4].

4. Conclusions

The artificial stone is made up of 85% w of quartzite, epoxy resin (AS-EP), and polyurethane resin (AS-PE), using the "vacuum, vibration and compression" method. From the experimental results, the following conclusions can be drawn:

- AS-EP and AS-PU developed with 85 wt% of quartzite waste are a lightweight coating with 2.35 and 2.22 g/cm^3 density, respectively;
- The porosity was lower than 0.5%, classifying both AS-EP and AS-PU as high-quality lining materials for civil construction. The 0.16% and 0.14% water absorption of AS-EP and AS-PU were also lower than $\leq 0.4\%$, a suggested requirement for indoor and outdoor environments, with frequent wetting and low to high traffic on floors;
- AS-EP artificial stone achieved a bend strength of 27.96 ± 1.86 MPa, classifying AS-EP as a high-quality material, due to its bend strength being higher than >20 MPa; compared to the other studied stones, AS-EP had a bend strength twice higher than NS and 50% lower than CS. AS-PU obtained a low bend strength, evidencing a probable inadequate cure or the excess of bubbles, as shown by the SEM; through the abrasive wear test, the low thickness (<1.5 mm) reductions observed in the artificial stone (1.20 mm for AS-EP and 1.21 mm for AS-PU) indicate that both stones can be used for intense traffic floor areas. The NS thickness reduction of 0.35 mm was the lowest, possibly owing to the presence of large mineral quartz crystals in the NS. The hard body impact resistance classified AS-EP as a material for low heights and AS-PU as a material for very high heights enabling its use in areas with a high risk of falling loads, such as supermarkets, airports, and industries in general;
- The petrographic slides analysis revealed that AS-EP has the best load distribution, ensuring greater compaction and presence of microcracked grains during the process of vibration and compression. As demonstrated, the development of artificial stones from wastes of ornamental stones productive chain associated with mining and processing, in terms of its technical feasibility, is proven.

The method adopted for the production of artificial rock demonstrates that it has the potential to be manufactured, which could reduce the negative environmental impacts caused by the natural rock industry. This method is currently the most adopted by researchers and the industry, and has shown better results, using vacuum, compression, and vibration (CVC). Higher mechanical properties and a more cohesive microstructure were exhibited by materials produced by VCV, with residues of marble, crushed stone, quartz, and glass.

The definition of the mixture with the highest packing factor, based on the variation of the particle size percentages, using the Simplex-lattice design (SLD) mathematical model, provided greater cohesion between the particles in studies aimed at artificial rocks. They calculated the minimum resin content, TMR (%), which allowed to reduce the resin content used, and they realized through their results that compacting the particles prevents the formation of pores in the samples, thus reducing the absorption of water and reinforcing the structure, compressed.

The development of artificial rock using materials from renewable sources has already been studied, obtaining satisfactory properties with the use of waste from the ornamental rock industry, with zero toxicity and without releasing gases during its production process—that is, a product totally ecological; it has been made possible through the substitution of resins derived from petroleum for resins of natural source.

Author Contributions: Methodology, C.P.A. and E.A.S.C.; writing—original draft preparation, E.A.S.C. and G.N.S.B.; writing—review and editing, E.A.S.C. and A.R.G.d.A.; resources, C.M.F.V., S.N.M. and M.C.B.G.; supervision, C.M.F.V. All authors have read and agreed to the published version of the manuscript.

Funding: This research was funded by the State University of the Northern Fluminense (UENF), partially financed by CAPES (Coordenação de Aperfeiçoamento de Pessoal de Nível Superior-Brazil), and provided additional financial by CNPq (Coordenação Nacional de Pesquisa). The participation of A.R.G.A. was sponsored by FAPERJ through the research fellowships proc. no: E-26/210.150/2019, E-26/211.194/2021, E-26/211.293/2021, and E-26/201.310/2021 and by CNPq through the research fellowship PQ2 307592/2021-9. The participation of M.C.B.G. was sponsored by FAPES through the research fellowships proc. n°: 80857019 and 84376732.

Institutional Review Board Statement: Not applicable.

Informed Consent Statement: Not applicable.

Data Availability Statement: Not applicable.

Acknowledgments: This work was made possible with the assistance of the Foundation of Support for Research in the State of Rio de Janeiro (E-26/200.847/2021) and (E-26/202.387/2021). They also thank UENF, FAPERJ, CNPq, CAPES, FAPES, CETEM AND IFES for the space and analyses carried out.

Conflicts of Interest: The authors declare no conflict of interest.

References

1. Mosaferi, M.; Dianat, I.; Khatibi, M.S.; Mansour, S.N.; Fahiminia, M.; Hashemi, A. Review of environmental aspects and waste management of stone cutting and fabrication industries. *J. Mater. Cycles Waste Manag.* **2014**, *16*, 721–730. [CrossRef]
2. Souza, L.R.; Ribeiro, R.C.C.; Carrisso, R.C.C.; Silva, L.P.; Pacheco, E.B.A.V.; Visconte, L.L.Y. Application of marble waste in the polymeric industry. *Environ. Technol. Ser.* **2009**, 1–36. (In Portuguese)
3. Vidal, F.W.H.; Azevedo, H.C.A.; Fernández Castro, N. *Ornamental Stone Technology: Research, Mining and Processing*, 4th ed.; Mineral Technology Center (CETEM): Rio de Janeiro, Brazil, 2014; pp. 1–684. (In Portuguese)
4. Lee, M.Y.; Ko, C.H.; Chang, F.C.; Lo, S.L.; Lin, J.D.; Shan, M.Y.; Lee, J.C. Artificial stone slab production using waste glass, stone fragments and vacuum vibratory compaction. *Cem. Concr. Compos.* **2008**, *30*, 583–587. [CrossRef]
5. ABIROCHAS. Balance of Brazilian Exports and Imports of Dimension Stones in 2019. Available online: https://abirochas.com. br/wp-content/uploads/2020/10/Informe-01_2020-Balanco_2019.pdf (accessed on 5 October 2021).
6. Souza, E.F.; Morais, C.R.S. Physical and mechanical characterization of artificial stone with sodium-calcium glass powder residue and unsaturated polyester resin by pressing molding. *Electron. Mag. Mater. Process.* **2018**, *13*, 190–195. (In Portuguese)
7. Demartini, T.J.C.; Rodriguez, R.J.S.; Silva, F.S. Physical and mechanical evaluation of artificial marble produced with dolomitic marble residue processed by diamond–plated bladed gang-saws. *J. Mater. Res. Technol.* **2018**, *7*, 308–313. [CrossRef]
8. Carvalho, E.A.S.; Vilela, N.F.; Monteiro, S.N.; Vieira, C.M.F.; Silva, L.C. Novel artificial ornamental stone developed with quarry waste in epoxy composite. *Mater. Res.* **2018**, *21*, 1–6. [CrossRef]

9. Gomes, M.L.; Carvalho, E.A.S.; Sobrinho, L.N.; Monteiro, S.N.; Rodriguez, R.J.S.; Vieira, C.M.F. Production and characterization of a novel artificial stone using brick residue and quarry dust in epoxy matrix. *J. Mater. Res. Technol.* **2018**, *7*, 492–498. [CrossRef]

10. Silva, F.S.; Ribeiro, C.E.; Rodriguez, R.J.S. Physical and mechanical characterization of artificial stone with marble calcite waste and epoxy resin. *Mater. Res.* **2018**, *21*, 1–6. [CrossRef]

11. Ribeiro, C.E. Development of an Alternative Artificial Marble with Residue from the Marble and Unsaturated Polyester Industry. Ph.D. Thesis, State University of the North Fluminense Darcy Ribeiro, Rio de Janeiro, Brazil. (In Portuguese).

12. *UNE–EN 14618*; Agglomerated Stone: Terminology and Classification. Spanish Association of Standards and Certification: Madri, Spain, 2009. (In Spanish)

13. Suta, S.; Wattanasiriwech, S.; Wattanasiriwech, D.; Duangphet, S.; Thanomsilp, C. Preparation of engineered stones. *Mater. Sci. Eng.* **2019**, *600*, 012009. [CrossRef]

14. Lee, D.J.; Shin, I.J. Effects of vacuum, molde temperature and cooling rate on mechanical properties of press consolidated glass fiber/PET composite. *Compos. Part A Appl. Sci. Manuf.* **2002**, *33*, 1107–1114. [CrossRef]

15. Gomes, M.A.P.M.; Carvalho, E.A.S.; Demartini, T.J.C.; Carvalho, E.A.; Colorado, H.A.; Vieira, C.M.F. Mechanical and physical investigation of an artificial stone produced with granite residue and epoxy resin. *J. Compos. Mater.* **2020**, *55*, 1247–1254. [CrossRef]

16. Ribeiro, C.E.G.; Rodriguez, R.J.S.; Vieira, C.M.F.; Carvalho, E.A.; Candido, V.S.; Monteiro, S.N. Production of synthetic ornamental marble as a marble waste added polyester composite. *Mater. Sci. Forum* **2014**, *775–776*, 341–345. [CrossRef]

17. Erfan, N.A. Recycling of marble calcite waste into useful artificial marble. *J. Adv. Eng. Trends* **2021**, *40*, 7–13. [CrossRef]

18. Gomes, M.L.P.M.; Carvalho, E.A.S.; Sobrinho, L.N.; Monteiro, S.N.; Rodriguez, R.J.S.; Vieira, C.M.F. Physical and Mechanical Properties of Artificial Stone Produced with Granite Waste and Vegetable Polyurethane. In *Green Materials Engineering*; The Minerals, Metals & Materials Series; Springer: Cham, Switzerland, 2019; pp. 23–29.

19. Çelik, M.Y.; Sabah, E. Geological and technical characterization of Iscehisar (Afyon-Turkey) marble deposits and the impact of marble waste on environmental pollution. *J. Environ. Manag.* **2008**, *87*, 106–116. [CrossRef]

20. Rodrigues, J.M.E.; Pereira, M.R.; Souza, A.G.; Carvalho, M.L.; Danta Neto, A.A.; Fonseca, J.L. DSC monitoring of the cure kinetics of a castor oil-based polyurethane. *Thermochim. Acta* **2005**, *427*, 31–36. [CrossRef]

21. Ribeiro, C.E.G.; Rodriguez, R.J.; Carvalho, E.A. Microstructure and mechanical properties of artificial marble. *Constr. Build. Mater.* **2017**, *149*, 149–155. [CrossRef]

22. Cornell, J.A. *Experiments with Mixtures: Designs, Models, and the Analysis of Mixture Data*, 3rd ed; John Wiley & Sons: New York, NY, USA, 2002.

23. *MB 3388*; Soil—Determination of the Minimum Void Index of Non-Cohesive Soils. Brazilian Association of Technical Standards: Rio de Janeiro, Brazil, 1991. (In Portuguese)

24. Shen, Y.; He, J.; Xie, Z.; Zhou, X.; Fang, C.; Zhang, C. Synthesis and characterization of vegetable oil based polyurethanes with tunable thermomechanical performance. *Ind. Crop. Prod.* **2019**, *140*, 111711. [CrossRef]

25. *ABN NBR–15845*; Rocks of Cladding—Test Methods. Brazilian Association of Technical Norms (ABNT): São Paulo, Brazil, 2010. (In Portuguese)

26. *ABN NBR 12042*; Inorganic Material-Determination of the Resistance to Abrasion. Brazilian Association of Technical Norms (ABNT): Rio de Janeiro, Brazil, 2002. (In Portuguese)

27. Chiodi, F.C.; Rodriguez, E.P. *Aplication Guide for Stone in Coating*; Abirochas: São Paulo, Brazil, 2009. (In Portuguese)

28. *ABNT NBR 15844*; Rocks for Coating—Requirements for Granite. Brazilian Association of Technical Standards (ABNT): Rio de Janeiro, Brazil, 2015. (In Portuguese)

29. Carvalho, E.A.S.; Marques, V.R.; Rodriguez, R.J.S.; Ribeiro, C.E.G.; Monteiro, S.N.; Vieira, C.M.F. Development of epoxy matrix artificial stone incorporated with sintering residue from steelmaking industry. *Mater. Res.* **2015**, *18*, 235–239. [CrossRef]

30. Frazão, E.B. *Rock Technology in Civil Construction*; Brazilian Association of Engineering and Environmental Geology: Rio de Janeiro, Brazil, 2002; 132p. (In Portuguese)

31. *ASTM C615*; Standard Specification for Granite Dimension Stone. American Society for Testing and Materials (ASTM): West Conshohocken, PA, USA, 2003.

32. Borselino, C.; Calabrese, L.; Di Bella, G. Effects of powder concentration and type of resin on the performance of marble composite structures. *Constr. Build. Mater.* **2019**, *23*, 1915–1921. [CrossRef]

33. Hamoush, S.; Abu-Lebdeh, T.; Picornell, M.; Amer, S. Development of sustainable engineered stone cladding for toughness, durability, and energy conservation. *Constr. Build. Mater.* **2011**, *25*, 4006–4016. [CrossRef]

34. Peng, L.; Qin, S. Mechanical behavior and microstructure of an artificial stone slab prepared using a SiO_2 waste crucible and quartz sand. *Constr. Build. Mater.* **2018**, *171*, 273–280. [CrossRef]

35. Barani, K.; Esmaili, H. Production of artificial stone slabs using waste granite and marble stone sludge samples. *J. Min. Environ.* **2016**, *7*, 135–141. [CrossRef]

36. Santos, A.M.; Neto, S.C.; Chierice, G.O. Influence of the NCO/OH ratio in the study of the thermal behavior of polyurethane derived from vegetable oil. *Braz. J. Therm. Anal.* **2014**, *3*, 1–4. (In Portuguese)

37. Miléo, P.C.; Baptista, C.A.R.P.; Mulinari, D.; Rocha, G.J.M. Mechanical Behaviour of Polyurethane from Castor Oil Reinforced Sugarcane Straw Cellulose Composites. *Procedia Eng.* **2011**, *10*, 2068–2073. [CrossRef]

38. Debnath, S.; Ranade, R.; Wunder, S.L.; Mccool, J.; Boberick, K.; Baran, G. Interface effects on mechanical properties of particle-reinforced composites. *Dent. Mater.* **2004**, *20*, 677–686. [CrossRef] [PubMed]

39. *ASTM C880*; Standard Test Method for Flexural Strength of Dimension Stone. American Society for Testing and Materials (ASTM): West Conshohocken, PA, USA, 1998.
40. Ribeiro, L.; Carvalho, E.; Gomes, M.L.; Borlini, M.; Monteiro, S.N.; Vieira, C.M.F. Reuse of the Iron Ore Residue through the Production of Coating. In *Green Materials Engineering*; The Minerals, Metals & Materials Series; Springer: Cham, Switzerland, 2019; pp. 23–29. [CrossRef]
41. Costa, F.P.; Fernandes, J.V.; Melo, L.R.L.; Rodrigues, A.M.; Menezes, R.R.; Neves, G.N. The Potential for Natural Stones from Northeastern Brazil to Be Used in Civil Construction. *Minerals* **2021**, *11*, 440. [CrossRef]
42. Marinho, N.P.; Nascimento, E.M.; Nisgoski, S.; Magalhães, W.L.E.; Neto, S.C.; Azevedo, E.C. Physical and Thermal Characterization of Polyurethane Composite Castor Oil Derivative Associated with Bamboo Particles. *Polímero* **2013**, *23*, 201–205.

sustainability

Review

Sustainability and Circular Economy Perspectives of Materials for Thermoelectric Modules

Manuela Castañeda [1], Elkin I. Gutiérrez-Velásquez [1], Claudio E. Aguilar [2], Sergio Neves Monteiro [3], Andrés A. Amell [4] and Henry A. Colorado [1,*]

1. CCComposites Laboratory, Universidad de Antioquia UdeA, Calle 70 No. 52-21, Medellín 050010, Colombia; manuela.castanedam@udea.edu.co (M.C.); elkin.gutierrez@udea.edu.co (E.I.G.-V.)
2. Department of Metallurgical Engineering and Materials, Universidad Técnica Federico Santa María, Valparaíso 2340000, Chile; claudio.aguilar@usm.cl
3. Departamento de Ciência dos Materiais, Instituto Militar de Engenharia, Rio de Janeiro 22290-270, Brazil; snevesmonteiro@gmail.com
4. Grupo de Ciencia y Tecnología del Gas y Uso Racional de la Energía, Facultad de Ingeniería, Universidad de Antioquia UdeA, Calle 70 No. 52-21, Medellín 050010, Colombia; andres.amell@udea.edu.co
* Correspondence: henry.colorado@udea.edu.co

Abstract: The growing demand for energy and the environmental problems derived from this problem are arousing interest throughout the world in the development of clean and efficient alternative energy sources, which involve ecological processes and materials. The materials used in the processes associated with thermoelectric generation technology will provide solutions to this situation. Materials related to energy make it possible to generate energy from waste heat residues, which are derived from various industrial processes in which significant fractions of residual energy are deposited into the environment. However, despite the fact that thermoelectric technology represents some relative advantages in relation to other energy generation processes, it in turn faces some technical limitations such as its low efficiency with respect to the high costs that its implementation demands today, and this has been the subject of intense research in recent years. On the other hand, the sustainability of the processes when analyzed from a circular economy perspective must be taken into account for the implementation of this technology, particularly when considering its large-scale implementation. In this article, a systematic search focused on the sustainability of thermoelectric modules is carried out as a step towards a circular economy model. The review aims to examine recent developments and trends in the development of thermoelectric systems in order to promote initiatives in favor of the environment. The aim of this study is to present a current overview, including trends and limitations, in research related to thermoelectric materials. As a result of this analysis, it was found that aspects related to costs and initiatives related to circular economy models have been little explored, which represents not only an opportunity for the development of new approaches in the conception of thermoelectric systems, but also for the conception of optimized designs that address the current limitations of this technology.

Keywords: thermoelectric modules; cost-efficiency ratio; sustainability; life cycle; circular economy

Citation: Castañeda, M.; Gutiérrez-Velásquez, E.I.; Aguilar, C.E.; Neves Monteiro, S.; Amell, A.A.; Colorado, H.A. Sustainability and Circular Economy Perspectives of Materials for Thermoelectric Modules. *Sustainability* **2022**, *14*, 5987. https://doi.org/10.3390/su14105987

Academic Editors: Luisa F. Cabeza, José Ignacio Alvarez and Asterios Bakolas

Received: 8 February 2022
Accepted: 30 April 2022
Published: 15 May 2022

Publisher's Note: MDPI stays neutral with regard to jurisdictional claims in published maps and institutional affiliations.

1. Introduction

The progressive deterioration of the global ecosystem, as well as the decline of fossil fuel reserves, are some of the consequences of the current global energy problem. This is primarily due to the excessive use of conventional energy sources. In recent years, this has led to a great developments in the use of alternative energy sources that are more environmentally friendly and that in turn have renewable characteristics, which would guarantee that are exhausted in the short or medium term. Among these alternative sources of energy, the sources based on direct solar radiation stand out [1], as well as the sources of use of indirect solar energy, such as hydraulic [2], wind [3], and tidal sources [4]. Other sources are based on biomass [5] and on the use of energy from waves [6]. Likewise, there is potential for geothermal energy [7] and nuclear energy [8]. However, a significant energy source is based on the use of energy residues from industrial processes, and recently there has been a notable boom in the use of thermoelectric systems, which are based on the use of energy waste in the form of heat that is regularly discarded into the atmosphere. In recent years there have been various advances in technologies that have made it possible to make better use of this energy waste [9], since it is not enough to achieve high efficiency, but it is also important to consider the useful life of thermoelectric systems, especially for systems that operate in low temperature ranges. This enables both long-term reliability and the recyclability of the system's components, in order to certify as sustainable development and conserve natural resources for the benefit of present and future generations. However, based on a recent literature review, it has been found that recent developments in these areas are not paying enough attention to the use of sustainable processes and materials, which are fundamental aspects in the development of current technology. This is why the present study focuses on reviewing the main thermoelectric materials used today, and also analyzes the types of modules used and the main applications of these thermoelectric systems. Therefore, the main objective of this analysis is to establish which of the current technologies are designed in a way that favors recycling processes, and therefore, to determine the degree of sustainability of these technologies. In this sense, the present review seeks to characterize, using recent publications from the last decade, which technologies are being implemented, and to what degree they are contributing to a more sustainable production system.

These developments result from a growing demand from the community to produce cleaner manufacturing processes and materials in sectors such as manufacturing, construction and transport that represent around 75% of the energy consumed throughout the world [10].

Recently, considerable attention has been given to the use of thermoelectric (TE) materials as a complement to enhance the efficiency of renewable energy systems based on solar and biomass energy sources [1]. In addition, other applications in industrial sectors are making progress with these systems to become more efficient and reduce the impact on the environment, which translates to a significant recovery of wasted heat normally disposed into the environment. According to various reports, the energy consumption levels of 2019 will significantly increase by 2040 in sectors such as transport and construction, and the use of renewable energy and natural gas use will grow faster than the use of coal and oil [11].

These increases in the energy consumption of the different sectors and in renewable energies result in a promising outlook for TE materials. Studies have reported that about 60% of the energy consumed worldwide is dissipated as waste heat in the transport sector, with the manufacturing and construction sectors also making large contributions of residual heat [12–14]. The recovery of this residual heat, even partially, would represent a significant saving in world energy consumption, as well as significant reductions in greenhouse gas emissions [15], and would be a major contribution to a better environment and to a better circular economy model [16].

Residual heat can be classified into three categories according to the temperatures of the heat sources [11]. In this classification, low temperature ranges are temperatures below 100 °C; average temperatures when the temperatures are between 100 °C and 300 °C; and high temperatures when they are above 300 °C, such as in calcining, forming, heat treating, thermal oxidation and metal reheating processes [11]. Considering these ranges, the operating temperatures of commercial thermoelectric generation modules (TEG) show a great potential for the implementation of heat recovery systems using TEG systems in many different processes [17–20]. It is known that TE materials do not allow 100% heat loss recovery due to their current low efficiency. Furthermore, the high cost of the elements used to construct commercial TEG modules is another factor preventing their widespread implementation in many systems and economies [21–23]. Currently, one of the main goals is to improve the performance of TE materials so that they are competitive and efficient, which is can be obtained with a high Seebeck coefficient and electrical conductivity, while the thermal conductivity has to be low. Factors such as profitability and complexity of processing have limited this efficiency increase [24].

In the recent literature, there is a wide variety of studies that examine the development of sustainable alternatives and promote the development of alternative generation systems to reduce dependence on traditional fossil fuels and to reduce the current rates of emissions. Furthermore, numerous works aim to improve environmental conditions based on the development of clean generation systems. Clearly, the generation alternatives based on thermoelectric systems are viable and reliable; however, the present study goes beyond ensuring the implementation of cleaner generation alternatives to demonstrate that those same proposed solutions are in themselves sustainable. Therefore, in this analysis we intend to present some of the work that is being carried out today to contribute to the construction of a better future for future generations.

This work presents a systematic search on TE materials focused on the sustainability and profitability of TEG modules, and is divided in the following topics: materials, TEG modules, costs, efficiency, comparisons with other types of energy collection or cooling systems, and modeling of properties and construction. Finally, the recycling and end disposal of this type of TE material is reviewed (see Figure 1), with a particular focus on the circular economy for thermoelectric modules.

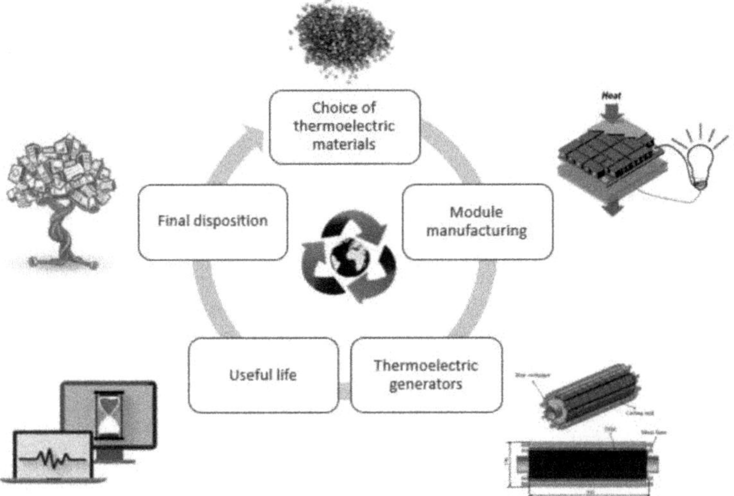

Figure 1. Themes of the article related to the circular economy of thermoelectric modules.

2. Methodology

The methodology proposed for this study was based on a comprehensive systematic review in which the most relevant publications related to the development of thermoelectric generation systems were taken into consideration. Based on this preliminary investigation, a relationship was established between the most relevant sources in order to establish comparisons between them in order to critically analyze the information collected in relation to the sustainability of these systems and to subsequently classify the documentation collected.

In this review, the most relevant publications were selected by the authors and summarized to identify their most notable aspects, and tables and graphs were developed, seeking to synthesize the collected evidence in the best way.

Based on the proposed methodology, a systematic search was initially made in the Scopus database using the keywords "thermoelectric AND modules" with the AND logical connector. In this initial search, a total of 4676 results were obtained. Then, two filters were used to refine the search. In the first filter, only selected review type documents were selected, and in the second filter, the publication date was restricted to the range from 2015 to 2020, resulting in a final total of 61 reviews, as shown in Table 1. These filters were based on reviews by Alam and Ramakrishna [25] and Zheng et al. [26], who provided an overview of research in the area of thermoelectric materials incorporating semiconductor-doped materials and highlighted efforts to date to increase thermoelectric efficiency from different perspectives and the potential for economic and environmental benefits through the improvement of thermoelectric systems.

Table 1. Summary of the initial systematic search for the subject of the review.

Database	Number of Articles Found		
	Unfiltered	Filter 1	Filter 2
Scopus	4676	81	61

Figure 2 presents the classification according to content of the works selected in the initial bibliographic review using four categories: TE materials, TEG modules, applications and sustainability. The classification was applied by grouping the articles into different themes and subtopics after reading the titles, abstract and keywords. From this selection, it was found that 10 articles were not relevant to the topic of TE materials and, out of the 51 relevant articles, some were classified in two of the categories.

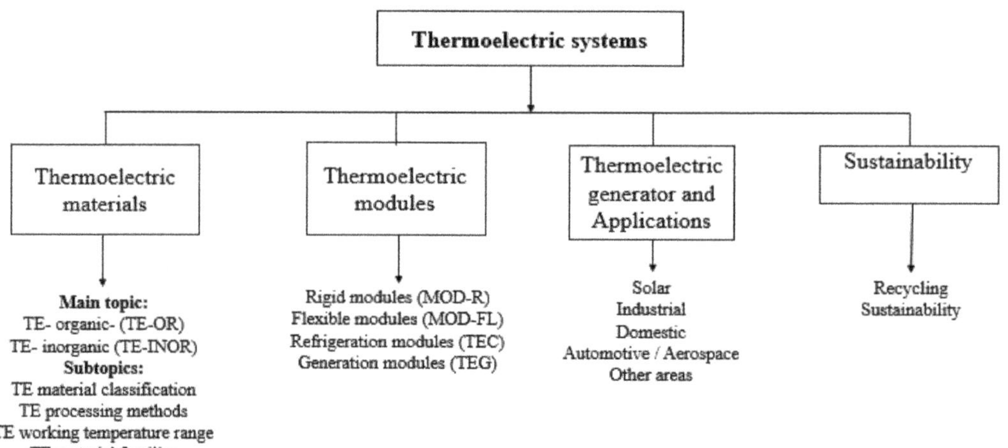

Figure 2. Topics identified in the articles of the initial systematic search.

In this initial systematic search, a knowledge gap was found in the sustainability category in which themes of profitability, cost/efficiency ratio and recycling, among others, were developed. For this reason, it was decided to focus on this area in the present analysis.

Regarding the development of this review, six basic themes related to the sustainability of these types of materials were considered: efficiency, cost, recycling, sustainability, life cycle and profitability. For this, after the removal of duplication in articles, four filters were applied. The first filter was the year of publication in which only the articles published from 2010 to 2020 were considered. The second filter was the type of document; in this case only the research articles were considered. The third filter was based on reading the title and abstract, and finally, a thorough reading of the selected articles was carried out, which resulted in a total of 36 relevant articles, from which the analysis presented below was developed. Table 2 shows the search strings used and the results of their systematic debugging. Likewise, Figure 3 shows the results of the search carried out, where

Figure 3a shows the data obtained by percentage for each search string. Figure 3b shows a classification of the publications analyzed by year, revealing the growing interest in TEG. In Table 3, the publications and topics covered by each of these strings are presented according to the topics to be addressed in this study.

Table 2. Search strings and filters used in the systematic search.

Search Strings	Number of Articles Found				
	Unfiltered	Filter 1	Filter 2	Filter 3	Filter 4
"Thermoelectric" AND "Modules" AND "Recycling"	37	23	22	6	5
"Thermoelectric" AND " Modules" AND "Cost" AND "Efficiency"	211	115	105	25	12
"Thermoelectric generators" AND "Cost" AND "Efficiency"	298	249	143	9	5
"Thermoelectric generators" AND "Sustainability"	25	24	13	5	4
"Thermoelectric" AND "Modules" AND "Life cycle"	25	19	11	3	2
"Thermoelectric modules" AND "demand"	43	38	24	5	1
"Thermoelectric" AND "Materials" AND "Modules" AND "Cost" AND "efficiency"	92	77	45	22	7

Filter 1: year of publication of the article from 2010 to 2020, Filter 2: type of document "article type", Filter 3: reading of title and summary, Filter 4: reading the article.

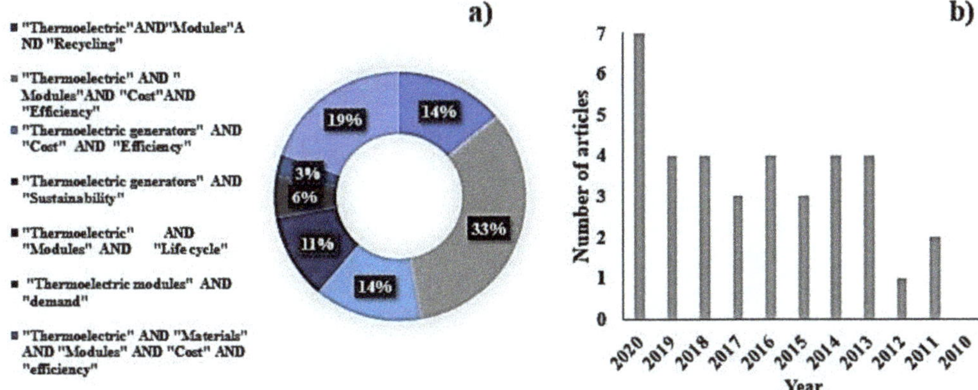

Figure 3. Systematic search results for (**a**) number of articles found per search string, and (**b**) number of articles per year.

Table 3. Articles found in the systematic search with their classification by subject matter.

Ref.	Year	MAT	MOD	TEG	COST	EFF	MDL	RE
[27]	2020	X	X			X		
[28]	2020		X			X	X	
[22]	2020		X		X	X		
[21]	2020		X	X	X	X	X	
[16]	2020			X	X	X	X	

Table 3. *Cont.*

Ref.	Year	MAT	MOD	TEG	COST	EFF	MDL	RE
[29]	2020		X	X		X		
[18]	2020		X					X
[30]	2019	X				X	X	
[31]	2019		X		X	X	X	
[32]	2019		X			X	X	
[23]	2019		X					
[33]	2018	X			X	X		
[34]	2018		X			X	X	
[35]	2018		X			X	X	
[36]	2018	X	X		X	X	X	
[37]	2017		X					X
[38]	2017		X					X
[39]	2017	X	X			X		
[40]	2016	X	X		X	X	X	
[41]	2016		X			X	X	
[42]	2016			X	X		X	
[43]	2016			X	X	X	X	
[44]	2015	X	X			X		
[45]	2015		X	X		X	X	
[46]	2015	X	X			X	X	
[47]	2014		X	X		X	X	
[48]	2014		X			X		
[49]	2014		X				X	
[50]	2014		X					X
[51]	2013		X		X	X		
[52]	2013	X	X	X		X		
[53]	2013		X					X
[54]	2013			X	X	X	X	
[55]	2012		X		X		X	
[56]	2011	X	X			X		
[57]	2011					X		

MAT: TE materials, MOD: thermoelectric modules, TEG: thermoelectric generators, EFF: efficiency, MDL: modeling, and RE: recycling.

3. Results and Discussion

3.1. Thermoelectric Materials

It is estimated that between 50% and 80% of the studies related to thermoelectric generation systems (TEG) focus on the TE materials themselves. Therefore, the development of materials that have a low cost and high efficiency is fundamental to achieving commercial profitability in these systems [17,18].

In the systematic search carried out, 10 articles were found in which the TE material was studied in depth in order to increase its efficiency, decrease costs or develop viable materials in both environmental and sustainability aspects by analyzing factors such as the material's abundance and its degree of toxicity among others. Table 4 presents a classification of these items according to the cost, efficiency and sustainability of the TEG modules. In this table, the cost column expresses the materials cost reduction with a minus sign ($-$), and the materials cost increase with a plus sign ($+$). In this aspect, the elements that compose in some cases the type of processing are taken into account. Note that in most of the articles reviewed, the cost of the material was not stated explicitly. For efficiency, the minus sign refers to materials with low efficiencies (ZT < 1) and the plus sign ($+$) for materials with significant efficiencies (ZT > 1). Finally, the sustainability of the materials is assessed using measures such as the availability, toxicity and useful life; therefore, a minus sign indicates that the material is unsustainable and a plus sign that it is sustainable.

Table 4. Classification of the articles found according to the results obtained in the development of thermoelectric materials.

Ref.	Cost	Efficiency	Sustainability	Year
[27]	$-$	$+$	$-$	2020
[30]	$-$	$-$	$+$	2019
[33]	$-$	$+$	$-$	2018
[36]	$+$	$+$	$+$	2018
[39]	$-$	$-$	$+$	2017
[40]	$-$	$+$	$+$	2016
[44]	$-$	$+$	$-$	2015
[46]	$-$	$-$	$+$	2015
[52]	$-$	$+$	$+$	2013
[56]	$-$	$+$	$+$	2011

Cost: $-$, Low cost; $+$, high cost. Efficiency: $+$, high efficiency; $-$, low efficiency. Environmental aspects and sustainability: $-$, unsustainable; $+$, more sustainable.

Among the articles shown in Table 3, three case studies were found [19,58,59] that examined cost of TE materials, including their cost effectiveness and the processing technique used to develop them.

Among the cost-effective materials, the use of oxides as raw materials stands out. Hung et al. [43] studied different oxides (Na_2CoO_4, $Ca_3Co_4O_9$, ZnO, $SrTiO_3$ and $CaMnO_3$) in order to decrease costs in the manufacture of TEG modules since the cost of manufacturing the oxides is approximately 1.1 \$/kg, which is equivalent to only a quarter of composite materials composed of metals and rare earths. On the other hand, Lee et al. [39] studied the potential of TiO_{2-x} for TE materials manufactured by plasma deposition. Ozturk et al. [30] studied two types of oxides: $Ca_{2.5}Ag_{0.3}X_{0.2}Co_4O_9$ type n and $Zn_{0.96}Al_{0.02}Y_{0.02}O$ type n, where X and Y are different dopants manufactured using the sol-gel method. In these works, the benefits of using oxides as raw materials are highlighted. Among these benefits are low cost, abundance, resistance to high temperatures, as well as simplified manufacturing processes not requiring controlled atmospheres. However, it can be seen that the purpose of these studies is to improve the efficiency of materials. In Table 5 it can be seen that the oxides present the least merit (efficiency). According to the review, the modules' oxide-based TEGs can increase their efficiency through the use of special processing [37] or doping techniques [28], which makes their use more viable.

Table 5. Thermoelectric materials type, merit rating, and temperature.

Material	Type	ZT	T (°C)	Ref.	Material	Type	ZT	T (°C)	Ref.
SnSe	p	2.5	627	[36]	$Pb_{0.93}Sb_{0.05}S_{0.5}Se_{0.5}$	n	1.7	627	[5]
FeNbSb	p	1.5	927	[44]	$Si_{80}Ge_{20}$	n	1.5	727	[36]
Bi_2Te_3	p	0.9	100	[35]	$AgPbmSbTe_2+m$	n	2	527	[36]
Na_2CoO_4	p	0.75	527	[46]	$Cu_xBi_2Te_{2.7}Se_{0.3}$	n	1	127	[36]
$Ca_{2.5}Ag_{0.3}Eu_{0.2}Co_4O_9$	p	0.57	800	[30]	$Mg_2(Si_{0.4}Sn_{0.6})_{0.99}Sb_{0.01}$	n	0.8	327	[29]
$Ca_{2.5}Ag_{0.3}Er_{0.2}Co_4O_9$	p	0.54	800	[30]	$Mg_2(Si_{0.53}Sn_{0.4}Ge_{0.05}Bi_{0.02})$	n	1.4	527	[29]
$Ca_{2.5}Ag_{0.3}Nb_{0.2}Co_4O_9$	p	0.52	800	[30]	ZnO	n	0.6	902	[46]
$Ca_{2.5}Ag_{0.3}Sm_{0.2}Co_4O_9$	p	0.51	800	[30]	$SrTiO_3$	n	0.4	827	[46]
$Ca_{2.5}Ag_{0.3}Lu_{0.2}Co_4O_9$	p	0.50	800	[30]	$CaMnO_3$	n	0.3	902	[46]
$Ca_3Co_4O_9$	p	0.5	827	[46]	$Zn_{0.96}Al_{0.02}Ge_{0.02}O$	n	0.04	800	[30]
$Mm_{0.28}Fe_{1.52}Co_{2.48}Sb_{1.2}$	p	0.5	477	[41]	$Zn_{0.96}Al_{0.02}Ga_{0.02}O$	n	0.17	800	[30]
Ca2.5Ag0.3Yb0.2Co4O9	p	0.47	800	[30]	$Zn_{0.96}Al_{0.02}In_{0.02}O$	n	0.12	800	[30]
$MnSi_{1.75}Ge_{0.01}$	p	0.4	527	[29]	$Mg_2Si_{0.4}Sn_{0.6}$	n	1.2	450	[35]
$MnSi_{1.81}$	p	0.3	400	[35]	Bi_2Te_3	n	0.8	100	[35]
$Ca_3Co_2O_6$	p	0.25	877	[35]	TiO_{2-x}	n	0.132	477	[28]
H-SnSe	p	2	427	[36]	Ni	n	0.020	477	[28]
BiBaCuSeO	p	1	527	[35]	$Yb_{0.09}Ba_{0.09}La_{0.05}Co_4Sb_{12}$	n	1.2	477	[41]
PbTe-SrTe+Te 2%	p	1	350	[33]	$Pb_{0.93}Sb_{0.05}S_{0.5}Se_{0.5}$	n	1.7	627	[27]

On the other hand, the use of cheap and abundant materials such as lead-based materials or silicides has also been the subject of recent studies. Han et al. [33] studied the feasibility of PbTe-SrTe base materials doped with 2% Te. They concluded there was a cost reduction through a low-cost processing method such as stable screen printing, although one of the base materials and the tellurium are high cost and low abundance elements. Jiang et al. [27] proposed PbS as an alternative to the base material PbTe, arguing that by doping with Sb and Se, efficiency can be considerably improved, in addition to them being abundant and low-cost materials.

Fu et al. [44] and Salvador et al. [52] developed materials such as FeNbSb (Half-Heuslers) and $Yb_{0.09}Ba_{0.09}La_{0.05}Co_4Sb_{12}$ (skutterudite), respectively, which are composed of low-cost and abundant elements, and by doping techniques are able to increase their efficiency. However, Ouyang et al. [36] evaluated some of the latest generation materials and recommended that materials such as skutterudites and half-heuslers could only be used in applications where cost is not of concern, due to the high manufacturing costs of these materials. On the other hand, Skomedal et al. [40] suggested the use of magnesium silicides as a favorable TE material, due to their low cost, abundance and low toxicity, despite their low efficiency when doping with elements such as Sn and Sb. They concluded that materials based on magnesium silicides are recommended for applications where low cost or low weight are more important than efficiency.

In addition, Homm et al. [56] analyzed some TE materials such as SiGe, PbTe, Bi_2Te_3, $FeSi_2$ and ZnO. The authors classified them according to selection criteria for different applications that required certain specifications for temperature, efficiency and cost, but taking into account the environmental aspects that each one presented.

According to the present review, it is observed that there is a conflict between the aspects of cost, efficiency and sustainability. Figure 4 presents a classification based on these aspects of recent studies addressed in this analysis. Three articles were found involving costs and efficiency in zone A, [27,29,44]; three articles between costs and sustainability in zone B [30,39,46]; one article involving efficiency and sustainability in zone C [36]; and tree articles involving all aspects, costs, efficiency and sustainability in zone U [40,52,56]. From this classification it is concluded that the oxides are inexpensive TE materials with important advantages. In particular, they are abundant, do not require high-cost processing, and resist high temperatures, which prevents premature degradation of the TE material. Moreover, they enable the formation of robust materials with a longer useful life, and have a good cost-sustainability ratio. However, their efficiency is reduced with respect to the commercially used TE materials, which prompts us to think about the different research approaches to improve them, such as nano-structuring, electronic band engineering, quantum confinement, as well as strategies such as crystal electron glass phonon, doping, and introduction of defects, among others. However, the use of any of these techniques requires specialized and complex processes, which would be reflected in the final cost of the product and would probably mean that the cost-efficiency ratio is not viable for developments in a commercial environment. Therefore, the development of this research is of utmost importance for providing not only a better future perspective of TE materials, but also because there are few investigations that specifically address the economic component of these materials.

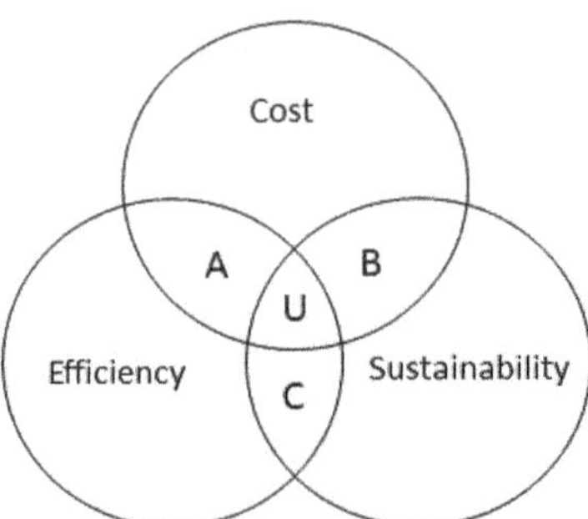

Figure 4. Classification of the articles found according to the results obtained for TE materials.

Furthermore, from a sustainability perspective, little information is available on commonly used TE materials. An example of this is the use of toxic materials such as lead, tellurium and bismuth in their fabrication. Therefore, an important aim of research is to explore is the environmental risks that these materials can present at different stages of the useful life of TEG modules and to search for abundant and low-cost elements.

Interest in certain thermoelectric materials is based on a combination of their characteristics and performance. Figure 5 shows the trends in the number of publications in recent years in relation to some representative thermoelectric materials, according to a survey carried out in the Scopus database. The figure shows the growing research interest in these thermoelectric materials.

Recent Development of New Materials for Thermoelectric Applications

The increases in the ZT values are produced especially by the decrease in the thermal conductivity of crystal lattices, and the recent advances in the development of new TE materials are based on the search for mechanisms that make it possible to minimize the thermal conductivity in the crystal structures of TE materials. Advances in TE materials provide measurable improvements in ZT values through the use of nanotechnology-based techniques. Nanophonon metamaterials provide special local resonance states in semiconductor materials for suppression of thermal conductivity. According to Ouyang et al. [60], nascent theories are being forged in the field of TE materials. Among the most promising are the coherent phonon theories (https://www.nature.com/articles/nmat3826, aacessed on 23 August 2021), the nanophonon metamaterial [61], the rattling effect [62], the topological phonon [63,64] and the topological electron [65]. Likewise, the synthesis of low-dimensional materials would allow the separation of related thermoelectric parameters to optimize thermoelectric performance. Among the advances in this field, the 1D Nanowires stand out [66,67], as well as the 2D Materials [68,69] and the Nanomesh Structures [70,71]. Finally, it should be noted that given the recent advances in computing, artificial intelligence and machine learning in combination with atomic simulation techniques, the development of new tools to predict new structures and characteristics of novel materials is envisioned, and these will provide accurate forecasts of the inherent properties of TE materials.

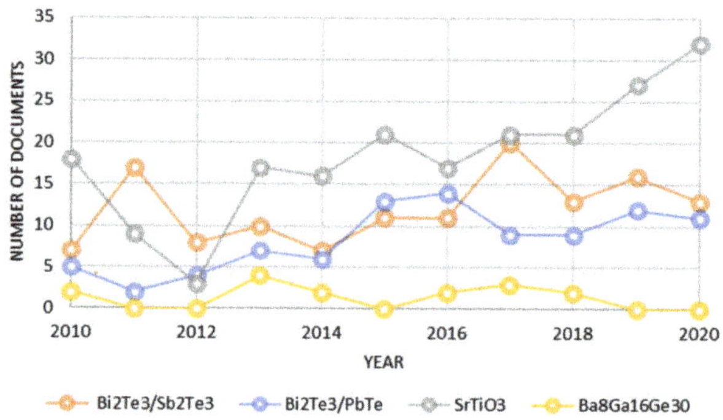

Figure 5. Research in thermoelectric materials in recent years.

3.2. TEG Modules

In the systematic search carried out, 29 articles related to TEG modules were found. We observed that one of the most commonly addressed topics is the efficiency/cost relationship presented by TE materials. This aspect is usually approached from several points of view, such as cost reduction, varying the TE material, or through a design of the TEG module that preserves its efficiency. However, another important factor that must be taken into account is the sustainability of the modules, which ranges from the analysis of their useful service life to the study of their final disposal. In Figure 6, the classification of the articles according to their content can be observed. These were classified by costs, sustainability, efficiency and modeling. In this graph, one article as found classified involving costs [42]; seven involving efficiency [27,39,44,49,52,56,57]; one in modelling [48]; and six in sustainability [18,23,37,38,50,53]. In the common areas D, E, F, G, H, it was not found papers between costs and sustainability and costs and modelling in zones D and G respectively. Between costs and efficiency, zone E, three articles were found [18,22,55]; while for zone F, between efficiency and modelling, seven were found [28,34,35,41,45–47]. Finally, ten references were found in thermoelectric modules, zone H [16,21,30–32,36,40,41,43,55].

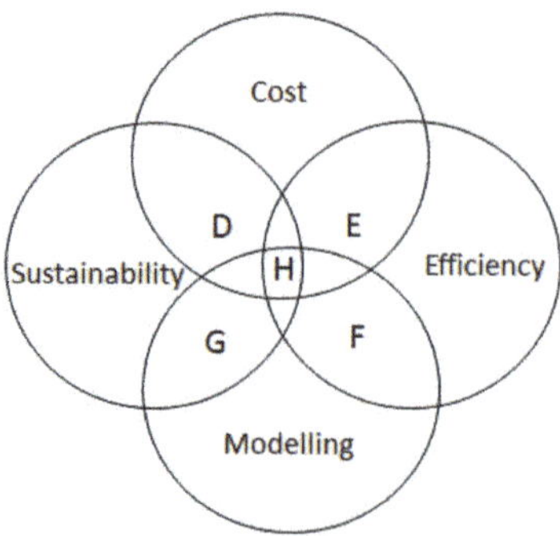

Figure 6. Classification of the articles found according to the results obtained for thermoelectric modules.

3.3. Cost/Efficiency

Currently, TEG modules are not in such high demand as could be expected as they are a clean source of energy that does not require exhaustive maintenance. However, the high cost/efficiency ratio that TEG modules present means that their introduction into the industry is difficult and still limits their viability.

3.4. Module Manufacturing

With the aim of decreasing the production costs without affecting the efficiency of the modules, some authors focus their research on the design of TE modules, the search for new materials in order to decrease costs, or increasing their efficiency. The use of oxides, half-heusler materials, skutterudites, and composite materials (organic/inorganic) can be a solution to overcome the limitations of the TEG modules. It is important, however, that these studies be developed in a holistic context oriented to applicability.

Studies such as those by Salvador et al. [52], on skutterudite encapsulated modules with a ZT of 1, enable this type of module to compete directly with the efficiencies of commercial PbTe modules. However, to date, commercial modules with this type of materials are scarce and costly due to their current manufacturing methods. Fu et al. [44] evaluated the potential of doped half-heusler materials with an efficiency of 6.2% and a power density of 2.2 W/cm^2. These modules can resist high temperatures (~927 °C) and are an economical alternative to commercially used TE materials.

In order to reduce costs in TEG modules and to improve their coupling to any surface, Lee et al. [39] studied the manufacture of TEG devices and basic electronics using titanium oxides (which operate at temperatures ~500 °C), and deposited them by plasma sintering, thus obtaining an assembly of thermocouples connected in series and in parallel with an efficiency of 0.85% and an electrical power of 2.43 mW at 450 °C. The authors did not conduct an economic analysis on the assembly, only referring to its easy and low manufacturing cost through the elimination of many of the parts that make up the commercial TEG modules. On the other hand, Yazawa et al. [54] proposed the use of flexible modules that incorporate organic/inorganic composite materials and reduce costs but with a reduced performance (ZT between 0.01 and 0.25), a performance that is relatively low compared with commercial TE materials.

Anderson et al. [23] carried out a techno-economic analysis on the total cost of TEG devices, finding that the use of impure TE materials such as oxides or other types of cheaper TE materials are not the most feasible option at a cost level, since TE material only represents 15% of the total value of a TEG module.

Table 6 presents some characteristics of commercial modules such as base material, dimensions, power output, open circuit voltage, operating temperature range and cost. Currently, the most common TEG modules are those manufactured from bismuth telluride. Among these, a great variety can be found in which characteristics such as their configuration, output power and circuit voltage might vary. Most modules can work at maximum temperatures between 320 °C and 350 °C, but their optimal operating conditions are around 250 °C. These types of modules can be found in the market with prices ranging between $10 and $28 US.

Likewise, commercially it is possible to find modules that resist higher temperature ranges, such as TEG PbTe-BiTe modules. These TEG modules can work at maximum temperatures of around 360 °C, and commercially they can be found with better output powers than the BiTe based ones. Naturally, the improvement of these characteristics is reflected in their cost.

Table 850 °C reaching 6% efficiency, being attractive for the recovery of residual heat at high temperatures. However, this type of component is up to seven times the cost of traditional BiTe-based modules. This makes them less attractive due to their cost-efficiency ratio.

Table 6. Technical characteristics and cost of commercial modules.

Commercial Modules	Materials	Dimensions (W × L × H) (mm^3)	Output Power (W)	Open Circuit Voltage (V)	Operating Temperature Range	Cost ($ US)	Ref
Hz-1	BiTe based	29.21 × 29.21 × 5.08	2.3	6.1	50° to 250°	$10.00	[72]
Hz-9	BiTe based	62.74 × 62.74 × 6.63	9	6.1	50° to 250°	$15.00	[72]
Hz-14	BiTe based	62.74 × 62.74 × 6.63	14	3.1	50° to 250°	$25.00	[36]
Hz-14HV	BiTe based	61.05 × 71.05 × 7.87	14	8	50° to 250°	$25.00	[17]
Hz-20	BiTe based	74.68 × 74.68 × 5.08	20	4.5	50° to 250°	$50.00	[13]
Hz-20HV	BiTe based	74.50 × 68.00 × 5.00	20	10.8	50° to 250°	$50.00	[72]
TEP1-1264-3.4	BiTe based	40 × 40 × 5	5.4	10.8	30° to 300°	$40.89	[11]
TEG1-126610-5.1	BiTe based	40 × 40 × 4	5.1	7.8	30° to 300°	$28.41	[73]
TEM 070-6006	SnTe based	40 × 40 × 4	16.8	-	70° to 600°	-	[12]
TEG1-PB-12611-60	PbTe-BiTe	56 × 56 × 7.7	21.7	9.2	30° to 350°	$69.00	[14]
CMO-32-62S	BiTe cold site-Calcium	64.5 × 64.5 × 8.3	12.30	12.8	50° to 800°	$375.00	[73]
CMO-25-42S	manganese oxide hot site	42 × 42 × 7	7.5	10.0	50° to 800°	$330.00	[73]

3.5. Module Design

Commercial modules, such as those from Table 6, generally have a pre-designed configuration from the supplier, which means that for some applications, they are not suitable or do not show their optimal performance. On the other hand, the design is an important factor since by means of the design parameters, the efficiency, the cost and the useful life of the modules can be improved. Likewise, most of the studies related to TEG components or systems are carried out using simulation tools [15], due to the complexity of their manufacture, as well as the cost that these would generate for their development if they were carried out exclusively by experimental means. Some of the most relevant studies in this regard are listed below.

Segmented modules represent one of the most viable alternatives from the view point of design. In these, various TE materials are used to manufacture the module legs, seeking to increase the working temperature, minimize the thermal effects, and increase the efficiency of the TEG modules. Recent work related to the manufacture of segmented modules includes the work of Hung et al. [46], who implemented commercial TE materials in cold areas of the cell and oxides in hot areas, in order to increase the working temperature of the modules. The viability of these TEG modules was analyzed using numerical modeling, which found that the oxide-segmented modules have an efficiency of around 10%. Ouyang et al. [36] carried out a study in which high-ZT TE materials were evaluated by finite element analysis. A systematic model was achieved for the segmented modules, finding a cost-performance ratio of ~0.86 $/W with an efficiency of 17.8%. Similarly, Jiang et al. [27] evaluated a TEG module composed of np Bi2Te3 and np PbS/PbTe, which exhibited an efficiency of 11.2% with a $\Delta T = 317\,°C$. In addition, they optimized the ratio of the legs at low and high temperatures, determining the optimal ratio to be 7:17. The maximum power obtained in this TEG module was 0.53 W for a $\Delta T = 312\,°C$.

During the design of TEG modules, the optimization of parameters such as the length of the legs or the number of thermocouples of the TE materials helps to reduce the amount of material used, without compromising the efficiency of the modules. According to Rezania et al. [49], the temperature differences at the n- and p-type junctions of TE elements are not identical. Such temperature differences are lower in n-type TE elements, compared with those in p-type TE elements, due to the higher thermal conductivity in the n-type material. Consequently, the footprint size of the n-type element must be larger than the footprint of the p-type TE element, due to the higher thermal conductivity in the n-type material. Therefore, the optimal ratio of footprint areas to achieve the maximum generation and the best cost-efficiency ratio in thermoelectric modules must satisfy that An/Ap < 1, where An and Ap are the footprint areas of the the n-type and p-type junctions, respectively. Brito et al. [41], found that when the thickness of the TE elements is smaller, the electrical resistance is reduced, but this will impact the ΔT of the TE module, because a lower thermal resistance will increase the thermal output and attenuate the temperature difference between the hot and cold sources. However, this will only occur if the usable hot source is low and the other thermal resistances are high enough to significantly affect the ΔT of the TE module. In their study, Dongxu et al. [31] found that the thickness of the TE legs can be reduced to 1.1 mm, which is 4 mm less than commercial modules, while still achieving the same efficiency. In all these works, simulation tools were successfully used to find the relationships between the different parts of the modules and their respective powers.

3.6. Useful Life

The evolution in the design of TEG modules has also contributed to increasing their useful life. Moreover, the thermal stresses to which the TEG modules are subjected, affect them adversely by the formation of microcracks and the expansion and contraction of TE materials. To address this problem, Skomedal et al. [40] incorporated spring-supported contacts in the legs of TEG modules to dampen thermal expansion and contraction in the modules; these authors concluded that good diffusion barriers and possible coatings can reduce oxidation of the hot side electrodes and interconnections. Likewise, they noted that the TEG module could generate 1 to 3 W/cm^2. Furthermore, Ming et al. [45] studied, via numerical analysis, how the non-uniform flow causes the junctions of the TEG modules to be damaged and also decreases their output power. In addition, Merienne et al. [32] investigated the effect of thermal cycles on commercial TEG modules, finding that when rapid temperature changes are applied, their output power can decrease by up to 61%, compared with a module in which the heat flow is constant and the temperature changes are minimal. Therefore, it is of utmost importance to analyze in detail the design of TEGs and the operating conditions to which they will be subjected in order to establish applications that allow them to extend their useful life.

The current commercial TEG modules have still not reached the economic feasibility, nor the efficiency required for thermal energy recovery applications. Therefore, as mentioned above, the characterization of commercial TEG modules, in the specific conditions to which they are going to be subjected, requires the use of expensive control equipment, or manual processes, which leads to difficulty in making long-term measurements. Therefore, Elzalik et al. [28] developed a characterization method that allows precise and inexpensive estimation of the maximum power point and the dynamic parameters of the TEG module. The proposed procedure can be used with different sources of residual thermal energy and under different operating conditions.

In relation to the sustainability of TEG modules, authors such as Khanmohammadi et al. [22] and Heber et al. [21] state that the amortization period and the power ratio of commercial TEGs make them not viable to be used exclusively as a generation method. Therefore, they recommend using them as a complement to other methods of recovering residual thermal energy, thus generating an integrated system, or the development of segmented TEG modules that allow greater efficiency and profitability.

3.7. Thermoelectric Generators

As previously mentioned, the main limitation for the use of TEG is its cost-efficiency ratio; therefore, this type of generator still does not compete with conventional mechanical thermodynamic systems or the electrochemical systems of Rankine, Stirling, Brayton, expansion devices or fuel cells. These types of conventional generation systems have a wide application ranging from heat utilization with a lower output power range to greater than 10 kW. For their part, TE materials are usually utilized for applications with an output power of less than 10 kW [51], which makes them suitable for specific power requirements.

Yazawa et al. [40], focused their research on the economic viability that TEGs can achieve with respect to other types of electricity generation systems. The authors found that TEG modules with a ZT of 0.8 can achieve a cost-performance ratio of 0.86 \$/W. This makes TEG systems competitive with other power generation systems. It was also found that TEG systems have commercial profitability if they have a cost-performance ratio of less than $1/W [34].

In the manufacture of TEG systems, the main component is the TEG module, and the other components such as heat sinks, and supports help the system to perform better. Mori et al. [42] estimated that the TEG modules only represent 60% of the cost of whole systems and focused the design of the system on heat concentration structures that would help to double the efficiency of power generation and that could reduce the total cost in the system by half. Likewise, Hendricks et al. [43] warned that the cost of the heat exchanger, which is the element of the system that most frequently increases the total costs of TEG modules, should be further investigated to achieve profitability of the system.

Lately, the automotive industry has initiated investigations into TEG systems due to the considerable losses of caloric energy that arise from the combustion process, which could be used to power other vehicle systems. Indeed, the incorporation of these energy recovery systems, could significantly reduce CO_2 emissions into the atmosphere.

Arsie et al. [47] proposed the incorporation of a TEG system in the exhaust of a car using commercial 14 Hz TEG modules. In their research, the temperature gradient was guaranteed using refrigerant on the cold side of the module; the system was connected directly to the battery and alternator of a vehicle, and using a longitudinal model, it was possible to determine that the system displaces the energy of the alternator by between 15 and 20%, having an average saving of ~1 g/km of CO_2 in standard driving cycles.

Fernández et al. [34] used commercial Bi2Te3 TEG modules for heat recovery from light duty diesel engines as they produce ~386 W of recoverable power under common vehicle driving conditions. In their research they found that in commercial TEG modules, it is only possible to recover about 37.6 W, if no additional improvements, such as advanced heat exchangers, are applied. The authors also found that when a cooling system is able to maintain the cold side of the TEG module at 50 °C (less than the engine system coolant temperature), up to 75 W of power can be obtained.

Heber et al. [21] in 2020, manufactured a TEG system for natural gas heavy vehicles, in which they used 168 commercial modules based on SnTe. The cost of the TEG was EUR 1811, and a maximum power of 1507 W was achieved with a power density 50 W/kg, a reduction in CO_2 emissions of 4.9 (9.4) g (CO_2)/km, and a cost-efficiency ratio of 1.2 EUR/W, which suggests that the system is profitable.

Likewise, the use of TEGs as complementary systems has been investigated to compensate the cost-efficiency ratio in different applications to reduce CO_2 emissions. Thus, Bellos et al. [74] investigated the efficiency of a solar energy-induced TEG using commercial Bi_2Te_3 TEG modules, and carrying out a financial analysis, found that the cost of the investment would be 1 EUR/W, with a payback period of 4.55 years and a leveled cost of electricity of 0.0441 EUR/kWh, indicating that this system would be unprofitable.

3.8. Sustainability (Circular Economy)

As noted above, commercial TE materials are not yet sufficiently cheap, and high efficiency materials are not yet mass produced. Until now, the most commonly used commercial TE materials are Bi2Te3-based alloys because they have advantages such as easy bulk processing. However, although they are precursors, high energy expenditure and expensive techniques are used in their processing both for power generation and for cooling at temperatures close to ambient levels. On the other hand, TE materials use elements such as bismuth, tellurium, antimony, selenium, and lead, among others, which are expensive, scarce, and sometimes toxic. As mentioned earlier, from the point of view of the circular economy, the recycling of TEG modules could generate great economic benefits since it would allow obtaining raw materials for the manufacture of new TEG modules or other electronic devices, generating a reduction in the consumption of scarce elements. Moreover, they also generate environmental advantages because the improper disposal of these materials is avoided, which can benefit both the environment and human health.

At the time of writing this paper, the scientific publications on recycling TEG modules are still quite sparse. However, currently there are different ways to recycle TEG modules based on tellurium bismuth, from which three approaches can be differentiated based on the separation techniques: (i) chemical, (ii) thermal, and (iii) bacterial methods. On the other hand, in some cases only some parts of the TEG modules are recycled or only the elements of the semiconductors are recovered. According to the bibliographic review carried out, approaches have been proposed for the recycling of commercial TEG modules based on bismuth tellurium, by taking advantage of the differences in melting temperature of the constituent materials. In this way, the separation of the different constituents of commercial modules (plastics, Cu, Bi_2Te_3 and Al_2O_3) in an efficient way might be achieved by mechanical processing which relies on the entropy changes of these materials [27].

The TE materials have been separated by thermal processes followed by chemical separation processes, in which the characterization of the materials of the TE modules was conducted by techniques such as differential scanning calorimetry (DSC), X-ray diffraction (DRX) and field emission scanning electron microscopy (FESEM), which give information on the material types, melting temperatures and the distribution of the materials. This allows their separation based on the differences in thermal and chemical properties. This type of separation is initially carried out by means of thermal treatments such as hot oil baths at 250 °C for the removal of solder from the -n (Bi_2Te_3) and the p-type ($Bi_{0.5}Sb_{1.5}Te_3$) semiconductors. Later they are subjected to a mixed acid solution (HCl and HNO3 in a 3:1 ratio) at room temperature. At this point, the Sb of the semiconductors precipitates. Then, the solution is then filtered, washed and sintered in order to obtain nano-powders of Bi_2Te_3 -n type with a particle size of ~15 nm purity [25,36].

None of the previous works reports a characterization of the thermoelectric properties of the recovered TE materials. The characterization would be of great importance in order to know if the processes used for their separation in any way affect the properties of the recovered products and their possible use in future applications. Table 7 lists some recent works in relation to the final disposal of thermoelectric modules.

Table 7. Bibliographic review of the final disposal of thermoelectric modules in recent years.

Year	Ref	Number of Articles	Country
2020	[9]	1	Finland
2017	[26]	2	France
	[27]		Korea
2014	[39]	1	Korea
2013	[42]	1	Korea

It is clear then that there is a global need for sustainable technologies [75–77], and the circularity of materials and processes are areas where TE can have a significant impact as the main, partner, or complementary technology since it is a particularly adaptable technology [78].

4. Remarks and Conclusions

Between 2010 and 2020, there have been few studies concerning thermoelectric generator (TEG) modules that examine the larger holistic picture from mass manufacturing, profitability (from the economic point of view), efficiency, life cycle, and competitiveness to final disposal. Despite this, in the systematic search, an approach towards circular economy and sustainability has been considered indirectly since the articles analyzed covered topics ranging from raw thermoelectric TE materials, manufacturing of TEG modules, energy efficiency, and also the applicability and recycling of materials and modules.

The cost of manufacturing the modules is a little explored topic since in most works, only the relationships concerning the cost of the constituent TE materials are analyzed, and other factors that contribute to the high commercial cost of these modules are not examined.

In the present review it is highlighted that the efficiency of TEG modules depends, in addition to the TE materials, on external agents such as refrigeration systems, connections between the TEG modules, and the environment in which they will be operated. Therefore, these are important research topics to improve the cost-efficiency ratio of TEG systems.

Modeling is a valuable tool in predicting the efficiency of TEG modules at different stages of their development and implementation.

Additionally, the present research evidences the lack of work oriented towards the circular economy and sustainability of TEG modules, highlighting this aspect as an extremely important area for the development of TEG systems.

Taking into account the limitations that current TEG modules present in relation to the efficiency-cost relationship and the lack of research from circular economy and sustainability perspectives, this review proposes an approach for new studies that would allow an improved understanding of the processes used and their useful life. The research approach would aim to balance the cost-efficiency ratio from a sustainability perspective, not only addressing the base material, but also provide a more general approach to understanding the entire life-cycle of the modules from their manufacture to their final disposal. This would provide more information on the economic viability and sustainability of TE materials and modules.

5. Implications and Final Reflection

The circular economy represents a development opportunity that would lay the foundations for a sustainable recovery. Responding to the problem of the large accumulation of waste will in turn allow progress in solving the climate crisis and thus avoid the loss of valuable habitats.

The present study was developed with the aim of promoting the development of alternative generation systems to reduce dependence on the consumption of oil and other minerals. This will lead to a reduction in the generation of high levels of emissions, which will help the restoration of the battered ecosystems of our planet. Here is an opportunity that the authors believe should be seized as soon as possible.

As shown in this study, there are few development initiatives, particularly in the field of thermoelectric materials, that implement sustainable models. This is striking since what has sparked interest in the development of this technology has been precisely the need to counteract an existing problem. However, there is no awareness of the implications of not adopting an environmental perspective and transforming our way of thinking in order to build a better world for all.

Author Contributions: Conceptualization, H.A.C., E.I.G.-V. and A.A.A.; methodology, H.A.C. and C.E.A.; validation, C.E.A. and S.N.M.; investigation, M.C., E.I.G.-V. and H.A.C.; writing—original draft preparation, H.A.C., M.C. and E.I.G.-V.; writing—review and editing, H.A.C., S.N.M. and E.I.G.-V.; supervision, H.A.C.; project administration, H.A.C. and A.A.A.; funding acquisition, H.A.C. and A.A.A. All authors have read and agreed to the published version of the manuscript.

Funding: This research received financial support from the Colombia Scientific Program within the framework of the call Ecosistema Científico (Contract No. FP44842-218-2018).

Institutional Review Board Statement: Not applicable.

Informed Consent Statement: Not applicable.

Data Availability Statement: No new data were created or analyzed in this study. Data sharing is not applicable to this article.

Acknowledgments: The authors gratefully acknowledge Universidad de Antioquia for the support in this project.

Conflicts of Interest: The authors declare no conflict of interest.

References

1. Nazir, R.; Laksono, H.D.; Waldi, E.P.; Ekaputra, E.; Coveria, P. Renewable energy sources optimization: A micro-grid model design. *Energy Procedia* **2014**, *52*, 316–327. [CrossRef]
2. Tan, L.; He, X.; Xiao, G.; Jiang, M.; Yuan, Y. Design and energy analysis of novel hydraulic regenerative potential energy systems. *Energy* **2022**, *249*, 123780. [CrossRef]
3. Maradin, D.; Cerović, L.; Šegota, A. The efficiency of wind power companies in electricity generation. *Energy Strategy Rev.* **2021**, *37*, 100708. [CrossRef]
4. Lande-Sudall, D.; Stallard, T.; Stansby, P. Co-located deployment of offshore wind turbines with tidal stream turbine arrays for improved cost of electricity generation. *Renew. Sustain. Energy Rev.* **2019**, *104*, 492–503. [CrossRef]
5. Rey, J.R.C.; Pio, D.T.; Tarelho, L.A.C. Biomass direct gasification for electricity generation and natural gas replacement in the lime kilns of the pulp and paper industry: A techno-economic analysis. *Energy* **2021**, *237*, 121562. [CrossRef]
6. de Oliveira, L.; dos Santos, I.F.S.; Schmidt, N.L.; Tiago Filho, G.L.; Camacho, R.G.R.; Barros, R.M. Economic feasibility study of ocean wave electricity generation in Brazil. *Renew. Energy* **2021**, *178*, 1279–1290. [CrossRef]
7. Barasa Kabeyi, M.J.; Olanrewaju, O.A. Geothermal wellhead technology power plants in grid electricity generation: A review. *Energy Strategy Rev.* **2022**, *39*, 100735. [CrossRef]
8. Ahmad, S.; Nadeem, A.; Akhanova, G.; Houghton, T.; Muhammad-Sukki, F. Multi-criteria evaluation of renewable and nuclear resources for electricity generation in Kazakhstan. *Energy* **2017**, *141*, 1880–1891. [CrossRef]
9. Shoeibi, S.; Kargarsharifabad, H.; Sadi, M.; Arabkoohsar, A.; Mirjalily, S.A.A. A review on using thermoelectric cooling, heating, and electricity generators in Solar energy applications. *Sustain. Energy Technol. Assess.* **2022**, *52*, 102105. [CrossRef]
10. BP. *BP Energy Outlook*, 2019th ed.; BP: London, UK, 2019.
11. United States Department of Energy. *Improvent Process Heating Sytem Performace: A Sourcebook for Industry*, 2nd ed.; Energy Government: Washington, DC, USA, 2007.
12. Tan, G.; Ohta, M.; Kanatzidis, M.G. Thermoelectric power generation: From new materials to devices. *Philos. Trans. R. Soc. A* **2019**, *377*. [CrossRef]
13. Kishore, R.A.; Nozariasbmarz, A.; Poudel, B.; Sanghadasa, M.; Priya, S. Ultra-high performance wearable thermoelectric coolers with less materials. *Nat. Commun.* **2019**, *10*, 1765. [CrossRef] [PubMed]
14. Kishore, R.A.; Nozariasbmarz, A.; Poudel, B.; Priya, S. High-performance thermoelectric generators for field deployments. *ACS Appl. Mater. Interfaces* **2020**, *12*, 10389–10401. [CrossRef] [PubMed]
15. Prieto, A.; Knaack, U.; Auer, T.; Klein, T. COOLFACADE: State-of-the-Art review and evaluation of solar cooling technologies on their potential for Façade integration. *Renew. Sustain. Energy Rev.* **2019**, *101*, 395–414. [CrossRef]
16. Halli, P.; Wilson, B.P.; Hailemariam, T.; Latostenmaa, P.; Yliniemi, K.; Lundström, M. Electrochemical recovery of tellurium from metallurgical industrial waste. *J. Appl. Electrochem.* **2020**, *50*, 1–14. [CrossRef]
17. Babu, C.; Ponnambalam, P. The Role of Thermoelectric generators in the Hybrid PV/T Systems: A review. *Energy Convers. Manag.* **2017**, *151*, 368–385. [CrossRef]
18. Velázquez-Martinez, O.; Kontomichalou, A.; Santasalo-Aarnio, A.; Reuter, M.; Karttunen, A.J.; Karppinen, M.; Serna-Guerrero, R. A recycling process for thermoelectric devices developed with the support of statistical entropy analysis. *Resour. Conserv. Recycl.* **2020**, *159*, 104843. [CrossRef]
19. Papapetrou, M.; Kosmadakis, G.; Cipollina, A.; la Commare, U.; Micale, G. Industrial waste heat: Estimation of the technically available resource in the EU per industrial sector, temperature level and country. *Appl. Therm. Eng.* **2018**, *138*, 207–216. [CrossRef]
20. Alghoul, M.A.; Shahahmadi, S.A.; Yeganeh, B.; Asim, N.; Elbreki, A.M.; Sopian, K.; Tiong, S.K.; Amin, N. A review of thermoelectric power generation systems: Roles of existing test rigs/prototypes and their associated cooling units on output performance. *Energy Convers. Manag.* **2018**, *174*, 138–156. [CrossRef]
21. Heber, L.; Schwab, J. Modelling of a thermoelectric generator for heavy-duty natural gas vehicles: Techno-economic approach and experimental investigation. *Appl. Therm. Eng.* **2020**, *174*, 115156. [CrossRef]
22. Khanmohammadi, S.; Saadat-Targhi, M.; Ahmed, F.W.; Afrand, M. Potential of thermoelectric waste heat recovery in a combined geothermal, fuel cell and organic rankine flash cycle (thermodynamic and economic evaluation). *Int. J. Hydrog. Energy* **2020**, *45*, 6934–6948. [CrossRef]
23. Anderson, K.; Brandon, N. Techno-economic analysis of thermoelectrics for waste heat recovery. *Energy Sources Part B* **2019**, *14*, 147–157. [CrossRef]
24. Twaha, S.; Zhu, J.; Yan, Y.; Li, B. A comprehensive review of thermoelectric technology: Materials, applications, modelling and performance improvement. *Renew. Sustain. Energy Rev.* **2016**, *65*, 698–726. [CrossRef]
25. Alam, H.; Ramakrishna, S. A review on the enhancement of figure of merit from bulk to nano-thermoelectric materials. *Nano Energy* **2013**, *2*, 190–212. [CrossRef]
26. Zheng, X.F.; Liu, C.X.; Yan, Y.Y.; Wang, Q. A review of thermoelectrics research—Recent developments and potentials for sustainable and renewable energy applications. *Renew. Sustain. Energy Rev.* **2014**, *32*, 486–503. [CrossRef]
27. Jiang, B.; Liu, X.; Wang, Q.; Cui, J.; Jia, B.; Zhu, Y.; Feng, J.; Qiu, Y.; Gu, M.; Ge, Z.; et al. Realizing high-efficiency power generation in low-cost pbs-based thermoelectric materials. *Energy Environ. Sci.* **2020**, *13*, 579–591. [CrossRef]

28. Elzalik, M.; Rezk, H.; Mostafa, R.; Thomas, J.; Shehata, E.G. An experimental investigation on electrical performance and characterization of thermoelectric generator. *Int. J. Energy Res.* **2020**, *44*, 128–143. [CrossRef]

29. Żelazna, A.; Gołębiowska, J. A PV-Powered TE cooling system with heat recovery: Energy balance and environmental impact indicators. *Energies* **2020**, *13*, 1701. [CrossRef]

30. Ozturk, T.; Kilinc, E.; Uysal, F.; Celik, E.; Kurt, H. Effects of electrical properties on determining materials for power generation enhancement in TEG modules. *J. Electron. Mater.* **2019**, *48*, 5409–5417. [CrossRef]

31. Dongxu, J.; Zhongbao, W.; Pou, J.; Mazzoni, S.; Rajoo, S.; Romagnoli, A. Geometry optimization of thermoelectric modules: Simulation and experimental study. *Energy Convers. Manag.* **2019**, *195*, 236–243. [CrossRef]

32. Merienne, R.; Lynn, J.; McSweeney, E.; O'Shaughnessy, S.M. Thermal cycling of thermoelectric generators: The effect of heating rate. *Appl. Energy* **2019**, *237*, 671–681. [CrossRef]

33. Han, C.; Tan, G.; Varghese, T.; Kanatzidis, M.G.; Zhang, Y. High-performance PbTe thermoelectric films by scalable and low-cost printing. *ACS Energy Lett.* **2018**, *3*, 818–822. [CrossRef]

34. Fernández-Yáñez, P.; Gómez, A.; García-Contreras, R.; Armas, O. Evaluating thermoelectric modules in diesel exhaust systems: Potential under urban and extra-urban driving conditions. *J. Clean. Prod.* **2018**, *182*, 1070–1079. [CrossRef]

35. Wilbrecht, S.; Beitelschmidt, M. The potential of a cascaded teg system for waste heat usage in railway vehicles. *J. Electron. Mater.* **2018**, *47*, 3358–3369. [CrossRef]

36. Ouyang, Z.; Li, D. Design of segmented high-performance thermoelectric generators with cost in consideration. *Appl. Energy* **2018**, *221*, 112–121. [CrossRef]

37. Balva, M.; Legeai, S.; Garoux, L.; Leclerc, N.; Meux, E. Dismantling and chemical characterization of spent peltier thermoelectric devices for antimony, bismuth and tellurium recovery. *Environ. Technol.* **2017**, *38*, 791–797. [CrossRef]

38. Swain, B.; Lee, K.-J. Chemical separation of P- and n-Type thermoelectric chips from waste thermoelectric module and valorization through synthesis of Bi 2 Te 3 nanopowder: A sustainable process for synthesis of thermoelectric materials. *J. Chem. Technol. Biotechnol.* **2017**, *92*, 614–622. [CrossRef]

39. Lee, H.; Chidambaram Seshadri, R.; Han, S.J.; Sampath, S. TiO 2−X based thermoelectric generators enabled by additive and layered manufacturing. *Appl. Energy* **2017**, *192*, 24–32. [CrossRef]

40. Skomedal, G.; Holmgren, L.; Middleton, H.; Eremin, I.S.; Isachenko, G.N.; Jaegle, M.; Tarantik, K.; Vlachos, N.; Manoli, M.; Kyratsi, T.; et al. Design, assembly and characterization of silicide-based thermoelectric modules. *Energy Convers. Manag.* **2016**, *110*, 13–21. [CrossRef]

41. Brito, F.P.; Figueiredo, L.; Rocha, L.A.; Cruz, A.P.; Goncalves, L.M.; Martins, J.; Hall, M.J. Analysis of the effect of module thickness reduction on thermoelectric generator output. *J. Electron. Mater.* **2016**, *45*, 1711–1729. [CrossRef]

42. Mori, M.; Matsumoto, M.; Ohtani, M. Concept for improving cost effectiveness of thermoelectric heat recovery systems. *SAE Int. J. Passeng. Cars-Mech. Syst.* **2016**, *9*, 17–25. [CrossRef]

43. Hendricks, T.J.; Yee, S.; LeBlanc, S. Cost scaling of a real-world exhaust waste heat recovery thermoelectric generator: A deeper dive. *J. Electron. Mater.* **2016**, *45*, 1751–1761. [CrossRef]

44. Fu, C.; Bai, S.; Liu, Y.; Tang, Y.; Chen, L.; Zhao, X.; Zhu, T. Realizing high figure of merit in heavy-band p-type half-heusler thermoelectric materials. *Nat. Commun.* **2015**, *6*, 8144. [CrossRef] [PubMed]

45. Ming, T.; Wang, Q.; Peng, K.; Cai, Z.; Yang, W.; Wu, Y.; Gong, T. The influence of non-uniform high heat flux on thermal stress of thermoelectric power generator. *Energies* **2015**, *8*, 12584–12602. [CrossRef]

46. Hung, L.T.; van Nong, N.; Linderoth, S.; Pryds, N. Segmentation of Low-Cost High Efficiency Oxide-Based Thermoelectric Materials. *Phys. Status Solidi (A)* **2015**, *212*, 767–774. [CrossRef]

47. Arsie, I.; Cricchio, A.; Marano, V.; Pianese, C.; de Cesare, M.; Nesci, W. Modeling analysis of waste heat recovery via thermo electric generators for fuel economy improvement and CO$_2$ reduction in small diesel engines. *SAE Int. J. Passeng. Cars-Electron. Electr. Syst.* **2014**, *7*, 246–255. [CrossRef]

48. Wang, H.; McCarty, R.; Salvador, J.R.; Yamamoto, A.; König, J. Determination of thermoelectric module efficiency: A survey. *J. Electron. Mater.* **2014**, *43*, 2274–2286. [CrossRef]

49. Rezania, A.; Rosendahl, L.A.; Yin, H. Parametric Optimization of thermoelectric elements footprint for maximum power generation. *J. Power Sources* **2014**, *255*, 151–156. [CrossRef]

50. Lee, K.-J.; Jin, Y.-H.; Kong, M.-S. Synthesis of the thermoelectric nanopowder recovered from the used thermoelectric modules. *J. Nanosci. Nanotechnol.* **2014**, *14*, 7919–7922. [CrossRef]

51. Patyk, A. Thermoelectric generators for efficiency improvement of power generation by motor generators—Environmental and economic perspectives. *Appl. Energy* **2013**, *102*, 1448–1457. [CrossRef]

52. Salvador, J.R.; Cho, J.Y.; Ye, Z.; Moczygemba, J.E.; Thompson, A.J.; Sharp, J.W.; König, J.D.; Maloney, R.; Thompson, T.; Sakamoto, J.; et al. Thermal to electrical energy conversion of skutterudite-based thermoelectric modules. *J. Electron. Mater.* **2013**, *42*, 1389–1399. [CrossRef]

53. Kim, W.-B. Investigation of low-cost, simple recycling process of waste thermoelectric modules using chemical reduction. *Bull. Korean Chem. Soc.* **2013**, *34*, 2167–2170. [CrossRef]

54. Yazawa, K.; Shakouri, A. Cost–performance analysis and optimization of fuel-burning thermoelectric power generators. *J. Electron. Mater.* **2013**, *42*, 1946–1950. [CrossRef]

55. Yazawa, K.; Shakouri, A. Scalable cost/performance analysis for thermoelectric waste heat recovery systems. *J. Electron. Mater.* **2012**, *41*, 1845–1850. [CrossRef]
56. Homm, G.; Klar, P.J. Thermoelectric materials-compromising between high efficiency and materials abundance. *Phys. Status Solidi (RRL) Rapid Res. Lett.* **2011**, *5*, 324–331. [CrossRef]
57. Yazawa, K.; Shakouri, A. Cost-efficiency trade-off and the design of thermoelectric power generators. *Environ. Sci. Technol.* **2011**, *45*, 7548–7553. [CrossRef] [PubMed]
58. Liu, W.-D.; Chen, Z.-G.; Zou, J. Eco-friendly higher manganese silicide thermoelectric materials: Progress and future challenges. *Adv. Energy Mater.* **2018**, *8*, 1–18. [CrossRef]
59. Liu, W.-D.; Yang, L.; Chen, Z.-G.; Zou, J. Promising and eco-friendly Cu2X-Based thermoelectric materials: Progress and applications. *Adv. Mater.* **2020**, *32*, 87–92. [CrossRef]
60. Ouyang, Y.; Zhang, Z.; Li, D.; Chen, J.; Zhang, G. Emerging theory, materials, and screening methods: New opportunities for promoting thermoelectric performance. *Ann. Phys.* **2019**, *531*, 1800437. [CrossRef]
61. Ravichandran, J.; Yadav, A.K.; Cheaito, R.; Rossen, P.B.; Soukiassian, A.; Suresha, S.J.; Duda, J.C.; Foley, B.M.; Lee, C.H.; Zhu, Y.; et al. Crossover from incoherent to coherent phonon scattering in epitaxial oxide superlattices. *Nat. Mater.* **2013**, *13*, 168–172. [CrossRef]
62. Kleinke, H. New bulk materials for thermoelectric power generation: Clathrates and complex antimonides. *Chem. Mater.* **2010**, *22*, 604–611. [CrossRef]
63. Faílde, D.; Baldomir, D. Emergent topological fields and relativistic phonons within the thermoelectricity in topological insulators. *Sci. Rep.* **2021**, *11*, 1–9. [CrossRef] [PubMed]
64. Huber, S.D. Topological mechanics. *Nat. Phys.* **2016**, *12*, 621–623. [CrossRef]
65. Baldomir, D.; Faílde, D. On behind the physics of the thermoelectricity of topological insulators. *Sci. Rep.* **2019**, *9*, 1–8. [CrossRef] [PubMed]
66. Shi, L.; Chen, J.; Zhang, G.; Li, B. Thermoelectric figure of merit in Ga-Doped [0001] ZnO nanowires. *Phys. Lett. Sect. A* **2012**, *376*, 978–981. [CrossRef]
67. Qiu, B.; Chen, G.; Tian, Z. Effects of aperiodicity and roughness on coherent heat conduction in superlattices. *Nanoscale Microscale Thermophys. Eng.* **2015**, *19*, 272–278. [CrossRef]
68. Fei, R.; Faghaninia, A.; Soklaski, R.; Yan, J.A.; Lo, C.; Yang, L. Enhanced thermoelectric efficiency via orthogonal electrical and thermal conductances in phosphorene. *Nano Lett.* **2014**, *14*, 6393–6399. [CrossRef]
69. Roldán, R.; Silva-Guillén, J.A.; López-Sancho, M.P.; Guinea, F.; Cappelluti, E.; Ordejón, P. Electronic properties of single-layer and multilayer transition metal dichalcogenides MX2 (M = Mo, W and X = S, Se). *Ann. Phys.* **2014**, *526*, 347–357. [CrossRef]
70. Ahmad, A.; Singh, Y. In-plane behaviour of expanded polystyrene core reinforced concrete sandwich panels. *Constr. Build. Mater.* **2021**, *269*, 121804. [CrossRef]
71. Wolf, S.; Neophytou, N.; Kosina, H. Thermal conductivity of silicon nanomeshes: Effects of porosity and roughness. *J. Appl. Phys.* **2014**, *115*, 204306. [CrossRef]
72. Thermoelectric Modules Made by Hi-Z Technology-San Diego. Available online: https://hi-z.com/ (accessed on 23 August 2021).
73. TECTEG MFR. Available online: https://tecteg.com/ (accessed on 23 August 2021).
74. Bellos, E.; Tzivanidis, C. Energy and financial analysis of a solar driven thermoelectric generator. *J. Clean. Prod.* **2020**, *264*, 121534. [CrossRef]
75. Vergragt, P.J. *How Technology Could Contribute to a Sustainable World*; GTI Paper Series: Frontiers of a Great Transition; Tellus Institute: Boston, MA, USA, 2006.
76. Cardona-Vivas, N.; Correa, M.A.; Colorado, H.A. Multifunctional composites obtained from the combination of a conductive polymer with different contents of primary battery waste Powders. *Sustain. Mater. Technol.* **2021**, *28*, e00281. [CrossRef]
77. Revelo, C.F.; Correa, M.; Aguilar, C.; Colorado, H.A. Composite materials made of waste tires and polyurethane resin: A case study of flexible tiles successfully applied in industry. *Case Stud. Constr. Mater.* **2021**, *15*, e00681. [CrossRef]
78. Colorado, H.A.; Mendoza, D.E.; Lin, H.T.; Gutierrez-Velasquez, E. Additive manufacturing against the Covid-19 pandemic: A technological model for the adaptability and networking. *J. Mater. Res. Technol.* **2021**, *16*, 1150–1164. [CrossRef]

 *sustainability*

Article

Ornamental Stone Processing Waste Incorporated in the Production of Mortars: Technological Influence and Environmental Performance Analysis

Pamella Inácio Moreira [1], Josinaldo de Oliveira Dias [2], Gustavo de Castro Xavier [1], Carlos Maurício Vieira [3], Jonas Alexandre [1], Sergio Neves Monteiro [4], Rogério Pinto Ribeiro [5] and Afonso Rangel Garcez de Azevedo [1,*]

[1] Civil Engineering Laboratory, State University of the Northern Rio de Janeiro, Rio de Janeiro 28013-602, Brazil; pamella_moreira@hotmail.com (P.I.M.); gxavier@uenf.br (G.d.C.X.); jonas@uenf.br (J.A.)
[2] Forest and Wood Sciences Department, Federal University of Espírito Santo, Victoria 29075-910, Brazil; josinaldo.engproducao@gmail.com
[3] Laboratory of Advanced Materials, State University of the Northern Rio de Janeiro, Rio de Janeiro 28013-602, Brazil; vieira@uenf.br
[4] Department of Material Sciences, Military Engineering Institute, Rio de Janeiro 22290-270, Brazil; snevesmonteiro@gmail.com
[5] São Carlos School of Engineering, University of São Paulo, São Paulo 13566-590, Brazil; rogerioprx@sc.usp.br
* Correspondence: afonso@uenf.br

Abstract: The technological performance and environmental advantages of replacing sand by ornamental stone processing waste (OSPW) in the production of mortars for civil construction were studied. Technological properties associated with the standard consistency index, squeeze flow and bulk densities as well as the determination of water retention and calorimetry analysis were evaluated in the mortars' fresh state, whereas capillarity tests as well as mechanical resistance by flexural and compression tests were determined in the hardened state for mortars incorporated with 10, 30 and 60 wt.% of OSPW substituting sand. Three different types of Portland Cements were considered in the incorporated mortars production. For these mortars environmental analysis, their corresponding life cycle assessment results were compared to that of conventional waste-free (0% OSPW) control mortar. It was found that the OPSW incorporation acts as nucleation sites favoring a hydration process, which culminates after 28 days of curing in the formation of more stable phases identified as hydrated calcium silicates by X-ray diffraction (XRD) amorphous halo. It was also revealed that both flexural and compression improved resistance for the incorporated mortars after 28 curing days. In particular, the calorimetry and XRD results explained the better mechanical resistance (12 MPa) of the 30 wt.% OSPW incorporated mortar, hardened with Portland Cement V, compared not only to the control, but also to the other incorporated mortars. As for the environmental analysis, the replacement of sand by OSPW contributed to the reduction in associated impacts in the categories of land use (−5%); freshwater eutrophication (−9%); marine eutrophication (−6%) and global warming (−5%).

Keywords: life cycle assessment; squeeze flow; cement

Citation: Moreira, P.I.; de Oliveira Dias, J.; de Castro Xavier, G.; Vieira, C.M.; Alexandre, J.; Monteiro, S.N.; Ribeiro, R.P.; de Azevedo, A.R.G. Ornamental Stone Processing Waste Incorporated in the Production of Mortars: Technological Influence and Environmental Performance Analysis. *Sustainability* **2022**, *14*, 5904. https://doi.org/10.3390/su14105904

Academic Editor: Antonio Caggiano

Received: 30 March 2022
Accepted: 6 May 2022
Published: 13 May 2022

Publisher's Note: MDPI stays neutral with regard to jurisdictional claims in published maps and institutional affiliations.

1. Introduction

Brazil stands out for having large mineral deposits, such as ornamental stones, which are transformed into products that can replace ceramic coverings and floors. In the Brazilian national territory, the state of Espírito Santo is known as a major producer of ornamental stones, such as marble and granite [1].

The ornamental stone industry is responsible for moving significant volumes in the national and international markets [2,3]. The Brazilian exports of natural stone materials totaled USD 1.34 billion and 2.40 Mt in 2021, an improvement of 35.5% and 11.4%, respectively, compared to 2020. Revenue and annual variation surpassed the historical records recorded in 2013 (USD 1.30 billion and 22.8%). The processed stone export activities were responsible for USD 1.077 billion, about 80% of total revenue in 2021 [4]

The observed growth consumption of the ornamental stone industry, in the national and international marketS, is directly linked to the amount of waste from the processing stages of this material. It is estimated that of the total natural stones available for processing, about 41% becomes tailing. Thus, in 2021, internal processing activities in Brazil generated 3.276 million tons of tailing [4]. The tailings obtained in the beneficiation phase form a sludge.

When drying, this sludge turns into non-biodegradable solid ornamental stone processing waste (OSPW) classified as class IIA waste—non-inert and non-hazardous [5]—which often receives inadequate final disposal, having its impacts reflected on human health and depletion of the environment [6]. The OSPW can be classified as non-biodegradable, from the particle's point of view. Due to the strong presence of quartz in its crystalline phase, it resembles the waste glass previously studied [7], which corroborates that this class of waste, produced in increasing volumes, owing to the lack of adequate landfills for packaging, contributes to the intensification of pollution and ecosystem degradation, impacting factors such as water and soil. It is noteworthy that the glass residue, in its X-ray diffraction analysis, presents a quartz peak, and the remainder in the form of an amorphous halo [7].

The expressive volume of OSPW has a great potential for reuse in civil construction, being compatible with cement matrices, and can be applied as a filler in mortars and concrete, positively influencing the stability of the material [8–11].

The processing of ornamental stone has the objective of transforming the blocks, extracted in the mining phase, into final or semi-finished products. In the mining phase, the blocks are extracted in large dimensions, while in the primary processing, the blocks are prepared and sawn into sheets of variable thickness, usually two or three centimeters. Thus, they are subjected to a surface finish, with or without resin, which can be a simple grinding, polishing, brushing and flaming [1,12].

For the production of concrete and mortars incorporated with ornamental stone waste obtained in the processing phase, it is relevant to study the waste chemical characterization, since elements such as magnesium can generate expandability in the hardened state, reflecting in fissures when used in coatings and paving. In addition, the fineness of the residue and the SiO_2 content associated with the production process might dislodge high reactivity, resulting in the occurrence of alkali-aggregate reactions [13,14]. Aiming at better technological performance of the mortar, in addition to prolonging its durability, some researchers evaluated the addition of OSPW, obtained in the processing phase, which is characterized as a non-plastic and non-biodegradable by-product. The study considered particles with a grain size between 2 μm and 60 μm, with chemical characteristics similar to the sand from the same region [15].

The OSPW, derived from granite and marble whose predominant minerals are calcite and dolomite, can be used as raw material in the production of mortar, presenting itself as an alternative that guarantees satisfactory technological performance, in addition to contributing to the reduction in the environmental impact associated with this industrial activity. Thus, investing this waste is mainly justified by the possibility of mitigating the environmental impact associated with the proper disposal of this waste [16,17].

Several studies of mortars incorporated with residues directed their objectives towards characterization and technological evaluation. However, issues from an environmental perspective and their impacts are still incipient [18]. According to Sánchez et al. [19], one of the main inputs for the production of mortar is cement, which, according to its composition and production process, contributes between 5 and 8% to CO_2 emissions in a global context.

The life cycle assessment (LCA) can be understood as a systematic methodology of analysis capable of contributing to the quantification and identification of environmental impacts and protection areas associated with productive activities. This applies to the construction industry, allowing characteristic and comparative analyses of conventional and alternative products. The principle of LCA is to evaluate the basic flows of processes and products, being able to consider such steps as extraction, production, transport, useful life and disposal [20,21].

Many studies direct their efforts to study the incorporation of waste in the elaboration of new materials, mainly evaluating their results based on technological tests in accordance with current standards. However, in addition to the technological feasibility provided by the standards, it is necessary to evaluate the environmental impacts obtained in the elaboration of these new materials.

Additionally, it should be noted that compliance with the standards represents the minimum parameter for the use of construction materials in the industry, which stimulates the development of new materials. In addition to mitigating the environmental impacts associated with their production, it also results in technological performance that is superior to that expected by the norm. Thus, the present research represents a starting point for studies related to the incorporation of OSPW in the production of mortar, aiming at reducing environmental impacts and improving the technological performance of these new materials, considering the chemical composition of this residue and its positive effects on the formation of mortar.

This study aims to analyze the technological and environmental effects of replacing sand with OSPW in the production process of mortars used in civil construction. Three types of cement were considered for the production of mortar, namely Portland Cement II (CPII), Portland Cement III (CP III), and Portland Cement V (CPV) according to ABNT NBR 16697 [22].

2. Materials and Methods

Figure 1 presents the methodological flow established for the execution of tests and analyses in this research.

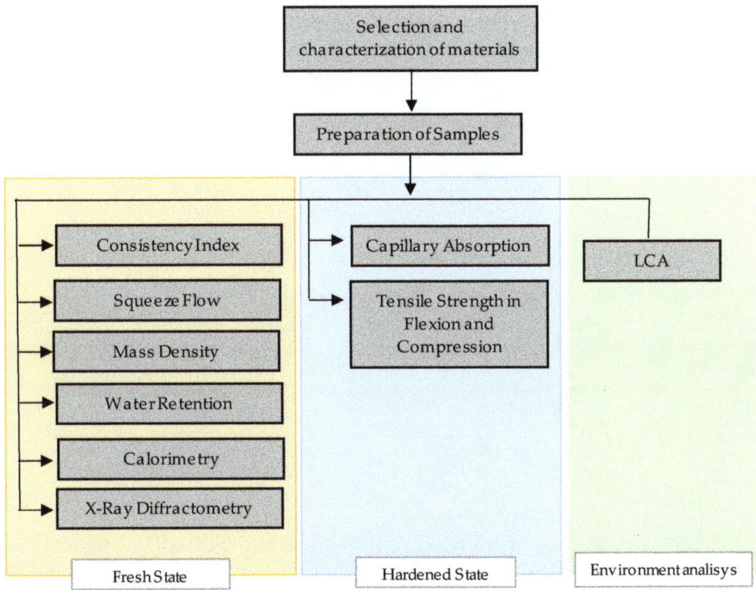

Figure 1. Schematic diagram of the methodological flow.

2.1. Selection and Characterization of Materials

To produce the mortar and carry out the technological tests, the selection stage consisted of obtaining the raw materials sand and OSPW (Figure 2). The sand extraction region was the city of Campos dos Goytacazes, Brazil, with the selection of 50 kg of sand. The OSPW, on the other hand, was obtained in the same amount, in Cachoeiro de Itapemirim, Brazil, the main ornamental stone producing city in the state of Espirito Santo.

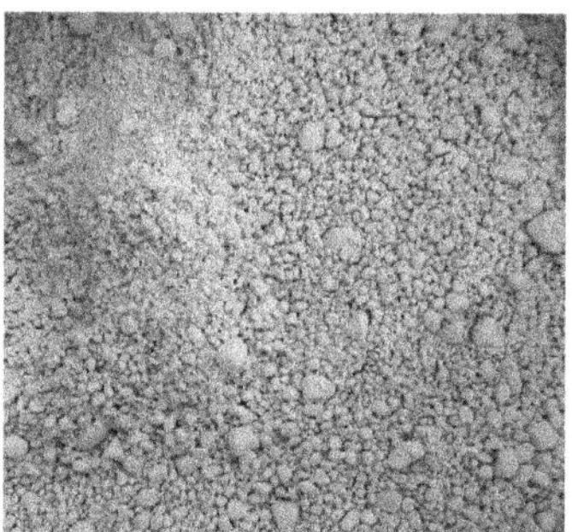

Figure 2. The ornamental stone processing waste sample.

The aforementioned materials were taken to an oven at a temperature of ±105 °C for a period of 24 h. After drying, the OSPW passed through a 30 min milling process, using a ball mill just to ensure the homogenization of the sample particle size. To finish the preparation of the materials the sand and the residue went through the sieving process in a 20 mesh sieve. The analysis of the particle size of the ornamental stone waste was determined following ABNT 7181 [23] by the sieving and sedimentation method.

2.2. Preparation of Samples

Together with the preparation of the materials, the composition of the samples according to the raw materials was adopted in the proportion 1:6 cement/sand. For its dosage, Portland cement types CP II, CP III and CP V (ABNT 16697) [22] were used, and the sand was replaced by 10%, 30% and 60% of OSPW. To compare the results, the reference trace was prepared with 0% substitution, as shown in Table 1.

Table 1. Samples composition.

OSPW (%)	Cement (Kg)	Sand (Kg)	OSPW (%)
0	0.214	1.280	0
10	0.214	1.152	0.128
30	0.214	0.896	0.384
60	0.214	0.512	0.768
Material	Cement (Kg)	Sand (Kg)	OSPW (%)
Density (Kg/m^3)	3.14	2.65	2.56
Unit Weight (Kg/m^3)	1.01	1.43	1.4
Cement/OSPW	Cement Consumption (Kg/m^3)	Sand Consumption (Kg/m^3)	OSPW Consumption (Kg/m^3)
CP II 0%	247	1035	0
CP II 10%	260	983	112
CP II 30%	259	762	333
CP II 60%	222	373	571
CP III 0%	250	1051	0
CP III 10%	260	983	112
CP III 30%	260	764	334
CP III 60%	254	427	654
CP V 0%	249	1043	0
CP V 10%	265	999	113
CP V 30%	255	748	327
CP V 60%	218	366	560

2.3. Technological Analysis—Fresh State

2.3.1. Consistency Index

According to the fresh state of the mortar, the consistency index was defined according to the ABNT NM 13276 [24]. Each mortar was produced following the methodology prescribed by the standard, being mechanically mixed for 90 s (speed 1) in the mortar (power of 0.3 CV at 62 rpm with a capacity of 5 L) and then placed in rest for 15 min covered by a damp cloth. After resting, the mixture was manually homogenized for 30 s with a spatula and placed in the mold on the flow table (MC—119), where it was subjected to 30 strokes with a socket. After the last fall from the table, the average value of the diameter measured in three different directions was calculated to verify the consistency of each mixture produced, being established by the aforementioned standard that the margin is 260 ± 5 mm. Measurements were performed in relation to the table plane, considering the vertical, horizontal and inclined readings.

2.3.2. Squeeze Flow

The technological analysis using the squeeze flow methodology was carried out according to ABNT NM 15839 [25], using a universal testing machine EMIC 26–30 with a capacity of 30 kN. However, for the execution of this study, a maximum load of 1 kN was used, as recommended. After preparation, the mortars were spread and poured into a cylindrical metal mold with a diameter of 100 mm and a height of 10 mm on an inflexible base made of stainless steel.

The squeeze flow test basically consists of the dynamic analysis of a material compressed by two horizontally parallel plates, thus obtaining a radial expansion of the compressed material.

The data obtained from the squeeze flow method can be interpreted according to the 3 phases of behavior that the material can present. The first phase corresponds to the linear elastic behavior (I), the second describes the plastic behavior, also defined as viscous flow (II) and the third phase is known as hardening (III) [26]. The three aforementioned phases, when analyzed according to the force versus displacement relationship, present their behaviors described by exponential curves, as illustrated in Figure 3.

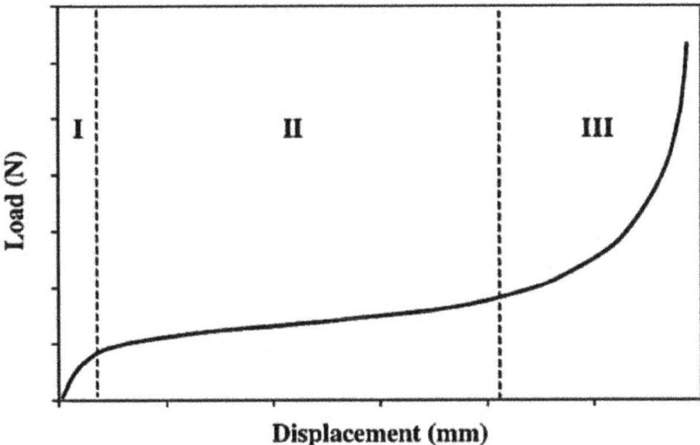

Figure 3. Observed phases according to the force versus displacement curve by the squeeze flow test [27].

2.3.3. Mass Density

To define the density of the mass, the mortar was prepared according to the ABNT NM 13276 [24] standard. Then, an empty rigid and cylindrical container was used, with a mass of 0.891 kg and a volume of 400 cm^3, continuing the test guided by the ABNT NM 13278 [28] standard.

2.3.4. Water Retention

First, considering the plate in the funnel, the filter paper positioned at the bottom of the plate was moistened. With the aid of a vacuum pump, a suction of 51 mm of mercury was applied to the set, which was weighed, and its mass (m_v) recorded. After this procedure, the set was filled with the mortar, thickened with 37 socket strokes, flattened, weighed, and its mass (m_a) was recorded.

Thus, with the set filled in the device, a suction of 51 mm of mercury was applied to the set, which was weighed, and its mass (m_s) recorded. The entire continuation of the test was guided by the ABNT NM 13277 [29] standard. In addition to the three values (m_v, m_a, m_s) recorded, the total mass of water (m_w) and the mass of the anhydrous components (m) were also recorded.

2.3.5. Calorimetry

According to the squeeze flow test, the samples with cement type CPV were prioritized. To evaluate the rate of hydration reactions of the mortars, the calorimetry technique (ASTM C1679-17) [30] was used, being tested in two dosages, only in the CPV, with the substitutions of 0% and 30%, subjected first to consistency and later to the calorimeter, in amounts of 111.9 g and 111.8 g, respectively. The cement type CPV was used because it does not contain added blast furnace slag and/or pozzolana in the composition, and it contains calcium sulfate and carbonate material in the limit of up to 10% ABNT 16697 [22]. This makes it easier to identify the changes caused by the residue next to the folder. In addition, it was easily acquired in the local market. This cement is used for prefabricated parts when high initial strength is required.

2.3.6. X-ray Diffractometry Analysis—XRD

For the XRD analysis, the residue and paste samples were sieved through a 200 mesh (0.074 mm) opening sieve and taken to an oven at 110 °C for 24 h. The pastes were composed of CP V cement and water (CH_2O), CP V cement with 30% of OSPW (COSPW) and water, and CP V cement with 30% sand and water (CS). For the XRD analysis, the CPV was used due to its composition, which does not register the presence of blast furnace slag, which is not in the CPII and CPIII types. The absence of blast furnace slag granulate contributes to the observation of phase formations and the presence of the main components when the sample is incorporated with OSPW [31].

The samples were mixed with isopropyl alcohol for 3 m in a 1:10 volumetric ratio (cement and alcohol), in a glass container with a glass rod. It was filtered and left to dry in a desiccator at 20 °C until it reached the consistency of the masses [30], and it was analyzed at the age of 28 days. Qualitative crystalline phases were obtained by XRD, in Proto Manufacturing equipment, XRD Powder Diffraction System: the generator of 30 kV and 2 mA, Cu-Kα1 radiation, angular step of 0.0149°, time interval of 0.5 s, sweep of 47 min and 2θ ranging from 5° to 60°. The crystalline phases in the residue were identified with reference to COD.

2.4. Technological Analysis—Hardened State

2.4.1. Capillary Absorption

To carry out this test, it was necessary to make 3 prismatic parts for each sample, using Portland cement types CP II, CP III and CP V, with cures of 7, 14 and 28 days. For this, it was based on the ABNT NM 13279 [32] standard, which describes the necessary procedures for the preparation of prismatic parts.

After finishing the curing days, the capillary absorption test was carried out, guided by the ABNT NM 15259 [33] standard. It was necessary to use a tray with water and an individual platform made of rubber material to lift the prismatic parts, allowing minimal contact with the water.

2.4.2. Analysis of Tensile Strength in Flexion and Compression

For this test, guided by the ABNT NM 13279 [32] standard, it was necessary to prepare 3 prismatic parts for each mix, using Portland cement types CP II, CP III and CP V, with cures of 7, 14 and 28 days, in the different substitutions of sand by OSPW (10%, 30% and 60%).

After completing the curing days, the prismatic parts were subjected to flexural and compressive strength tests, using an EMIC 26–30 universal testing equipment with a maximum force of 30 kN. First, the flexural tensile strength test was performed, where the prismatic parts were positioned in the machine in order to avoid contact between the flat surface and the supporting and loading components.

Then, to run the compressive strength test, the halves of each part of the flexural test were used and also positioned in the machine in order to avoid contact between the flat surface and the supporting and loading components.

2.5. Life Cycle Assessment

The LCA methodology was conducted considering the production process of mortar, incorporated with OSPW, according to the international standards ISO 14040 [34] and ISO 14044 [35]. Indeed, the development of this study aims to assess the environmental performance of mortar incorporated with OSPW, compared to conventional mortar. Both were evaluated according to the LCIA methods ReCiPe 2016 at midpoint, considering the impact categories, global warming (GW), stratospheric ozone depletion (SOD), ionizing radiation (IR), ozone formation, human health (OFHH), fine particulate matter formation (FPMF), ozone formation, terrestrial ecosystems (OFTE), terrestrial acidification (TA), freshwater eutrophication (FE), marine eutrophication (ME), terrestrial ecotoxicity (TE), freshwater ecotoxicity (FEC), marine ecotoxicity (MEC), human carcinogenic toxicity (HCT), human non-carcinogenic toxicity (HNCT), land use (LU), mineral resource scarcity (MRS), fossil resource scarcity (FRS), and water consumption (WC).

The functional unit considered in this study corresponds to the production of 1 kg of mortar. To define the phases of the process to be studied, the limits of the system were delimited, considering the productive activities from the extraction of raw material to the final process (Figure 4).

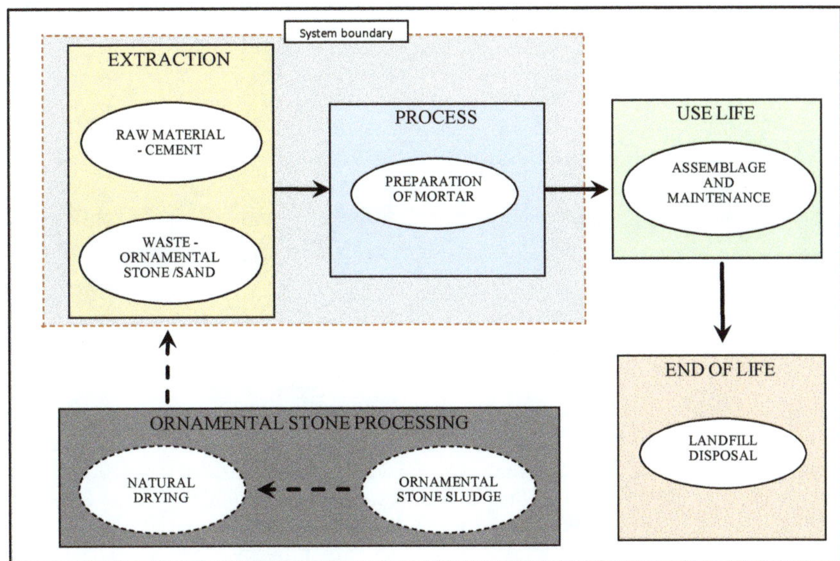

Figure 4. Mortar system boundary.

The life cycle inventory (LCI) report (Table 2) was prepared considering the production process from the extraction and obtaining of raw material to the finished product phase (cradle to gate). In this case, the life cycle inventory consists of detailed compilations of all inputs considered in the study (materials and natural resources). For the ornamental stone sludge, this research considered the natural drying to obtain the OSPW.

Table 2. Life cycle inventory data.

Description
CPV Cement, Portland {RoW}	production	APOS, U
Conveyor belt {GLO}	market for	APOS, U
Electricity, medium voltage {BR}	market for	APOS, U
Industrial machine, heavy, unspecified {GLO}	market for	APOS, U
Tap Water
Packing, cement {GLO}	market for	APOS, U
Silica sand {GLO}	market for	APOS, U
Ornamental Stone Processing Waste [1]

[1] Embedded in the 30% of OSPW mortar sample.

The data used for the model were extracted from the Ecoinvent 3.3 database, according to database allocations, as shown in Table 2. The OSPW data were obtained in studies by Rebello et al. [6]. The modeling, assessment of life cycle impacts (LCI) and interpretation of results were aided by the SimaPro 9.0 software. For the LCI assessment, the method used was ReCiPe 2016, at midpoint.

3. Results and Discussion

3.1. Technological Analysis—Fresh State

3.1.1. Consistency Index

According to the ABNT NM 13276 [24,36] standard, it is recommended to adopt the amount of water required for a consistency index of 260 ± 5 mm. Table 3 shows the results obtained by the related test.

Table 3. Consistency index test results.

Sample	HA [1] (mm)	VA[2] (mm)	TA[3] (mm)	Average (mm)	Water (g)	W/C [4]
0% OSPW—CP II	262	254	252	256	315	1.47
10% OSPW—CP II	253	270	272	265	268	1.25
30% OSPW—CP II	250	262	260	257	276	1.29
60% OSPW—CP II	257	264	250	257	400	1.87
0% OSPW—CP III	258	255	264	259	301	1.41
10% OSPW—CP III	250	259	269	259	265	1.24
30% OSPW—CP III	258	267	269	265	278	1.30
60% OSPW—CP III	256	254	255	255	415	1.94
0% OSPW—CP V	260	257	269	262	308	1.44
10% OSPW—CP V	248	259	257	255	255	1.19
30% OSPW—CP V	251	255	258	256	282	1.32
60% OSPW—CP V	259	264	265	263	420	1.96

[1] HA—horizontal average; [2] VA—vertical average; [3] TA—tilted average; [4] W/C—water/cement.

Thus, following the margin established by the aforementioned standard, the averages of all replacements (0%, 10%, 30% and 60%) of sand by OSPW in the three different types of cement (CP II, CP III and CP V) presented results within the acceptable limit.

Analyzing the water/cement ratio of the mortars produced, it was found that all the percentages of 10%, in the three different types of cement, had the lowest ratio, with the percentage of 10% in Portland cement type CP V being the lowest, with 1.19.

Taking into account that the lower the water/cement ratio, the greater the resistance, the lower the permeability and the greater the durability [37], the mortars produced with 10% substitution in cement type CP V have an advantage over the others.

3.1.2. Squeeze Flow

In view of the data obtained by the squeeze flow method, it is possible to elaborate the curves of the behavior of the material, guided by ABNT NM 15839 [25], which characterizes the workability level of the studied mortar. It is possible to analyze three workability phases: (i) the high workability being represented by the occurrence of the extension of the plastic deformation stage when the material is subjected to very low loads, registering transition to the hardening stage only in larger displacements; (ii) the low workability is characterized by the absence of the plastic deformation stage; and (iii) the medium workability is characterized by presenting variation of the load level as the plastic deformation occurs. It is noteworthy that, in mortars with medium workability characteristics, the material tends to flow and then interrupt the flow, being subjected to greater loads, and when they return flow, the load tends to decrease [26].

According to the dynamic squeeze flow test, the studied mortars were placed in an equipment composed of two parallel plates, where according to the variation of the time and speed parameters, the equipment recorded the data referring to force versus deformation, as shown in Figure 3.

Figures 5–7 shows the results of the squeeze flow test for the mortars obtained from the diversifications regarding the type of cement (CP II, CPIII and CP V) and OSPW/sand replacement fractions (10%, 30% and 60% of OSPW), also considering the speed parameters (0.1 mm/s and 3.0 mm/s) of material compression. Observing the curve described by the results of Portland cement mortar types CP II, CP III and CP V, considering the mixtures with 10% of OSPW tested at speeds of 0.1 mm/s and 3.0 mm/s, they obtained similar results.

Thus, the squeeze flow test was performed in three different groups. The first one considered the CPII type cement, the second one considered the CPIII type cement and the third one considered the CP V type cement. The three groups were evaluated according to the incorporation of 0%, 10%, 30% and 60% of OSPW, under the variation of speed parameters (0.1 mm/s and 3 mm/s) and maximum force of 1 kN.

The definition of the groups occurred based on the parameters foreseen by the standard, in addition to the composition presented by each type of cement and also the effects observed in the variation of the incorporation of residue in the materials studied.

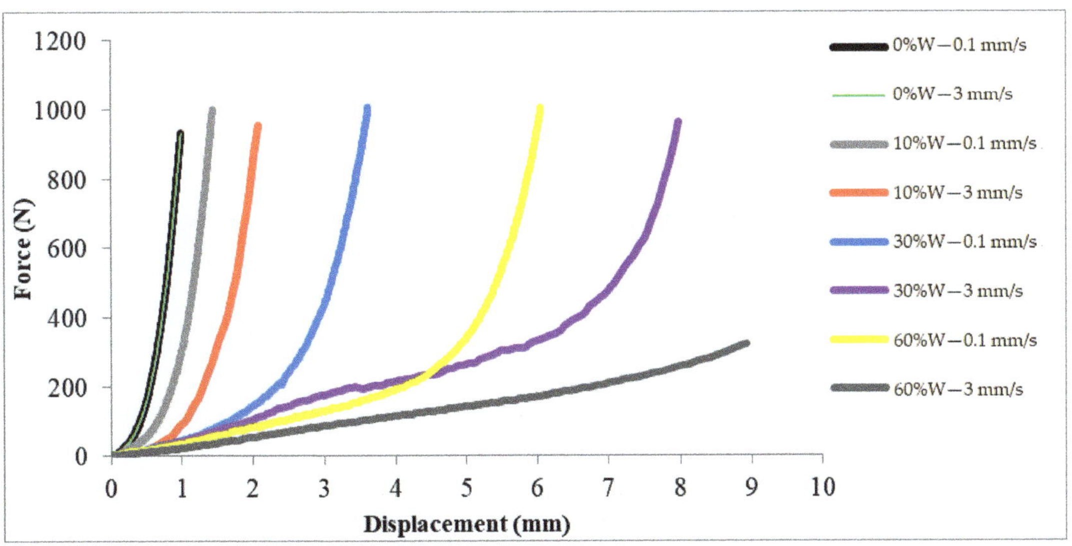

Figure 5. Squeeze flow test results—mortar with cement Portland type CPII (%W represents the OSPW grade incorporated).

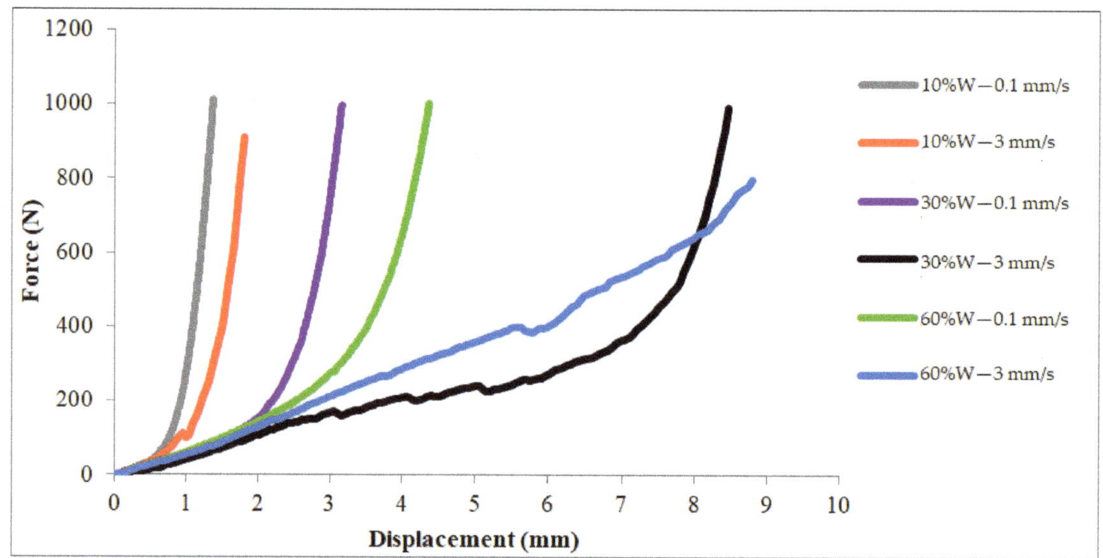

Figure 6. Squeeze flow test results—mortar with cement Portland type CPIII (%W represents the OSPW grade incorporated).

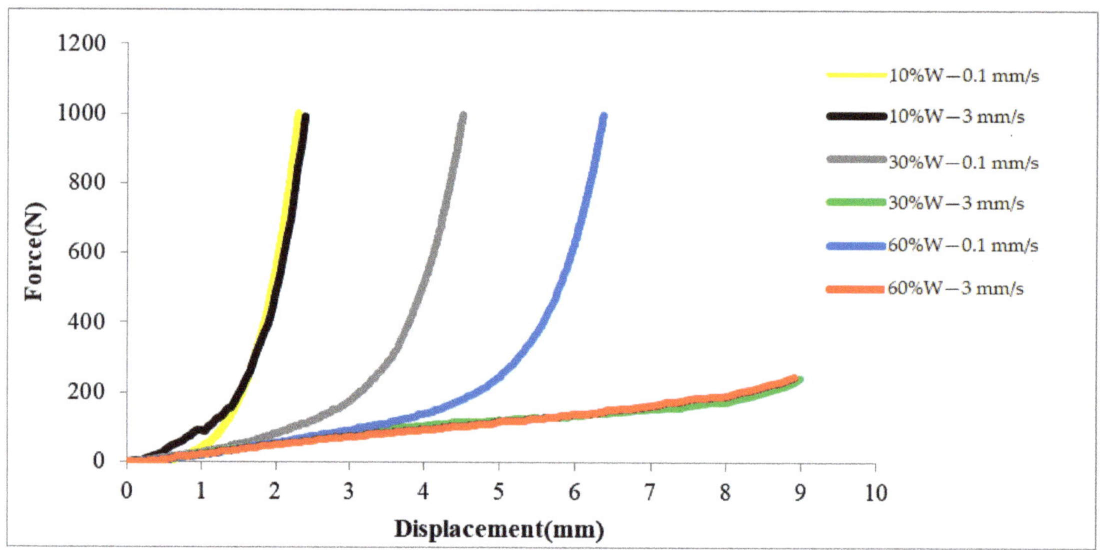

Figure 7. Squeeze flow test results—mortar with cement Portland type CPV (%W represents the OSPW grade incorporated).

The materials composed of cement type CPII, with a variation of 10, 30 and 60% OSPW/sand, subjected to the test with a velocity of 0.1 mm/s, presented different behaviors in the phase transitions. In particular, the CPII cement mortar, incorporated with 10% OSPW and 90% sand, registered a phase transition of 0.5 mm in the displacement axis.

This behavior may be associated with the increase in the presence of OSPW in the samples, which contributed to the material having its plastic characteristics accentuated according to the increase in the incorporation of the waste. According to the variation of the incorporation rate, from 30% to 60% of OSPW, the phase transition doubled.

It is noteworthy that the curves responded to a maximum load of 1 kN; therefore, the behavior of the materials tested, considering the different ranges of OSPW incorporation, was directly influenced by the displacement speed (0.1 mm/s), emphasizing sensitivity to segregation.

Considering the materials formed by cement type CP III, with variations of 10, 30 and 60% of OSPW incorporation, subjected to a test speed of 0.1 mm/s, it can be seen that the mortar with the incorporation at 10% of OSPW presented its plastic phase up to the mark of 0.80 mm of displacement. The mortar incorporated with 30% of OSPW behaved plastically until the record of 1.80 mm of displacement. The mortar with 60% OSPW incorporation presented the highest displacement value, registering 2.60 mm.

According to the change in the speed parameter to 3.0 mm/s, the behavior of the curves was significantly sensitive, as seen in Figure 6. Thus, in a comparative analysis between the curves obtained at 0.1 mm/s and 3.0 mm/s, it was observed that the materials subjected to the lowest velocity parameter reached the maximum displacement in the system with a load of 1 kN, varying between 1.0 and 4.50 mm. On the other hand, the materials tested under a speed of 3.0 mm/s showed punctual behavior, being the loads to reach the maximum displacement, conditioned to the composition of each mortar. Taking into account the average load of 1 kN, the variation of the maximum displacement of the samples varied between 1.80 and 9.0 mm.

The curve obtained from the squeeze flow test with mortar incorporated with cement of the CPV type generated a different result from the others, showing only the viscous flow phase. Comparing the curves of the CP II type mortars, with the CP III type and CP V, in both the percentage of 60% reached maximum displacement, reaching the mark of 9.0 mm.

The tested materials formed by cement of the CPV type, with the incorporation of OSPW in 10, 30 and 60%, subjected to the parameter of speed of 0.1 mm/s, and of maximum force of 1 kN, presented phase changes punctually in displacements 1.0 mm, 2.90 mm and 4.30 mm respectively. The largest displacement recorded was 6.30 mm, presented by the product with 60% OSPW incorporation. When the speed parameter was changed to 3.0 mm/s, the product with 10% OSPW incorporation presented a similar curve to the same product subjected at a speed of 0.1 mm/s. Additionally, the CPV-type cement mortar incorporated with 10% OSPW, subjected to a speed parameter of 3 mm/s, registered a variation in the phase transition range.

Considering the speed of 3.0 mm/s, it was noticed that the products with 30 and 60% of OSPW incorporation described similar curves in the squeeze flow test, behaving plastically up to a maximum displacement of 9.0 mm.

In general, it appears that the curves mainly present the stages of plastic deformation and hardening stage. The materials subjected to the squeeze flow test started their curves in stage 2, known as plastic deformation or viscous flow, subjected to small loads. According to the increase in the load, the displacement responded with its gradual growth. In the transition from stage 2 to stage 3, some mortars presented the transition phase in a short displacement space. On the other hand, other mortars showed the transition phase in the long displacement space. As observed in Table 3, as the amount of waste increases, the amount of water and the w/c ratio increase, converging the results of the curves in Figures 5–7.

From the point of view of viscosity, regardless of the type of cement, the curves are similar, despite the different values. It is noteworthy that the CPII curves with 30% of OSPW at 3 mm/s and 60% of OSPW and CPV with 60% of OSPW, both at 1 mm/s and at 3 mm/s, differ mathematically from the other curves, having similarities between them. Through linear regression, the equations presented in Tables 4–6 were obtained, which suggest that the fluid mortar behaves like a grade 2 Herschel Bulkley fluid, unlike the grade 1 Bingham fluid, according to [38], showing that the fluid was influenced by the addition of waste.

It is observed that the linear coefficient of all equations is different from zero, that is, the initial yield stress occurred, differentiating these mortars from Newtonian fluids.

Table 4. Linear regression and coefficient of determination—mortar sample with CPII.

Equation	R^2	Trend Line
$y = 742.61x^2 - 487.13x + 86.24$	0.99	10% OSWP—0.1 mm/s
$y = 380.39x^2 - 400.32x + 88.49$	0.98	10% OSPW—3.0 mm/s
$y = 106.02x^2 - 164.94x + 59.18$	0.97	30% OSPW—0.1 mm/s
$y = 32.44x^2 - 76.478x + 49.95$	0.94	30% OSPW—3.0 mm/s
$y = 12.95x^2 - 20.44x + 61.47$	0.91	60% OSPW—0.1 mm/s
$y = 1.36x^2 + 21.02x + 4.16$	0.99	60% OSPW—3.0 mm/s

Table 5. Linear regression and coefficient of determination—mortar sample with CPIII.

Equation	R^2	Trend Line
$y = 982.46x^2 - 765.47x + 128.08$	0.98	10% OSWP—0.1 mm/s
$y = 550.26x^2 - 569.95x + 122.42$	0.96	10% OSPW—3.0 mm/s
$y = 145.15x^2 - 222.47x + 73.56$	0.95	30% OSPW—0.1 mm/s
$y = 60x^2 - 81.77x + 46.55$	0.97	30% OSPW—3.0 mm/s
$y = 3.44x^2 + 52.82x + 6.56$	0.99	60% OSPW—0.1 mm/s
$y = 12.42x^2 - 29.94x + 77.30$	0.86	60% OSPW—3.0 mm/s

Table 6. Linear regression and coefficient of determination—mortar sample with CPV.

Equation	R^2	Trend Line
$y = 388.74x - 203.41$	0.7996	10% OSWP—0.1 mm/s
$y = 349.73x - 157$	0.7795	10% OSPW—3.0 mm/s
$y = 158.27x - 100.46$	0.7941	30% OSPW—0.1 mm/s
$y = 23.203x + 4.4478$	0.9728	30% OSPW—3.0 mm/s
$y = 105.6x - 84.594$	0.7785	60% OSPW—0.1 mm/s
$y = 24.778x - 4.0944$	0.989	60% OSPW—3.0 mm/s

Another observed fact is that the yield stress decreases with the increase in the amount of waste, regardless of the load speed and the type of cement, and the mortar tends to be more fluid, that is, to have a greater workability. Ying et al., [39] studied the incorporation of waste glass processing in the mortar, noting that its increased incorporation in the mortar also contributed to a better workability.

Considering the value obtained in the linear regression analysis for R^2, it is possible to analyze the linear correlation of the variables force and displacement. For values above 95%, it is possible to infer that the variables have a positive linear correlation, i.e., the data are accurate.

3.1.3. Mass Density

The density test was carried out in accordance with ABNT NM 13278 [28], and the mass of the full container was recorded. Then, the mass density of each mixture was calculated in the different substitutions (0%, 10%, 30% and 60%) in the three types of Portland cement (CP II, CP III and CP V). Tables 7–9 show the values obtained, respectively, by CP II, CP III and CP V cements.

Table 7. Mass density of Portland cement type CP II.

	Incorporation of OSPW			
	0%	10%	30%	60%
Container mass (g)	1643.5	1654.2	1697.1	1654.4
Mass density (g/cm^3)	1.88	1.91	2.02	1.91

Table 8. Mass density of Portland cement type CP III.

	Incorporation of OSPW			
	0%	10%	30%	60%
Container mass (g)	1666.4	1609.9	1540.6	1637.6
Mass density (g/cm^3)	1.94	1.80	1.62	1.87

Table 9. Mass density of Portland cement type CP V.

	Incorporation of OSPW (%)			
	0	10	30	60
Container mass (g)	1649.2	1673.4	1703.1	1654.5
Mass density (g/cm^3)	1.90	1.96	2.03	1.91

Analyzing the mass density in Portland cement type CP II, it appears that the mortar with the lowest density was the one with 0% replacement of sand by the residue with 1.88 g/cm^3, while the mortar with the highest density was the one with 30% replacement with 2.01 g/cm^3. In the case of cement type CP III, the opposite occurs, where the mortar produced with 30% replacement has the lowest density with 1.62 g/cm^3 and the mortar with 0% replacement has the highest density with 1.94 g/cm^3.

In Portland cement type CP V, the same occurs as for CP II, the mortar with the lowest density is the one with 0% replacement with 1.90 g/cm^3, and the one with the highest density is the one with 30% replacement with 2.03 g/cm^3. Analyzing the densities in the three types of cement, the one that presents the highest density is the Portland cement type CP V, while the type CP III presents the lowest density.

Although the mortars tested in the three types of cements in percentages of different replacements show variations in densities, most are around 2 g/cm^3, and densities lower than those mentioned are also obtained, bringing advantage to the mixture, as in the case of Portland cement type CP III with 30% replacement [40].

3.1.4. Water Retention

According to the registration of the five values (m_v, m_a, m_s, m_w, and m) of the mass of the set, the water/fresh mortar factor (AF) and the water retention (Ra) were then calculated, considering each mixture in the different substitutions (0%, 10%, 30% and 60%) in the three types of Portland cement (CP II, CP III and CP V), as shown in Tables 10–12.

Table 10. Water retention in mortar with Portland cement type CP II.

OSPW (%)	m_v (g)	m_a (g)	m_s (g)	m_w (g)	m (g)	AF	Ra (%)
10	605.80	1698.70	1688.70	268.0	1494	0.22	95.81
30	601.80	1735.60	1733.10	276.0	1494	0.23	99.03
60	602.40	1597.20	1593.80	400.0	1494	0.37	99.07
0	622.98	1746.19	1735.06	315.2	1494	0.27	96.29

Table 11. Water retention in mortar with Portland cement type CP III.

OSPW (10%)	m_v (g)	m_a (g)	m_s (g)	m_w (g)	m (g)	AF	Ra (%)
10	622.8	1799.2	1793.4	265	1494	0.22	97.71
30	621.1	1924.8	1920.1	278	1494	0.23	98.42
60	623.5	1684.9	1681.9	415	1494	0.38	99.27
0	623.3	1715.8	1712.3	301	1494	0.25	98.73

Table 12. Water retention in mortar with Portland cement type CP V.

OSPW (%)	m_v (g)	m_a (g)	m_s (g)	m_w (g)	m (g)	AF	Ra (%)
10	604.3	1664.4	1662.0	255	1494	0.21	98.90
30	605.5	1631.9	1628.2	282	1494	0.23	98.45
60	603.7	1589.5	1585.4	420	1494	0.39	98.94
0	622.0	1715.9	1711.9	308	1494	0.26	98.59

Analyzing the water retention in Portland cement type CP II, it appears that the mortar with the lowest retention was the one with 10% of sand replacement by OSPW with 95.81%, while the mortar with the highest retention was the one with 60% of replacement with 99.07%. The same characteristics are found in cement type CP III; the mortar produced with 10% replacement has the lowest retention with 97.71% and the mortar with 60% replacement has the highest retention with 99.27%.

Considering the Portland cement type CP V, there is a difference from the others. The mortar with the lowest retention is the one with 30% replacement with 98.45% instead of 10%, as in the CP II and CP III cements. The one with the highest retention is the 60% replacement with 98.94%, as in the other cements.

Examining the control mortar (0% OSPW), it is noted that the lowest value is found in CP II cement with 96.29%, while the highest is found in CP III with 98.73% retention. At 10% replacement, the lowest value is also found in CP II cement with 95.81%, while the highest is for CP V with 98.90% retention.

As for the analysis of the mortars with 30% replacement, it is noted that the lowest value is found in the CP III cement with 98.42%, while the highest is found in the CP II with 99.03% of retention. At 60% replacement, the lowest value is found in CP V cement with 98.94%, while the highest is found in CP III with 99.27% retention.

3.1.5. Calorimetry

Regarding the calorimetry analysis the registration of the hydration reactions of the mortars, through the software connected to the calorimeter device, it was then possible to plot the heat and thermal energy curves per unit weight of cement. Figures 8 and 9 present the values obtained.

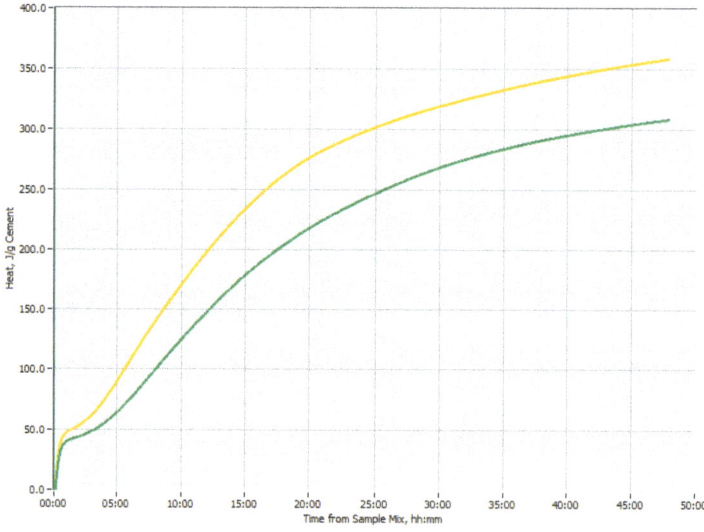

Figure 8. Heat per unit weight of cement: Yellow curve with 30% and green curve with 0%.

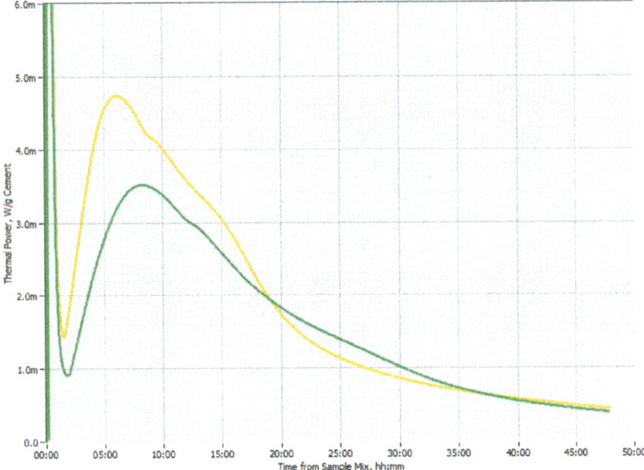

Figure 9. Thermal energy per unit weight of cement: Yellow curve with 30% and green curve with 0%.

By means of the thermal power curves, it was possible to observe in the sample with 30% substitution an increase in the curve in the acceleration period from 1.5 h to 5 h approximately, due to the faster heat release. This is directly associated with a nucleation process and growth of hydration products faster, a behavior also shown in previous studies [13,41].

On the other hand, in the heat of hydration curves, it was possible to observe that the 30% sample presented higher heat of hydration accumulated since the beginning of the test, as a consequence of a greater formation of hydration products, a similar result to that obtained by Topçu et al. [42].

3.1.6. X-ray Diffractometry Analysis—XRD

Figure 10 shows that the particle size distribution of the OSPW particles dimensions is 88% between 2 μm and 60 μm. This class is considered as a filler. It was observed 10% of particles smaller than 2 μm, represented by D10 referring to fine quartz particles found by XRD. The particle size of the sand fraction is 2%, which is the coarser particle of the raw material. The residue is classified as non-plastic. These results are similar to those of Xavier et al. [5].

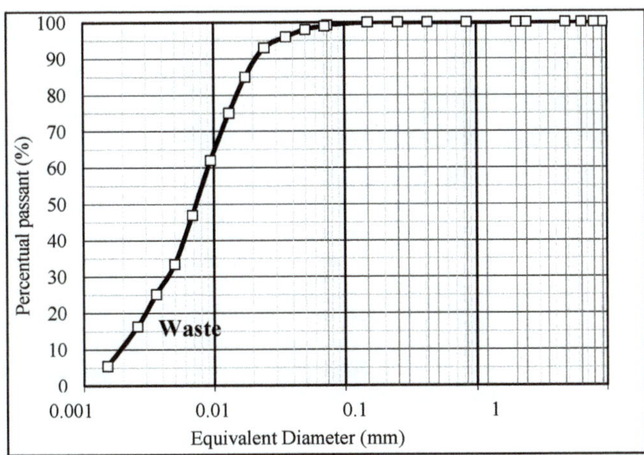

Figure 10. Particle size distribution of the ornamental stone processing waste.

Figure 11a presents the XRD pattern of the OSPW. Note that it is basically formed by quartz, identified by well-defined peaks. Figure 11b shows the presence of ettringite, portlandite, quartz and calcium carbonate. There is also an amorphous halo of C–S–H between 20° and 28° in the COSPW and CH_2O samples. An amorphous halo of C–S–H is found between 22° and 26° (X-ray angle of incidence) in the CS sample, which shows a smaller amount of this stable and resistant phase. This shows that the waste contributes to the formation of this more resistant phase, as observed in Figure 8 with the increase in the heat of hydration being faster with the paste containing 30% of OSPW with the nucleation of the smaller particles, these being in amounts of 10% of OSPW fraction of 2 μm (Figure 10—OSPW particle size distribution). There is no difference between the COSPW and CH_2O paste when compared to each other.

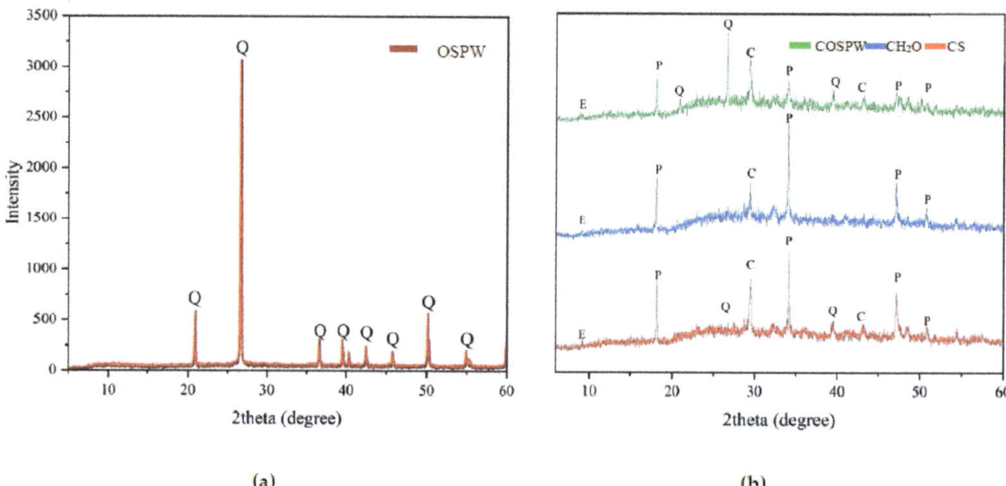

(a)   (b)

Figure 11. (a) X-ray diffraction patterns of the OSPW. (b) X-ray diffraction patterns of cement pastes. Legend: E (ettringite), P (portlandite), Q (quartz), C ($CaCO_3$).

The above-mentioned difference is observed in Figure 7, from the squeeze flow, in the fresh state, as it requires less load and has a greater possibility of displacement, which means that the mortar with OSPW has more workability than the control mortar with only sand. In addition, there is a difference in intensity in most portlandite peaks between the COSPW and CH_2O samples, with smaller peaks being observed in most of the COSPW samples. This may indicate a portlandite reduction and formation of stable phases, as discussed earlier.

3.2. Technological Analysis—Hardened State

3.2.1. Capillary Absorption

The capillary absorption test was carried out as per the ABNT NM 15259 [33] standard, from which two parameters can be taken to determine the phenomenon of capillarity in mortars: (i) the capillarity coefficient, and (ii) the water absorption by capillarity at 90 m [40].

Three prismatic parts were made for each mix, using Portland cement types CP II, CP III and CP V, with cures of 7, 14 and 28 days, in the different replacement of sand by OSPW (10%, 30% and 60%). Table 13 shows the average of each percentage of the capillarity coefficients of the mortars produced.

Table 13. Capillarity coefficients ($g/dm^2 \cdot min^{1/2}$).

		Capillarity Coefficients		
Cure (Days)	OSPW (%)	CP II	CP III	CP V
7 Days	10	0.85	0.44	0.35
	30	0.60	0.36	0.27
	60	1.07	0.85	0.25
14 Days	10	1.07	0.98	0.51
	30	0.78	0.6	0.46
	60	1.17	0.77	0.76
	10	0.92	0.78	0.48
28 Days	30	0.71	0.61	0.44
	60	1.30	1.06	0.70

Thus, analyzing the capillarity coefficient in the prismatic specimens, made with Portland cement type CP II, it appears that the 60% replacement mortars, from sand to OSPW, showed a higher capillarity coefficient regardless of the curing time. On the other hand, the mortars produced with 30% replacement showed the lowest capillarity coefficient also regardless of the curing time.

In the specimens made with Portland cement type CP III, the 30% replacement mortars, at all curing times, showed the lowest capillarity coefficient. In mortars with curing times of 7 and 28 days, the highest capillarity coefficient is at 60% replacement, while at 14 days of curing, it is at 10% replacement.

In Portland cement type CP V, mortars with curing times of 14 and 28 days presented the highest coefficient at 60%, while at 7 days of curing, the highest coefficient is at 10% replacement. As with mortars made with cement type CP II and CP III, the lowest coefficient is found at 30%, when analyzed at 14 and 28 days, while at 7 days of curing, the lowest value is at 60% of replacement.

3.2.2. Determination of Tensile Strength in Flexion and Compression

Table 14 presents the compressive strengths found in the mortars made with Portland cement types CP II, CP III and CP V, with curing of 7, 14 and 28 days, in the different replacement of sand by OSPW (10%, 30% and 60%).

Table 14. Flexural and compression strength test results.

OSPW (%)	Cure (Days)	CPII FS[1] ± SD[2] (MPa)	CPII CS[3] ± SD[2] (MPa)	CPIII FS[1] ± SD[2] (MPa)	CPIII CS[3] ± SD[2] (MPa)	CPV FS[1] ± SD[2] (MPa)	CPV CS[3] ± SD[2] (MPa)
0	7	1.10 ± 0.20	4.07 ± 0.90	1.07 ± 0.21	3.87 ± 0.65	2.00 ± 0.45	5.03 ± 0.60
	14	0.90 ± 0.19	3.55 ± 0.90	0.96 ± 0.21	3.09 ± 0.65	1.93 ± 0.45	6.00 ± 0.60
	28	1.11 ± 0.20	4.05 ± 0.90	1.32 ± 0.21	3.89 ± 0.65	2.05 ± 0.45	6.11 ± 0.60
10	7	1.19 ± 0.23	4.74 ± 0.45	1.14 ± 0.23	3.93 ± 0.76	2.01 ± 0.41	6.44 ± 0.51
	14	1.43 ± 0.23	3.83 ± 0.45	0.83 ± 0.23	2.78 ± 0.76	1.67 ± 0.41	7.08 ± 0.51
	28	0.97 ± 0.23	4.21 ± 0.45	1.29 ± 0.23	4.23 ± 0.76	2.49 ± 0.41	7.46 ± 0.51
30	7	2.17 ± 0.25	9.82 ± 1.33	1.19 ± 0.35	6.40 ± 1.41	2.38 ± 0.40	10.33 ± 1.15
	14	2.23 ± 0.25	7.34 ± 1.33	1.26 ± 0.35	6.34 ± 1.41	3.04 ± 0.40	10.95 ± 1.15
	28	2.63 ± 0.25	9.44 ± 1.33	1.83 ± 0.35	8.82 ± 1.41	3.12 ± 0.40	12.56 ± 1.15
60	7	1.61 ± 0.28	7.12 ± 0.46	0.79 ± 0.21	5.61 ± 1.08	1.36 ± 0.54	5.17 ± 1.69
	14	1.77 ± 0.28	6.66 ± 0.46	0.61 ± 0.21	4.26 ± 1.08	2.39 ± 0.54	6.47 ± 1.69
	28	1.21 ± 0.28	7.59 ± 0.46	1.04 ± 0.21	6.40 ± 1.08	2.16 ± 0.54	8.53 ± 1.69

[1] FS—flexural strength; [2] SD—standard deviation; [3] CS—compression strength.

Analyzing the compressive strength of mortars, it was found that in all three types of cement, the highest strengths were found in the replacement of 30% of OSPW, and these values were identified in CP III and CP V cements, at 28 days of curing, and in cement CP II, at 7 days of curing.

In Portland cement type CP II, the highest strength found was 9.82 MPa tested at 7 days of curing, in the percentage of 30% replacement, while the lowest strength was 3.83 MPa tested at 14 days of curing, in the 10% replacement percentage. Note that the strengths of the mortars tested at 7 and 28 days of curing present similar values in all replacements.

In mortars made with CP III type cement, the highest strengths, in all percentages, were at 28 days of curing, with the highest value being 8.82 MPa in 30% replacement. On the other hand, the lowest resistances found, regardless of residue replacements, were at 14 days of curing.

In Portland cement type CP V, the highest strengths, as well as in CP III, were at 28 days of curing, regardless of percentages, with the highest value of 12.56 MPa in 30% replacement. While the lowest resistances were at 7 days of curing in all percentages.

Table 14 also shows the flexural tensile strengths found in mortars made with Portland cement types CP II, CP III and CP V, with cures of 7, 14 and 28 days of curing, in the different replacements of sand by OSPW (10%, 30% and 60%).

Analyzing the flexural tensile strengths of mortars, it was found that in all three types of cement, the highest strengths were found in the replacement of 30% of OSPW, at 28 days of curing, with the highest value being 3, 12 MPa, found in CP V cement. In Portland cement type CP II, the mortars tested at 7 and 14 days of curing showed similar strengths, while those tested at 28 days showed greater discrepancies.

The mortars made with cement type CP III tested at 28 days showed higher strengths in all replacement percentages, when compared to the other curing times of the same cement. In contrast, the lowest resistance found was with 60% replacement tested at 14 days of curing. In Portland cement type CP V, the mortars tested at 28 days showed the highest strengths with 10% and 30% replacement, however, with 60% the mortars that showed higher values were those tested at 14 days of curing.

3.3. Life Cycle Assessment (LCA)

According to the results of the technological tests, LCA studies were directed on the environmental impacts associated with obtaining the mortar incorporated with 30% of OSPW and cement CPV type. For the analysis of the environmental impacts associated with the mortar production process, OSPW was considered as an alternative raw material to substitute the use of sand, which allows obtaining an avoided impact with the reduction of the extraction of this input.

The LCA study was compiled considering the geographic limitation of the states of Rio de Janeiro and Espírito Santo, in the Brazilian context, in addition to the chemical composition and morphological characteristics of the sand and OSPW found in the studied region. The study was conducted in a comparative way between the two products: the mortar produced in a conventional way (0% OSPW), and the mortar incorporated with 30% OSPW. For this study, the use of CPV-type cement was also considered.

The mortar production process considered at the system boundary consisted of traditional operations and equipment. Thus, studies that direct efforts to analyze the effects of new technologies are recommended. In addition, the present LCA study is dedicated only to analyzing the environmental impacts caused by the products of interest, recommending future analyses on economic and social impacts.

Figure 12 presents the environmental impact of the conventional mortar production process, without the incorporation of OSPW (the characterized analysis can be seen in Table S1 of the Supplementary Materials). Eighteen impact categories were analyzed at midpoint. In a global analysis, it is possible to realize that the cement and sand inputs significantly contribute to the impact associated with obtaining the mortar, as presented by Sánchez et al. [19]. Cement has an intensive impact on the global warming categories, accounting for about 75% of the CO_2 emissions considered in this production process, and also on the mineral resource scarcity category, corresponding to 91% of the impacts recorded in this category. Jiménez et al. [43] state that cement is the material with the greatest influence in all categories of impacts considered in their studies, in addition to being the main contributor to emissions to air.

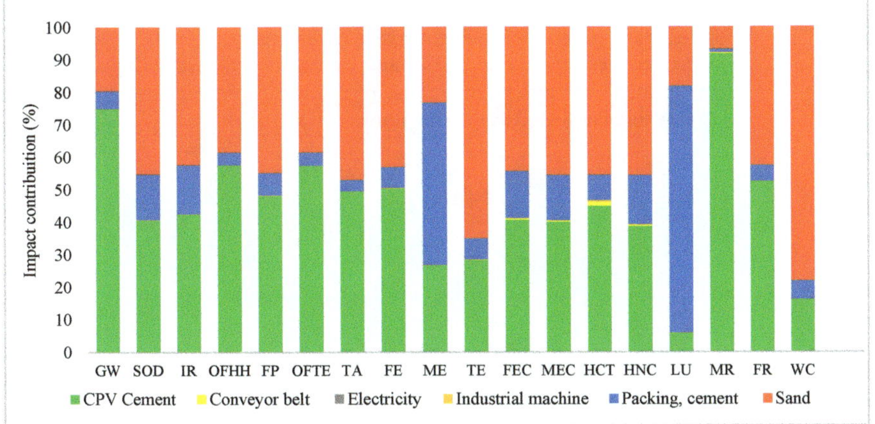

Figure 12. Impact life cycle assessment—mortar with 0% OSPW—CPV.

The sand has its largest impact contribution associated with the water consumption category, with over 77%. In addition, the cement used, identified as packing in the study, has considerable impacts on point categories, such as the land use category, reaching 76% of the relative impact. This can be attributed to the area needed to obtain wood that will supply the pulp industry for the manufacture of cement packages.

To highlight the main impacts associated with alternative mortar composed of cement type CPV, 30% of OSPW and sand, the LCA was analyzed (Figure 13). It was possible to identify that the replacement of sand by OSPW contributed to the reduction in the associated impacts in most categories, such as in the categories of land use −5%, freshwater eutrophication −9%, marine eutrophication −6% and global warming −5%. The most significant record of impacts associated with the use of OSPW is in the water consumption category, totaling 2.3% (the characterized analysis can be seen in Table S2 of the Supplementary Materials).

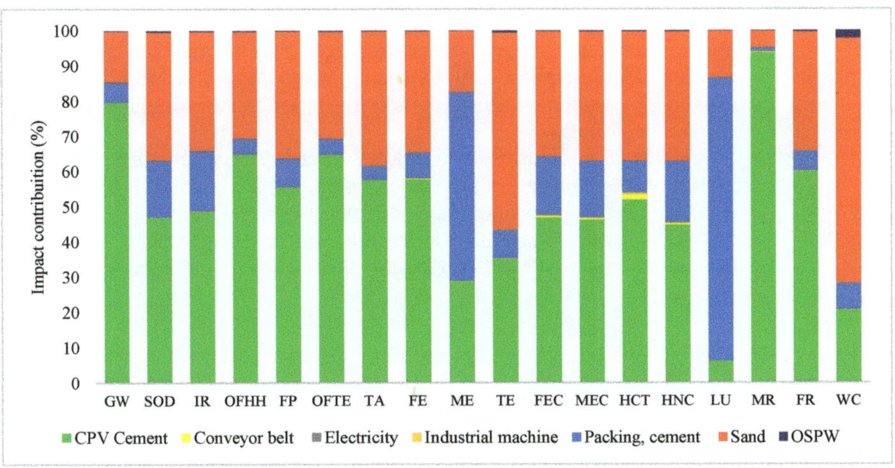

Figure 13. Impact life cycle assessment—mortar with 30% OSPW—CPV.

In a comparative way, the processes of obtaining the mortar with cement CPV, and the mortar with cement CPV incorporated with 30% of OSPW were analyzed. The normalized results show the reduction in global impacts when the mortar undergoes the replacement of sand by OSPW. The normalized scale also allows the identification of the main categories of impacts that are intensely affected by the industrial activity of obtaining the product under study, namely marine ecotoxicity, freshwater ecotoxicity, human carcinogenic toxicity, human non-carcinogenic toxicity and terrestrial ecotoxicity (Figure 14). These categories are allocated into two main areas of protection, those being human health and ecosystem quality, often assessed at the endpoint.

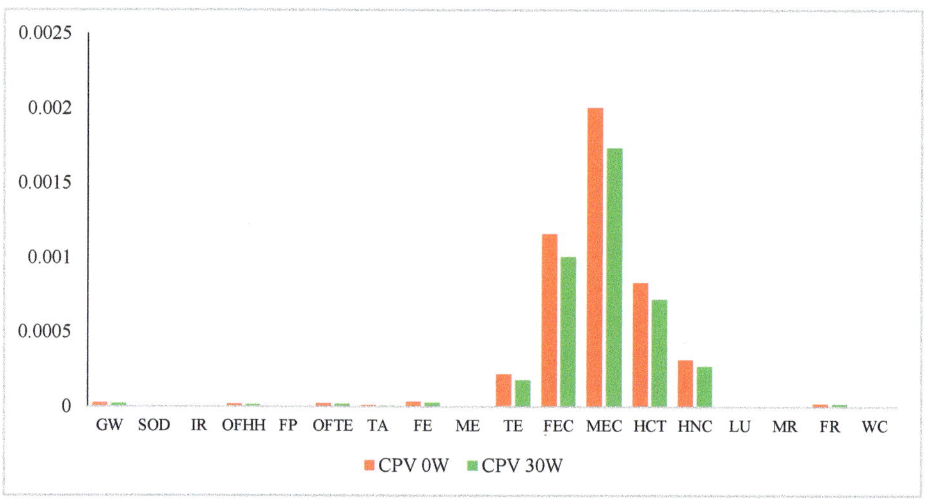

Figure 14. Impact life cycle assessment—normalized result of the comparison between conventional and 30% OSPW mortar.

The categories that estimate human health impacts aim to characterize human exposure to toxic substances, especially through ingestion and inhalation. Values referring to acute and chronic toxicological effects provide estimates of the toxicological risk and impacts associated with the mass (kilograms) of a substance emitted to the environment, such as dichlorobenzene, for example.

The categories related to ecosystem are related to the harmful, sometimes irreversible, action of substances toxic to the environment. This category can be defined for both water and soil, using chemical emissions to air, water and soil. Its indicator corresponds to the ecotoxicity potential of each emission in relation to the reference substance (for example, triethylene glycol).

Figure 15 presents the comparison in a characteristic mode between the conventional mortar and the mortar incorporation with OSPW. It is possible to notice the decrease in environmental impacts associated with all categories when the mortar is incorporated with OSPW. According to Rebello et al. [6], the incorporation of OSPW in the elaboration of civil construction materials presented better environmental performance compared to the use of sand.

This can be mainly related to the reuse of an inert residue, which implies the reduction in sand extraction for the composition of this mortar, being an avoided impact. The main reductions occurred in the water consumption and terrestrial ecotoxicity categories, recording reductions of 22% and 20%, respectively. According to Santos et al. [44], the incorporation of alternative aggregates in conventional mortars (with high environmental impact) requires less energy for production and transport. This mainly applies to the replacement of natural sand in mortars, which can also benefit the technological performance of the material.

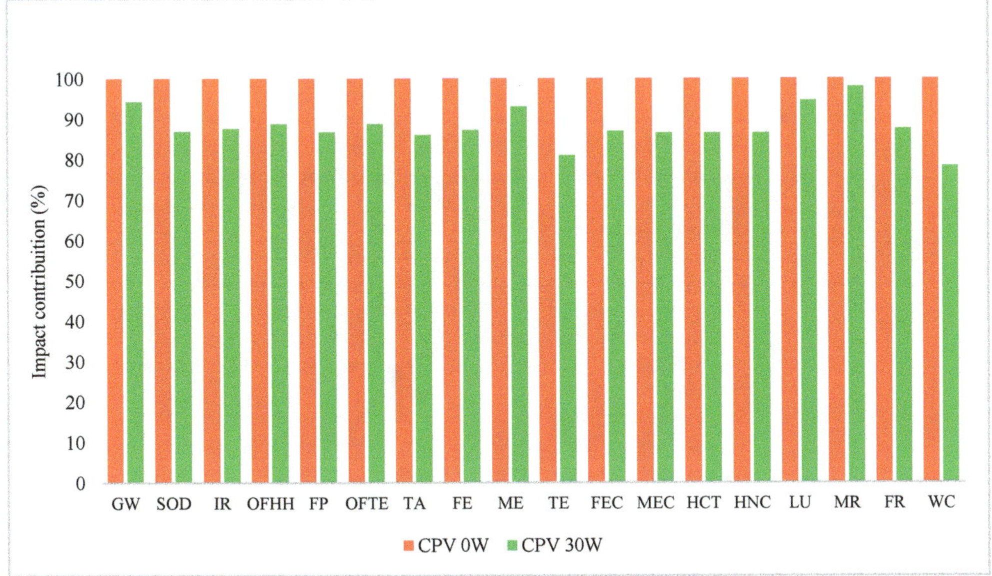

Figure 15. Impact life cycle assessment—characterized result of the comparison between conventional and 30% OSPW mortar.

In LCA studies oriented to the production of construction materials with incorporation of waste, the geographical distance to obtain the inputs stands out. Indeed, distance and transport have a direct relationship, that is, great distances can be associated with the increase in the environmental impacts considered [45].

The carbon footprints for both evaluated processes were obtained and are shown in Table 15. Thus, it was possible to identify the contribution to the environmental impacts of raw materials in the system, mainly associated with the global warming category. Predominantly, CPV cement presents itself as the main contribution to the carbon footprint in both scenarios, being responsible for 75.1% of the CO_2 equivalent generated in obtaining the mortar without incorporation of OSPW, and 79.7% of the CO_2 equivalent in the mortar with the incorporation of 30% of OSPW.

Table 15. Carbon footprint.

Inputs	CPV 0 W (CO_2 eq. %)	CPV 30 W (CO_2 eq. %)
CPV Cement	75.1	79.7
Sand	19.5	14.5
Packing, cement	5.28	5.61
OSPW	0	0.0885
Conveyor belt	0.0238	0.0252
Industrial machine	0.0088	0.00934
Electricity	0.000325	0.000345

4. Conclusions

- Considering the results found in the squeeze flow test, it can be concluded that the greater the displacement achieved by the material, the better its workability level, which was recorded in different mortar compositions incorporated by the ornamental stones processing waste (OSPW).
- According to the tests in the fresh state, it is possible to conclude that the mortar composed of cement of the CPV type, incorporated with 30% of OSPW, stands out in

comparison to the others analyzed, presenting higher density, which implies a material with lower porosity and good water retention index. The workability of this material also stands out, as well as the rapid release of heat that occurs in this material, which positively influences the nucleation process and rapid hydration.

- Through the calorimetry technique, it is possible to infer that the incorporation of the waste is acting as nucleation sites for the hydration products, favoring the cement hydration process, culminating at 28 days of curing, in the formation of more stable phases, according to the amorphous halo of the formation of C–S–H (hydrated calcium silicate).

- According to the tests in the hardened state, the mortar composed of cement of the CPV type, incorporated with 30% OSPW, stands out, which presents low capillarity and better performance in the resistance to flexural and compression strength, considering 7 days of curing when compared to other types of cement. There is still a better performance in the flexural and compression strength tests when the curing time increases to 28 days.

- In view of the life cycle assessment (LCA) for environmental analysis, it is concluded that the use of OSPW contributes to the reduction in the environmental impact associated with the process of obtaining the mortar. Indeed, this waste can replace the use of sand in the production mortar, which consequently reduces the sand extraction process and its impacts. Considering the LCA impacts generated by the production of mortars, the main areas of protection affected are human health and ecosystem quality.

Supplementary Materials: The following supporting information can be downloaded at: https://www.mdpi.com/article/10.3390/su14105904/s1, Table S1: Life Cycle Impact Assessment considering the mortar with 0% of OSPW and cement type CPV; Table S2: Life Cycle Impact Assessment considering the mortar with 30% of OSPW and cement type CPV.

Author Contributions: Conceptualization, P.I.M. and G.d.C.X.; methodology, G.d.C.X., C.M.V. and A.R.G.d.A.; software, J.d.O.D. and R.P.R.; validation, C.M.V., S.N.M. and J.A.; formal analysis, G.d.C.X. and R.P.R.; investigation, P.I.M. and G.d.C.X.; resources, G.d.C.X., S.N.M. and C.M.V.; data curation, A.R.G.d.A. and J.A.; writing—original draft preparation, G.d.C.X. and J.d.O.D.; writing—review and editing, R.P.R.; visualization, J.A.; supervision, G.d.C.X.; project administration, C.M.V.; funding acquisition, S.N.M. All authors have read and agreed to the published version of the manuscript.

Funding: This research was funded by State University of the Northern Fluminense (UENF), partial financed by CAPES (Coordenação de Aperfeiçoamento de Pessoal de Nível Superior—Brazil) and provided additional financial CNPq (Coordenação Nacional de Pesquisa) Code 309428/2020-3. The participation of A.R.G.A. was sponsored by FAPERJ through the research fellowships proc. no: E-26/210.150/2019, E-26/211.194/2021, E-26/211.293/2021, and E-26/201.310/2021 and by CNPq through the research fellowship PQ2 307592/2021-9.

Institutional Review Board Statement: Not applicable.

Data Availability Statement: Not applicable.

Acknowledgments: The authors acknowledge the Brazilian governmental research agencies CAPES (Coordenação de Aperfeiçoamento de Pessoal de Nível Superior), CNPq (Conselho Nacional de Desenvolvimento Científico e Tecnológico), and FAPERJ (Fundação de Amparo à Pesquisa do Estadodo Rio de Janeiro).

Conflicts of Interest: The authors declare no conflict of interest.

References

1. Azevedo, A.R.G.; Marvila, M.T.; Barroso, L.S.; Zanelato, E.B.; Alexandre, J.; Xavier, G.C.; Monteiro, S.N. Effect of Granite Waste Incorporation on the Behavior of Mortars. *Materials* **2019**, *12*, 1449. [CrossRef] [PubMed]
2. Montani, C. *Marble and Stones in the World XXVIII Report—International Situation Production and Interchange*, 18th ed.; Casa di Edizioni in Carrara: Carrara, Italy, 2017; pp. 1–132.
3. Chiodi, C. *The Brazilian Ornamental Stone Sector*; Brazilian Association of the Ornamental Stone Industry (ABIROCHAS): Vitoria, Brazil, 2018. (In Portuguese)

4. Chiodi, C. *Balanço das Exportações e Importações Brasileiras de Materiais Rochosos Naturais e Artificiais de Ornamentação e Revestimento em 2021*; Brazilian Association of the Ornamental Stone Industry (ABIROCHAS): Vitoria, Brazil, 2022.

5. Xavier, G.C.; Azevedo, A.R.G.; Alexandre, J.; Monteiro, S.N.; Pedroti, L.G. Determination of Useful Life of Red Ceramic Parts Incorporated with Ornamental Stone Waste. *J. Mater. Civ. Eng.* **2019**, *31*, 04018381. [CrossRef]

6. Rebello, T.A.; Zulcão, R.; Calmon, J.L.; Gonçalves, R.F. Comparative life cycle assessment of ornamental stone processing waste recycling, sand, clay and limestone filler. *Waste Manag. Res.* **2019**, *37*, 186–195. [CrossRef] [PubMed]

7. Xiao, R.; Polaczyk, P.; Zhang, M.; Jiang, X.; Zhang, Y.; Huang, B.; Hu, W. Evaluation of glass powder-based geopolymer stabilized road bases containing recycled waste glass aggregate. *Transp. Res. Rec. J. Transp. Res. Board* **2020**, *2674*, 22–32. [CrossRef]

8. Galetakis, M.; Soultana, A. A review on the utilisation of quarry and ornamental stone industry fine by-products in the construction sector. *Constr. Build. Mater.* **2016**, *102*, 769–781. [CrossRef]

9. Aydin, E.; Sahan Arel, H. High-volume marble substitution in cement-paste: Towards a better sustainability. *J. Clean. Prod.* **2019**, *237*, 117801. [CrossRef]

10. Baghel, R.; Pandel, U.; Vashistha, A. Manufacturing of sustainable bricks: Utilization of mill scale and marble slurry. *Mater. Today Proc.* **2020**, *26*, 2136–2139. [CrossRef]

11. Jain, A.; Gupta, R.; Chaudhary, S. Sustainable development of self-compacting concrete by using granite waste and fly ash. *Constr. Build. Mater.* **2020**, *262*, 120516. [CrossRef]

12. Awad, A.H.; El-Wahab, A.; El-Gamsy, R.; Abdel-Latif, H. A study of some thermal and mechanical properties of HDPE blend with marble and granite dust. *Ain Shams Eng. J.* **2019**, *10*, 353–358. [CrossRef]

13. Li, L.G.; Huang, Z.H.; Tan, Y.P.; Kwan, A.K.H.; Chen, H.Y. Recycling of marble dust as paste replacement for improving strength, microstructure and eco-friendliness of mortar. *J. Clean. Prod.* **2019**, *210*, 55–65. [CrossRef]

14. Messaouda, B.; Rahmouni, Z.E.A.; Tebbal, N. Influence of the addition of glass powder and marble powder on the physical and mechanical behavior of composite cement. *Procedia Comput. Sci.* **2019**, *158*, 366–375. [CrossRef]

15. Saboya, F.; Xavier, G.C.; Alexandre, J. The use of the powder marble by-product to enhance the properties of brick ceramic. *Constr. Build. Mater.* **2007**, *21*, 1950–1960. [CrossRef]

16. Dantas, A.P.A.; Acchar, W.; Leite, J.Y.P.; Araújo, F.S.D. Use of Ornamental Stone Residues in the Production of White Ceramics. *Holos* **2010**, *1*, 26.

17. Buyuksagis, I.S.; Uygunoglu, T.; Tatar, E. Investigation on the usage of waste marble powder in cement-based adhesive mortar. *Constr. Build. Mater.* **2017**, *154*, 734–742. [CrossRef]

18. Perera, A.G.; Dionisio, M.D.P.S.; Sancha, P. Alternativas de reducción de las emisiones de dióxido de carbono (CO_2) en la producción de cemento. Propuesta de un modelo de evaluación. *Innovar* **2016**, *26*, 51–66. [CrossRef]

19. Sánchez, A.R.; Ramos, V.C.; Polo, M.S.; Ramón, M.V.L.; Utrilla, J.R. Life Cycle Assessment of Cement Production with Marble Waste Sludges. *Int. J. Environ. Res. Public Health* **2021**, *18*, 10968. [CrossRef]

20. Malça, J.; Freire, F. Renewability and life-cycle energy efficiency of bioethanol and bio-ethyl tertiary butyl ether (bioETBE): Assessing the implications of allocation. *Energy* **2006**, *31*, 3362–3380. [CrossRef]

21. Ye, L.; Hong, J.; Ma, X.; Qi, C.; Yang, D. Life cycle environmental and economic assessment of ceramic tile production: A case study in China. *J. Clean. Prod.* **2018**, *189*, 432–441. [CrossRef]

22. *NBR 16697*; Portland Cement—Requirements. Brazilian Association of Technical Standards: Rio de Janeiro, Brazil, 2018. (In Portuguese)

23. *ABNT NBR 7181*; SOIL—Particle Size Analysis. Brazilian Association of Technical Standards: Rio de Janeiro, Brazil, 1984. (In Portuguese)

24. *NBR 13276*; Mortar for and Coating Walls and Ceilings—Mixing Preparation and Determination of Consistency Index. Brazilian Association of Technical Standards: Rio de Janeiro, Brazil, 2016. (In Portuguese)

25. *NBR 15839*; Mortars for Laying and Covering Walls and Ceilings—Rheological Characterization by the Squeeze Flow Method. Brazilian Association of Technical Standards: Rio de Janeiro, Brazil, 2010. (In Portuguese)

26. Cardoso, F.A.; John, V.M.; Pileggi, R.G.; Banfill, P.F.G. Characterization of rendering mortars by squeeze-flow and rotational rheometry. *Cem. Concr. Res.* **2014**, *57*, 79–87. [CrossRef]

27. Cardoso, F.A.; John, V.M.; Pileggi, R.G. Rheological behavior of mortars under different squeezing rates. *Cem. Concr. Res.* **2009**, *39*, 748–753. [CrossRef]

28. *NBR 13278*; Argamassas para Assentamento e Revestimento de Paredes e Tetos—Determinação da Densidade de Massa e Do teor de ar Incorporado. Associação Brasileira de Normas Técnicas: Rio de Janeiro, Brazil, 2005.

29. *NBR 13277*; Argamassas para Assentamento e Revestimento de Paredes e Tetos—Determinação da Retenção de Água. Associação Brasileira de Normas Técnicas: Rio de Janeiro, Brazil, 2005.

30. *C1679-17*; Standard Practice for Measuring Hydration Kinetics of Hydraulic Cementitious Mixtures Using Isothermal Calorimetry. American Society for Testing and Materials—ASTM: West Conshohocken, PA, USA, 2017.

31. Tambara, L.U.D.; Cheriaf, M.; Rocha, J.C.; Palomo, A.; Fernández-Jiménez, A. Effect of alkalis content on calcium sulfoaluminate (CSA) cement hydration. *Cem. Concr. Res.* **2020**, *128*, 105953. [CrossRef]

32. *NBR 13279*; Mortars for Laying and Covering Walls and Ceilings—Determination of Tensile Strength in Flexion and Compression. Brazilian Association of Technical Standards: Rio de Janeiro, Brazil, 2005. (In Portuguese)

33. *NBR 15259*; Mortars for Laying and Covering Walls and Ceilings—Determination of Water Absorption by Capillarity and Capillarity Coefficient. Brazilian Association of Technical Standards: Rio de Janeiro, Brazil, 2005. (In Portuguese)
34. *ISO 14040: 2006/Amd 1:2020*; Environmental Management Life Cycle Assessment—Principles and Framework. ISO: Geneva, Switzerland, 2006.
35. *ISO 14044: 2006/Amd 2:2020*; Environmental Management—Life Cycle Assessment—Requirements and Guidelines. ISO: Geneva, Switzerland, 2006.
36. de Oliveira Guimarães, C.A.; Delaqua, G.C.G.; de Azevedo, A.R.G.; Monteiro, S.N.; Amaral, L.F.; Souza, C.L.M.; Vieira, C.M.F. Heating rate effect during sintering on the technological properties of brazilian red ceramics. *Int. J. Adv. Man. Technol.* **2022**, *119*, 8125–8135. [CrossRef]
37. Ghasemipor, V.; Piroti, S. Experimental Evaluation of the Effect of Water-Cement Ratio on Compressive, Abrasion Strength, Hydraulic Conductivity Coefficient and Porosity of Nano-Silica Concretes. *J. Appl. Eng. Sci.* **2018**, *8*, 17–24. [CrossRef]
38. Baumert, C.; Garrecht, H. Minimization of the Influence of Shear-Induced Particle Migration in Determining the Rheological Characteristics of Self-Compacting Mortars and Concretes. *Materials* **2020**, *13*, 1542. [CrossRef] [PubMed]
39. Ying, J.; Huang, J.; Xiao, J. Test and theoretical prediction of chloride ion diffusion in recycled fine aggregate mortar under uniaxial compression. *Constr. Build. Mater.* **2022**, *321*, 126384. [CrossRef]
40. Marvila, M.T.; Alexandre, J.; Azevedo, A.R.G.; Zanelato, E.B.; Xavier, G.C.; Monteiro, S.N. Study on the replacement of the hydrated lime by kaolinitic clay in mortars. *Adv. Appl. Ceram.* **2019**, *118*, 373–380. [CrossRef]
41. Wadsö, L.; Winnerfeld, F.; Riding, K.; Sandeberg, P. Calorimetry. In *A Practical Guide to Microstructural Analisys of Cementitious Materials*, 1st ed.; Scrivener, K., Snellings, R., Lothenbach, B., Eds.; CRC Press: Boca Raton, FL, USA, 2016.
42. Topçu, I.B.; Bilir, T.; Uygunoglu, T. Effect of waste marble dust content as filler on properties of self-compacting concrete. *Constr. Build. Mater.* **2009**, *23*, 1947–1953. [CrossRef]
43. Jiménez, C.; Barra, M.; Josa, A.; Valls, S. LCA of recycled and conventional concretes designed using the Equivalent Mortar Volume and classic methods. *Constr. Build. Mater.* **2015**, *84*, 245–252. [CrossRef]
44. Santos, T.; Almeida, J.; Silvestre, J.D.; Faria, P. Life cycle assessment of mortars: A review on technical potential and drawbacks. *Constr. Build. Mater.* **2021**, *288*, 123069. [CrossRef]
45. Yazdanbakhsh, A.; Lagouin, M. The effect of geographic boundaries on the results of a regional life cycle assessment of using recycled aggregate in concrete. *Resour. Conserv. Recycl.* **2019**, *143*, 201–209. [CrossRef]

 *sustainability*

Article

Novel Sustainable Castor Oil-Based Polyurethane Biocomposites Reinforced with Piassava Fiber Powder Waste for High-Performance Coating Floor

Juliana Peixoto Rufino Gazem de Carvalho [1], Noan Tonini Simonassi [1,*], Felipe Perissé Duarte Lopes [1], Sergio Neves Monteiro [1,2] and Carlos Maurício Fontes Vieira [1]

[1] Advanced Materials Laboratory (LAMAV), State University of Northern Rio de Janeiro (UENF), Campos dos Goytacazes 28013-602, Rio de Janeiro, Brazil; julianarufino@pq.uenf.br (J.P.R.G.d.C.); felipeperisse@gmail.com (F.P.D.L.); snevesmonteiro@gmail.com (S.N.M.); vieira@uenf.br (C.M.F.V.)

[2] Materials Science Program, Military Institute of Engineering (IME), Praça General Tibúrcio 80, Urca 22290-270, Rio de Janeiro, Brazil

* Correspondence: noantoninisimonassi@gmail.com

Abstract: The search for new greener materials that contribute to a more sustainable world motivated the present study in which novel biocomposites with 10, 20 and 30 vol% of piassava fiber powder waste reinforcing castor oil-based polyurethane (COPU) intended for a high-performance coated floor (HPCF) were developed. The novel biocomposites were characterized by flexural, Izod impact and wear standard tests as well as Fourier transform infrared spectroscopy (FTIR) and fracture analysis using scanning electron microscopy (SEM). Both flexural modulus and strength displayed marked increases reaching more than 800 and 500%, respectively, compared to plain COPU for 30 vol% piassava powder incorporation. FTIR bands indicated the existence of interaction between the piassava constituents and COPU. However, SEM fractographs disclosed the presence of bubbles attributed to retained gases during the COPU curing. Consequently, the Izod impact resistance showed a 50% decrease while the wear was more than three times accentuated for 30 vol% piassava powder biocomposite. These results met the specified values of corresponding standards and revealed a promising new greener material for HPCFs.

Keywords: castor oil polyurethane; natural fiber biocomposite; piassava fiber powder; high-performance coating; high-performance coating floor

Citation: Carvalho, J.P.R.G.d.; Simonassi, N.T.; Lopes, F.P.D.; Monteiro, S.N.; Vieira, C.M.F. Novel Sustainable Castor Oil-Based Polyurethane Biocomposites Reinforced with Piassava Fiber Powder Waste for High-Performance Coating Floor. *Sustainability* **2022**, *14*, 5082. https://doi.org/10.3390/su14095082

Academic Editor: Mariateresa Lettieri

Received: 8 March 2022
Accepted: 7 April 2022
Published: 23 April 2022

Publisher's Note: MDPI stays neutral with regard to jurisdictional claims in published maps and institutional affiliations.

1. Introduction

In different kinds of activities, coatings are used to protect structures, goods and packaging from the aggressiveness of the environment mostly by providing special characteristics to the associated material. High-performance coatings is a generic term used to define coatings that provide great mechanical resistance along with other interesting features. These coatings must have particular properties in addition to mechanical resistance to ensure good protection against corrosion or to avoid the occurrence of cracks and fissures among other anomalies [1,2] without loss of performance throughout their life cycle. In particular, polymer resins are widely used as coatings in many industries from aerospace and automobile to pharmaceutical and food industries [3–5]. Resins normally used as high-performance coatings are actually polymer matrix composites reinforced by mineral aggregates, natural or synthetic fibers, as well as other fillers that provide great mechanical resistance along with high durability and waterproof conditions, together with any property of specific interest [6–8].

A particular case related to the civil construction common in Brazil is the high-performance coated floor (HPCF). Traditional floors are made of concrete plates with or without mortar and ceramic tiles. The joints in these floors pose problems such as biohazard contamination. This can be bothersome in places such as hospitals, and pharmaceutical and food processing buildings. In contrast, polymer-based HPCFs are easy to apply and do not need a previously prepared surface when compared to traditional ceramic tiles. Moreover, HPCFs do not have joints as the polymer is primarily applied in the form of liquid resin over the entire floor. In fact, not only do HPCFs improve the floor's durability, provide oil or waterproof characteristics, and maintain its cleanliness, but they also enhance sports performance because of their good adherence properties when applied on stadium floors [9].

Although it is common to find marketing information for HPCFs, to our knowledge, studies on this subject are scarce and no relevant work has been reported in the literature for their use in civil construction. Therefore, the main objective of this work was to evaluate the potential viability of using the piassava fiber waste in such an application. Although some nanocomposites have shown promising results for HPCFs [9], the reinforcements used are usually from synthetic, non-renewable origins. The use of biocomposites reinforced by natural and renewable piassava fibers for this application would be directly beneficial for the environment. The Brazilian standard NBR 14050 [10] regulates the use of HPCFs made with epoxy resins and provides the requirements for their application, as shown in Table 1. In this table, the "critical" requirements are properties that can be found in other kinds of polymers and their composites.

Table 1. Summarized requirements of the NBR 14050 [10] standard for a material to be used as HPCFs.

Requirements	Value Recommendations
Thickness	3 to 10 mm
Maximum Water Absorption	1.0%
Minimum Compressive Strength	45 MPa
Minimum Flexural Strength	20 MPa
Maximum Abrasion	2.30 mm/km

Epoxy resins are the most common choice not only for HPCFs but also for paints, car components and aerospace parts because of their well-known properties and reliable characteristics [11–13]. However, the production of epoxy-based HPCFs generates the evaporation of large amounts of volatile organic compounds [14]. Moreover, the use of non-renewable source materials poses a long-term problem. Both the academic and the industrial communities have been striving to develop new sustainable materials [15–18].

An alternative to the use of petroleum-derived resins such as epoxy is material from natural sources. One example is polyurethane (PU) synthesized from oilseed plants. The synthesis of PU occurs in stages of polyaddition and from hydroxyl compounds and isocyanates. Hydroxyl monomers can be obtained from vegetable oils [19]. This is the case for castor oil-based polyurethane (COPU). The oil is extracted from the fruit of the "castor oil plant" (*Ricinus communis* L.) a relatively tall shrub from tropical regions that has been successfully used to obtain the COPU polymer [19,20].

Studies have shown some interesting properties of COPU. Santan et al. [21] created an adhesive based on COPU and evaluated both the microstructural relationships and mechanical properties of the new material. Zeng et al. [22] modified the asphalt used for paving roads with COPU and noticed improvements in performance and a reduction in the deformations of these coatings. In fact, COPU is a bi-component polymer composed of prepolymer and polyol, both of them easily found commercially, and characterized by not emanating toxic substances [23,24].

In parallel, several researchers in past decades have suggested natural lignocellulosic fibers (NLFs) [25–32] as composite reinforcement material, mostly owing to its attractive features such as good mechanical properties, low density cost-effective production and sustainable motivation. The environmental appeal of these materials can be further enhanced by adding industrial, residential or agricultural wastes as reinforcement of these composites [32–36]. Usually, natural lignocellulosic fiber (NLF) waste is burned during its end-of-life cycle, and not only does the use of waste as a composite reinforcement or filler in natural source-derived resins directly reduce the cost associated with its disposal, but it also presents itself as a good alternative to materials with carbon neutral emissions.

One specific NLF commonly used in Brazil as a raw material in the manufacture of brooms for households is the piassava fiber. In the manufacturing process, the fibers are cut to a standard length, and those shorter are discarded and become disposable waste. These piassava fiber wastes have limited commercial use and are normally burned [37]. As an alternative, this fibrous waste can be further processed and ground to powder to be incorporated into polymer composites.

The use of piassava fiber powder, here referred as piassava powder for short, has shown some promising results. Borges et al. [38] initially ground the piassava fibers in a knife mill and sieved them at 50 mesh to incorporate as composite filler in a co-polypropylene matrix. They achieved satisfactory results of 35.5 MPa for the flexural strength, while those incorporated into a homopropylene matrix showed an even greater resistance of 47.7 MPa. These results revealed the possibility of using piassava powder as a reinforcement in polymer composites to be applied as an HPCF. Furthermore, a preliminary study [39] disclosed the great potential of the piassava fibers to be used as HPCFs. Indeed, the compressive strength obtained was around 50 MPa, and 0.8% of water absorption was found in 20 vol% of piassava powder-reinforced COPU.

Therefore, the present work aimed to continue the investigation into the use of piassava powder waste obtained as a processed material from a Brazilian broom factory, reinforcing a COPU biocomposite to be applied as an HPCF. In this study a comprehensive investigation was conducted not only to meet the NBR 14050 [10] standard recommendations, but also to perform a more extensive characterization of this HPCF material since, so far, there are no results to compare it to in the literature.

2. Materials and Methods

2.1. Materials

The piassava fibers used in this work were obtained as an industrial waste from a broom factory in the city of Campos dos Goytacazes, Brazil. After they were received, the fibers were washed in running water to eliminate any contaminant and put to dry in a stove for 24 h at 60 °C. For use as reinforcement material in HPCFs the piassava fibers were processed in a knife mill until the ground powder material passed through a 16-mesh sieve in accordance with the NBR 14050 standard [10]. Prior to the ground processing, however, a group of fibers was separated and went through preliminary characterization.

The polymeric matrix used in this study was the castor oil-based polyurethane (COPU) produced and commercially supplied by Imperveg, Brazil (Commercial tag AGT 1315). The COPU consisted of a bi-component. Both components—"A" (polyol) derived from castor oil and "B" (isocyanate)—were obtained as liquid. When the initiator (B) was added in A with 55% mass fraction, the polymerization process occurred and the resin hardened.

2.2. Methods

2.2.1. Characterization of the Piassava Fiber

The density, length and diameter distributions of the piassava fiber waste m reported in a previous work [39]. The density value of 1.42 g/cm^3 was used to calculate the volume fraction of reinforcement used on each composite in this work.

Initially, 100 fibers were randomly selected from the received batch (after the cleaning process) and had, one by one, their dimensions measured in a profile projector model PJ3150 made by Pantec (São Paulo, Brazil). These fibers were subjected to tensile tests in an Instron model 5582 universal machine with a 2.0 mm/min test speed at a controlled temperature of 20 °C. The test was conducted according to ASTM D3822-7 [40]. Adhesive tape was used on both ends of these fibers to avoid slipping as well as any damage that the machine grips may cause during the tensile test.

Images of the fiber surface morphology were obtained by scanning electron microscopy (SEM) on a microscope model SSX 550 made by Shimadzu (kyoto, Japan).

2.2.2. Biocomposites and Resin Processing

The biocomposites were made by adding the processed piassava powder in an already mixed component A and B resin, then leaving it at room temperature for 24 h in specific molds for the polymer cure to occur. Figure 1 illustrates the materials used in this study by showing the plain COPU Figure 1a and 30 vol% piassava powder Figure 1b,c biocomposites used for the abrasive wear tests as well as the material applied over a concrete surface. The piassava powder is black and the higher percentage of it added, the darker the biocomposites become. When applied over larger surfaces the material tends to form a smooth and even surface.

Figure 1. Plain COPU (**a**) and 30 vol% piassava powder biocomposite (**b**) samples used for abrasive wear tests, and 30 vol% piassava powder biocomposite applied over a concrete surface (**c**).

Biocomposite specimens were produced inside an open silicone mold. The molds were made according to the dimensions indicated by specific standards [40–42]. The piassava powder used in the preparation of the composites was dried in a stove for 24 h at 60 °C before the biocomposite preparation. To avoid excessive moisture absorption, the piassava powder was mixed with polymer, still heated to the polymer and finally poured inside silicone molds. After the 24 h period of curing, neat COPU, control with 0 vol% of piassava powder, and biocomposites made with 10, 20 and 30 vol% were demolded and the tests were carried out.

2.2.3. Tree Point Flexural Tests

The flexural tests were conducted according to ASTM C580 standard [40] on a universal machine model 5582 made by Instron (Norwood, MA, USA). For these tests, 10 rectangular cross-section samples with dimensions of 13 cm × 1.5 cm × 1.5 cm were used for each volume fraction of the biocomposites and the neat COPU resin, with a distance between the cleavers of 100 mm and a test speed of 1.3 mm/min.

2.2.4. Izod Impact Tests

For Izod tests, specimens were machined with dimensions of 150 mm $\times$ 120 mm $\times$ 10 mm. The specimens were then cut and polished to comply with the ASTM D256 standard [41] recommendations of 62 mm $\times$ 12 mm $\times$ 10 mm. Later a notch of 45° was made using a milling cutter and tests were conducted in a pendulum model XC-50 made by Pantec with a 22 J hammer. For each biocomposite studied a total of 10 samples were used.

2.2.5. Abrasive Wear Tests

Wear tests were conducted to determine the abrasion resistance of the material. Following the recommendations of the NBR 12042 standard [42], 70 mm $\times$ 70 mm $\times$ 30 mm samples were compressed with a constant and standard force against a circular-shaped track covered with sand, which ran at a specific speed. After a 500 and a 1000 m run the dimension difference was observed on the original 70 mm $\times$ 70 mm $\times$ 30 mm samples and the resistance was determined by the amount of material lost per distance.

2.2.6. Fourier Transform Infrared Spectroscopy (FTIR)

FTIR analyses were performed using attenuated total reflectance (ATR) on the piassava powder samples, COPU resin and its biocomposites. The equipment used was a spectrophotometer model IR-Affinity-1 made by Shimadzu, with a range interval of 4000–400 cm^{-1} and a resolution of 4 cm^{-1}, 32 scans. The preparation consisted of mixing the materials with KBr in order to increase the absorption signal.

2.2.7. Statistical Analysis: ANOVA and Tukey's Test

In some cases, the results obtained presented high statistical dispersion. In these cases, analysis of variance (ANOVA) was conducted with a 0.05 (5%) significance followed by Tukey's test if the results showed a p-value lower than the significance. Tukey's test is a complementary statistical analysis; the ANOVA results infer if there is difference between the analyzed groups (p-value < significance) while the Tukey's test shows which ones are different from each other.

2.2.8. Fracture Analysis

The fracture surfaces of the tested samples were analyzed by the scanning electron microscope (SEM) Shimadzu model SSX 550 (Shimadzu).

3. Results

3.1. Fiber Characterization

Well-known in the literature, piassava fibers, as seen in Figure 2, have oval-shaped cross-sections and a relatively high main diameter, as illustrated in Figure 2b. The surface morphology of the piassava fiber shows a rough surface and characteristic structures of spiny nodules made of pure silica (SiO$_2$) that occur in some spherical holes on the fiber surface [43,44]. These SiO$_2$ nodules are bonded on the fiber surface by weak hydrogen bridges. During the industrial processing steps, some of these nodules are lost, leaving a hollow cavity that provides good anchoring points for the polymeric resin [45]. This is an important feature since the hydrophilic characteristics of the piassava fiber, like those of many other NLFs, tend to form a week interface with hydrophobic polymer resins [25–36], thus creating defects that can accelerate the biocomposite degradation. All these structures were observed on the as-received fiber and are shown in Figure 2a. These structural morphologies were expected since there was no chemical treatment of the piassava fiber at any step of the broom production processes.

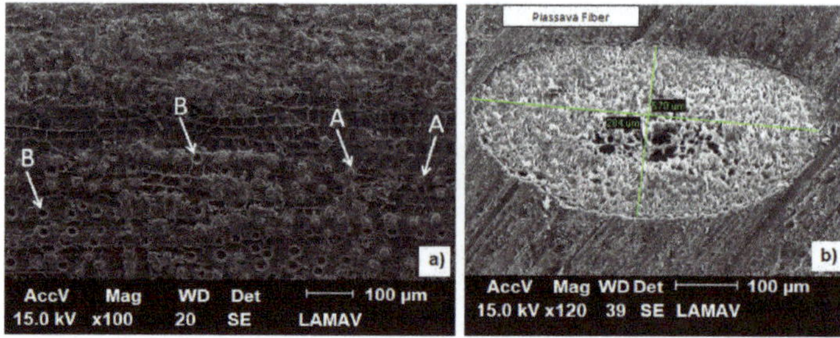

Figure 2. SEM photomicrography of the longitudinal surface of the piassava fiber (**a**) where the spiny silica spheres (indicated as "A") and some hollow holes (indicated as "B") and a cross-section of the piassava fiber (**b**) can be seen.

Regarding the mechanical behavior of the piassava fiber, the obtained tensile strength results of 87.9 ± 28.1 MPa and elastic modulus of 3.5 ± 1.2 GPa were similar to those reported in the literature [46], which corroborated the idea that, as expected, these fibers did not undergo significant changes during the broom manufacture process. Figure 3, obtained from SEM observations on the fracture surface, shows the characteristic rupture behavior of the piassava fibers. In this figure, the fiber cellulose microfibrils are ruptured in different regions, characterizing a non-uniform failure. Furthermore, Figure 3 displays a detailed image of the piassava fiber presenting columnar hollow lumen structures that justify its relatively low apparent density [39] and tensile strength with a great statistical dispersion.

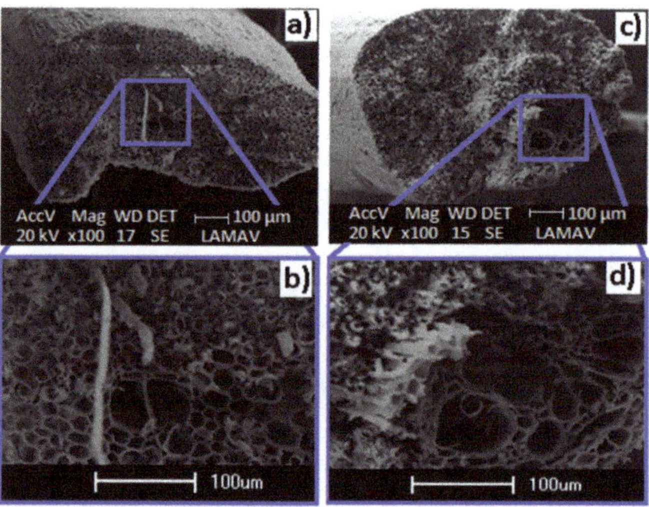

Figure 3. Surface of the fractures of piassava fibers at lower (**a**,**c**) and greater (**b**,**d**) magnifications.

Regarding the FTIR analysis of the piassava powder in Figure 4, the typical cellulose functional groups will appear around 3400 cm^{-1}—in this case, at 3317 cm^{-1}—and are related to OH groups that are linked in the main cellulose chain by a hydrogen bridge [47]. According to the literature [47,48], for CH aliphatic stretches of the methyl and methylene groups, the vibrations will be around 2936 and 2920 cm^{-1}, respectively, almost what was found for the piassava powder sample at the bands of 2930 and 2920 cm^{-1}. Furthermore, a band at 2116 cm^{-1} referred to a Si-H connection attributed to the spiny nodules on the fiber surface, already shown in Figure 2. This provided evidence that these structures were present even after the broom manufacture processing. The bands at 1424, 1512 and 1608 cm^{-1} were attributed to the vibration and elongations in the O-C-O of the aromatic ring present in the lignin. Finally, it was possible to identify the three distinct units that formed lignin molecules: guaiacyl (1278 cm^{-1}), syringyl (1332 cm^{-1}) and *p*-hydroxyphenyl (875 cm^{-1}).

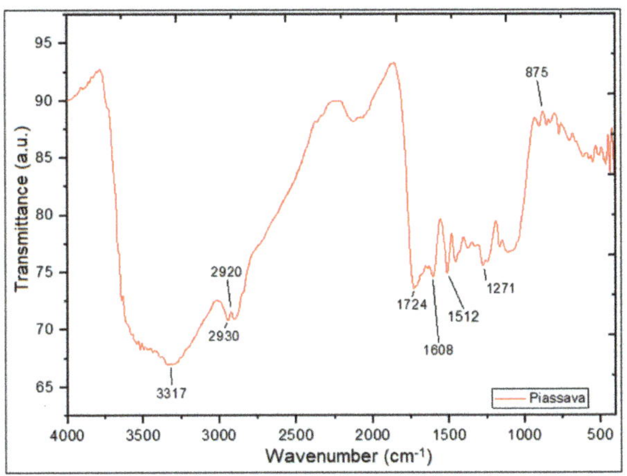

Figure 4. FTIR spectra of the piassava powder.

3.2. FTIR Analysis of the COPU Resin and Its Biocomposites

Figure 5 shows the FTIR spectra of plain COPU along with its piassava powder-incorporated biocomposites. As can be seen, the spectra obtained for the COPU biocomposites presented almost the same bands and were very similar to the COPU resin.

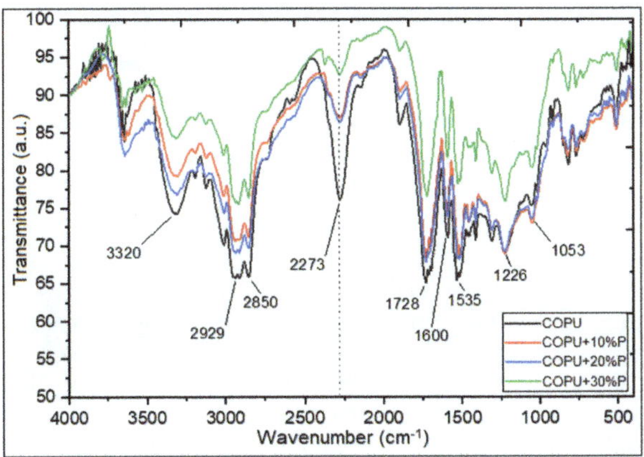

Figure 5. FTIR spectra for COPU and its composites reinforced by 10% (COPU + 10%P), 20% (COPU + 20%P) and 30% (COPU + 30%P) volume fraction of piassava powder.

It should be noticed in Figure 5 that the characteristic groups of the COPU resin were found, as reported by [49,50]. The bands at 3320, 2929 and 2850 cm^{-1} corresponded, respectively, to amine (NH), methyl (CH3) and methylene (CH2). The 1728 cm^{-1} band corresponded to the stretching of the pre-polymer carbonyl and, finally, the 1226 and 1053 bands corresponded to ether groups. The 2273 cm^{-1} band was attributed to groups of free isocyanates (NCO) in the COPU resin. Its presence indicated the excess of the NCO group in the polymeric structure. Moreover, as the amount of piassava powder increased in volume, the 3317 cm^{-1} band (OH cellulose group) on the composites had its intensity diminished along with the 2273 cm^{-1} band (free isocyanates in the COPU resin). Indeed, at 30% volume fraction of piassava powder, the 3317 cm^{-1} band almost disappeared and the 2273 cm^{-1} band had its intensity drastically lowered.

The interaction between the cellulose and the free isocyanates was expected [50,51]. The FTIR results in Figures 4 and 5 suggested that, not only was the interaction occurring, but the level of interaction was also greater with higher amounts of piassava powder on the biocomposites, compared to lower volume fractions.

3.3. Flexural Strength

Figure 6 shows the flexural strength and flexural elastic modulus obtained for the COPU resin (0%), and the 10, 20 and 30 vol% piassava powder-reinforced biocomposites along with a statistical analysis of the data. It was noticed that, in all cases, none of the biocomposite materials was able to meet the standard requirements presented in Table 1. However, it was evident that the use of piassava powder significantly enhanced both the flexural strength and elastic modulus of the COPU matrix. Indeed, the 30 vol% reinforced composite presented a resistance more than four times and stiffness more than seven times greater than the plain COPU. Moreover, the results showed almost irrelevant data dispersion, which is not a common behavior for composites reinforced by NLFs.

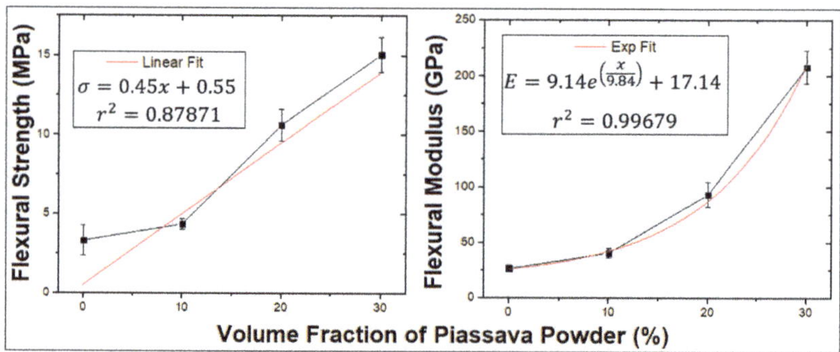

Figure 6. Flexural strength and elastic modulus for COPU resin and its composites along with linear fit of the results.

In fact, the COPU resin tended to present elastomeric characteristics, and the addition of reinforcement was enough, at least partially, to change this behavior [52]. Moreover, both good approximated linear fit (r^2 = 0.87871) and exponential fit (r^2 = 0.99679) indicated that the flexural strength and the stiffness, respectively, of the biocomposites could be further increased with higher amounts of piassava powder, enough to meet the standard criteria. In addition, the discussion of the FTIR results for the biocomposites indicated that not all the free NCO groups were completely bonded with the piassava cellulose OH, which indicated that more reinforcement could be added to the matrix without loss of strength of the interface between reinforcement and matrix. In contrast to what was observed in this work, piassava powder was reported to present a decrease of around 30% flexural resistance when incorporated with up to 30 vol% in other polymer matrixes [38], reaching around 30–35 MPa. This also suggested that the flexural strength of the composites can be enhanced, since the curves presented in Figure 6 showed a tendency to growth.

Figure 7 shows the fracture surface of the biocomposites. In this figure, we noticed the presence of bubbles on the matrix that were associated with existing retained gases during the polymer curing. On one hand, the use of the open mold tended to mimic the application of this composite as an HPCF. On the other hand, this also facilitated the appearance of defects such as the bubbles in Figure 7, which further explained the lower flexural strength than that required by the standard [10].

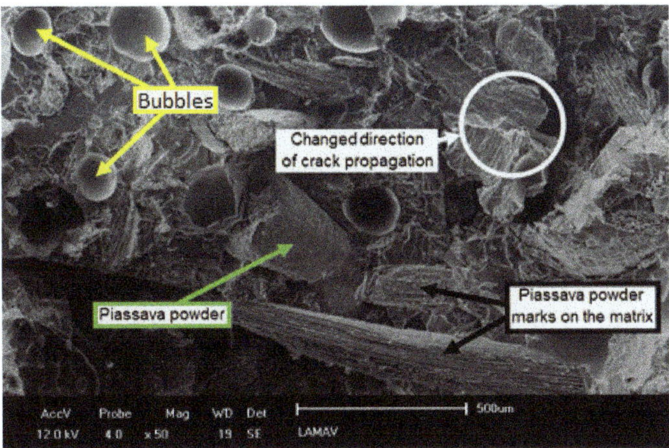

Figure 7. Surface of fracture for a 30% volume fraction reinforced by piassava powder COPU resin composites.

Regarding the crack propagation, also evidenced in Figure 7, it occurred on the matrix, as expected, by the change of direction when in contact with the reinforcement. The fiber powder marks presented on the matrix indicated a relatively stronger interface between reinforcement and matrix, which is not common on this kind of material, as already discussed.

3.4. Impact Resistance

As seen in Table 2, the addition of piassava powder to the COPU matrix tended to decrease the Izod impact resistance of the biocomposites. However, one may notice that, owing to greater data dispersion, almost all the conditions studied were statistically equal. Actually, with ANOVA tests, p-values of 0.0321 and 0.210 were obtained for notch resistance and impact resistance, respectively, which indicated that there was a statistical difference in at least one of the analyzed conditions for notch resistance and none for the impact resistance. In fact, the honest significant difference (HSD) of 14.59 was obtained on the Tukey's test for the notch resistance. This meant that only the 30 vol% reinforced COPU resin presented a significant HSD compared to the plain resin.

Table 2. COPU resin (0%) and its biocomposites' notch resistance and impact resistance obtained by Izod tests.

Volume Fraction of Piassava Powder	Notch Resistance (J/m)	Impact Resistance (kJ/m^2)
0%	69.0 ± 19.5	7.5 ± 2.7
10%	56.9 ± 7.1	5.6 ± 0.6
20%	52.8 ± 8.9	5.4 ± 0.9
30%	43.8 ± 4.6	4.0 ± 0.5

3.5. Abrasive Wear Resistance

Perhaps more important than the floor's ability to resist impact is its ability to withstand abrasion for long periods without wear loss. In this particular case, the biocomposites made with piassava powder reinforcing the COPU resin presented satisfactory results, since in all the cases studied the total material loss due to wear was lower than the standard requirements. The results in Table 3 show the amounts of material loss in wear tests after 500 and 1000 m on the abrasive track.

Table 3. Loss of material in the abrasive wear test for COPU resin (0%) and its piassava powder biocomposites after 500 and 1000 m on the abrasive track.

Volume Fraction of Piassava Powder	Wear (mm) after Distance on the Track	
	500 m Distance (mm)	1000 m Distance (mm)
0%	0.10 ± 0.06	0.28 ± 0.13
10%	0.47 ± 0.23	0.82 ± 0.30
20%	0.52 ± 0.05	0.97 ± 0.05
30%	0.59 ± 0.02	1.06 ± 0.11

The results in Table 3 show that the addition of piassava powder up to 30 vol% decreased by more than three times the wear resistance compared to the pure resin. This was somewhat expected for this kind of material since the interface between the two was relatively weak because of bubbles (Figure 7). Actually, by adding a material that creates a weak bond with the resin, the reinforcement should easily be pulled out from the matrix, leaving a hollow behind. In this scenario, the interface between the two phases should provide a weaker material. The fact that the biocomposites presented a total wear lower than half of the standard requirement from Table 1 revealed an interface that was moderately strong.

4. Discussions

From the results shown in this work the great potential of using piassava powder as a reinforcement in the COPU matrix for HPCF application was evident. Along with the results of a previous work [39], almost all standard requirements were attained. By comparing the results obtained for impact, abrasive wear and flexural strength with the results from FTIR analysis, there was space to improve the flexural strength of the material by further addition of piassava powder without greater loss of impact resistance and yet maintain the wear resistance at acceptable levels.

Finally, the use of the polyurethane material with approximately 60% volume fraction derived from castor oil (mass proportion use converted to the volume fractions) with a 30 vol% of piassava powder obtained as an industrial waste represented a biocomposite made almost 80% in volume from a renewable source. This alone represented a much greener solution compared to composites made with matrices or reinforcements from the non-renewable sources commonly applied in high-performance floors. Furthermore, the fact that the COPU resin used in this study was biodegradable, as well as the piassava powder, meant an environmentally friendly end-of-cycle alternative for natural materials.

To be used as HPCFs, according to the standard [10], the biocomposites are required to withstand a 10-year-long period. This can be a problem owing to the biodegradable characteristics of this biocomposite. However, the castor oil-derived polyol is reported to be antimicrobial, and the biocomposite itself can be designed to present a good resistance to biodegradability until its end of cycle [19]. Further studies should determine the life cycle times of these materials.

5. Conclusions

Novel sustainable castor oil-based polyurethane (COPU) biocomposites reinforced with piassava fiber powder waste were characterized for possible application as a high-performance coating floor (HPCF). The following conclusions were drawn:

FTIR results suggested interaction occurring between the piassava powder and COPU matrix.

Both flexural strength and modulus were substantially improved for the 30 vol% piassava powder biocomposite, with values reaching more than 500% the strength and more than 800% the stiffness of the plain COPU.

Izod notch impact resistance and absorbed energy displayed a tendency to decrease with the amount of piassava powder incorporated into the biocomposite. However, within the statistical precision, the values might be considered unchanged up to 20 vol% of incorporated piassava powder.

The loss of material in wear tests was accentuated in biocomposites with the incorporation of piassava powder. However, the maximum attained value of 1.06 mm/km for the 30 vol% biocomposite was significantly lower than the maximum abrasion of 2.30 mm/km specified by the standard.

Preliminary results from greener biocomposites made of a castor oil-based polyurethane matrix added with a natural piassava industrial waste revealed a promising material for high-performance coating floors. Ongoing research is being conducted aiming to improve the mechanical resistance of the biocomposites by enhancing the adhesion between the piassava and polyurethane matrix.

Author Contributions: Conceptualization, S.N.M. and C.M.F.V.; data curation, J.P.R.G.d.C., N.T.S. and F.P.D.L.; formal analysis, J.P.R.G.d.C., N.T.S., F.P.D.L. and C.M.F.V.; funding acquisition, C.M.F.V.; investigation, J.P.R.G.d.C., N.T.S. and F.P.D.L.; methodology, F.P.D.L., S.N.M. and C.M.F.V.; project administration, C.M.F.V.; resources, F.P.D.L. and C.M.F.V.; supervision, S.N.M. and C.M.F.V.; validation, S.N.M. and C.M.F.V.; visualization, S.N.M. and C.M.F.V.; writing—original draft, J.P.R.G.d.C. and N.T.S.; writing—review and editing, S.N.M. and C.M.F.V. All authors have read and agreed to the published version of the manuscript.

Funding: This research was funded by Fundação Carlos Chagas Filho de Amparo à Pesquisa do Estado do Rio de Janeiro (FAPERJ) grant numbers: E-26/202.773/2017, E-26/200.847/2021 and E-26/201.628/2021, and Conselho Nacional de Desenvolvimento Científico e Tecnológico (CNPq) grant number: 301634/2018-1.

Institutional Review Board Statement: Not applicable.

Informed Consent Statement: Not applicable.

Acknowledgments: The authors would like to thank FAPERJ and CNPQ as well as, Imperveg for the resin, the Laboratory of Advanced Materials—LAMAV/UENF for the support, Geovana Girondi (LAMAV/UENF), Rômulo Loiola (LAMAV/UENF), Gabriel Valeriolete (LAMAV/UENF) and Marcelo Mathias (LCQUI/UENF).

Conflicts of Interest: The authors declare no conflict of interest.

References

1. Katariya, M.N.; Jana, A.K.; Parikh, P.A. Corrosion inhibition effectiveness of zeolite ZSM-5 coating on mild steel against various organic acids and its antimicrobial activity. *J. Ind. Eng. Chem.* **2013**, *19*, 286–291. [CrossRef]
2. Gergely, A.; Bertóti, I.; Török, T.; Pfeifer, É.; Kálmán, E. Corrosion protection with zinc-rich epoxy paint coatings embedded with various amounts of highly dispersed polypyrrole-deposited alumina monohydrate particles. *Prog. Org. Coat.* **2013**, *76*, 17–32. [CrossRef]
3. Lamm, M.E.; Li, P.; Hankinson, S.; Zhu, T.; Tang, C. Plant oil-derived copolymers with remarkable post-polymerization induced mechanical enhancement for high performance coating applications. *Polymer* **2019**, *174*, 170–177. [CrossRef]
4. Jin, F.L.; Li, X.; Park, S.J. Synthesis and application of epoxy resins: A review. *J. Ind. Eng. Chem.* **2015**, *29*, 1–11. [CrossRef]
5. Kausar, A. Polymer coating technology for high performance applications: Fundamentals and advances. *J. Macromol. Sci. Part A* **2018**, *55*, 440–448. [CrossRef]
6. Hao, Y.; Liu, F.; Han, E.H. Protection of epoxy coatings containing polyaniline modified ultra-short glass fibers. *Prog. Org. Coat.* **2013**, *76*, 571–580. [CrossRef]
7. Feng, K.; Hong, G.; Liu, J.; Li, M.; Zhou, C.; Liu, M. Fabrication of high performance superhydrophobic coatings by spray-coating of polysiloxane modified halloysite nanotubes. *Chem. Eng. J.* **2018**, *331*, 744–754. [CrossRef]
8. Pistone, A.; Scolaro, C.; Visco, A. Mechanical properties of protective coatings against marine fouling: A review. *Polymers* **2021**, *13*, 173. [CrossRef]
9. Harifi, T.; Montazer, M. Application of nanotechnology in sports clothing and flooring for enhanced sport activities, performance, efficiency and comfort: A review. *J. Ind. Text.* **2017**, *46*, 1147–1169. [CrossRef]
10. *NBR 14050*; High Performance Mortars and Coatings Systems Epoxy Based Resins and Mineral Aggregates—Design, Application and Performance Evaluation—Procedure. Brazilian National Standards Organization: Sao Paulo, Brazil, 1998.
11. Martulli, L.M.; Creemers, T.; Schöberl, E.; Hale, N.; Kerschbaum, M.; Lomov, S.V.; Swolfs, Y. A thick-walled sheet moulding compound automotive component: Manufacturing and performance. *Compos. Part A* **2020**, *128*, 105688. [CrossRef]

12. Zah, R.; Hischier, R.; Le, A.L.; Braun, I. Curauá fibers in the automobile industry e a sustainability assessment. *J. Clean. Prod.* **2007**, *15*, 1032–1040. [CrossRef]

13. Siengchin, S. Editorial corner—A personal view. Potential use of 'green' composites in automotive applications. *Express Polym. Lett.* **2017**, *11*, 600. [CrossRef]

14. Chattopadhyay, D.K.; Raju, K.V.S.N. Structural engineering of polyurethane coatings for high performance applications. *Prog. Polym. Sci.* **2007**, *32*, 352–418. [CrossRef]

15. NASA. Climate Change: How Do We Know? 2020. Available online: https://climate.nasa.gov/evidence (accessed on 28 March 2022).

16. Sajeeb, A.M.; Babu, C.S.; Arif, M.M. Evaluation of Mechanical Properties of Natural Fiber Reinforced Melamine Urea Formaldehyde (MUF) Resin Composites. *Mater. Today Proc.* **2018**, *5*, 6764–6769. [CrossRef]

17. Ahmad, H.; Fan, M. Interfacial properties and structural performance of resin-coated natural fibre rebars within cementitious matrices. *Cem. Concr. Compos.* **2018**, *87*, 44–52. [CrossRef]

18. Guerra, G.N.; Albuquerque, E.C.M.C.; Campos, L.M.A.; Pontes, L.A.M. Chemical and Physicochemical Characterization of Alkali Pretreated and in Natura Sisal Solid Waste. *J. Nat. Fibers* **2021**, *18*, 203–212. [CrossRef]

19. Garrison, T.F.; Murawski, A.; Quirino, R.L. Bio-Based Polymers with Potential for Biodegradability. *Polymers* **2016**, *8*, 262. [CrossRef]

20. Ionescu, M.; Radojčić, D.; Wan, X.; Shrestha, M.L.; Petrović, Z.S.; Upshaw, T.A. Highly functional polyols from castor oil for rigid polyurethanes. *Eur. Polym. J.* **2016**, *84*, 736–749. [CrossRef]

21. Santan, H.D.; James, C.; Fratini, E.; Martínez, I.; Valencia, C.; Sánchez, M.C.; Franco, J.M. Structure-property relationships in solvent free adhesives derived from castor oil. *Ind. Crops Prod.* **2018**, *121*, 90–98. [CrossRef]

22. Zeng, M.; Li, J.; Zhu, W.; Xia, Y. Laboratory evaluation on residue in castor oil production as rejuvenator for aged paving asphalt binder. *Constr. Build. Mater.* **2018**, *193*, 276–285. [CrossRef]

23. Abdel-Hamid, S.M.S.; Al-Qabandi, O.A.; Elminshawy, N.A.S.; Bassyouni, M.; Zoromba, M.S.; Aziz, M.H.A.; Mira, H.; Elhenawy, Y. Fabrication and characterization of microcellular polyurethane sisal biocomposites. *Molecules* **2019**, *24*, 4585. [CrossRef] [PubMed]

24. Ahmad, A.; Jamil, S.N.A.M.; Choong, T.S.Y.; Abdullah, A.H.; Mastuli, M.S.; Othman, N.; Jiman, N.N. Green flexible polyurethane foam as a potent support for Fe-Si adsorbent. *Polymers* **2019**, *11*, 2011. [CrossRef] [PubMed]

25. Faruk, O.; Bledski, A.K.; Fink, H.P.; Sain, M. Progress report on natural fiber reinforced composites. *Macromol. Mater. Eng.* **2014**, *299*, 9–26. [CrossRef]

26. Mohamed, L.; Ansari, M.N.M.; Pua, G.; Jawaid, M.; Islam, M.S. A review in natural fiber reinforced polymer composites and its applications. *Int. J. Polym. Sci.* **2015**, *2015*, 243947. [CrossRef]

27. Güven, O.; Monteiro, S.N.; Moura, E.A.B.; Drelich, J.W. Re-emerging field of lignocellulosic fiber—polymer composites and ionizing radiation technology in their formulation. *Polym. Rev.* **2016**, *56*, 706–736. [CrossRef]

28. Dixit, S.; Goel, R.; Dubey, A.; Shivhare, P.R.; Bhalavi, T. Natural fibre reinforced polymer composite materials—A review. *Polym. Renew. Resour.* **2017**, *8*, 71–78. [CrossRef]

29. Kumar, R.; Haq, M.I.U.; Raima, A.; Anand, A. Industrial applications of natural fibre-reinforced polymer composites—Challenges and opportunities. *Int. J. Sustain. Eng.* **2018**, *12*, 212–220. [CrossRef]

30. Vigneshwaran, S.; Sundarerannan, R.; Jonh, K.; Johnson, R.D.J.; Drasath, K.A.; Agith, S.; Arumugaprabu, V.; Uthayakumar, M. Recent advancement in the natural fiber polymer composites: A comprehensive review. *J. Clean. Prod.* **2020**, *277*, 124109. [CrossRef]

31. Karimah, A.; Ridho, M.R.; Munawar, S.S.; Adi, D.S.; Damayant, R.; Subiyanto, B.; Fatriasari, W.; Fudkoli, A. A review on natural fibers for development of eco-friendly bio-composites: Characteristics and utilizations. *J. Mater. Res. Technol.* **2021**, *13*, 2442–2458. [CrossRef]

32. Chakraborty, I.; Chatterjee, K. Polymers and Composites Derived from Castor Oil as Sustainable Materials and Degradable Biomaterials: Current Status and Emerging Trends. *Biomacromolecules* **2020**, *21*, 4639–4662. [CrossRef]

33. Kuram, E. Advances in development of green composites based on natural fibers: A review. *Emergent Mater.* **2021**. [CrossRef]

34. Kumar, R.; Singh, T.; Singh, H. Solid waste-based hybrid natural fiber polymeric composites. *J. Reinf. Plast. Compos.* **2015**, *34*, 1979–1985. [CrossRef]

35. Cabrera, F.C. Eco-friendly polymer composites: A review of suitable methods for waste management. *Polym. Compos.* **2021**, *42*, 2653–2677. [CrossRef]

36. Mahmoud, M.A.; Alzebdeh, N.H.; Pervez, T.; Al-Hinai, N.; Munam, A. Progress and challenges in sustainability, compatibility, and production of eco-composites: A state-of-art Review. *J. Appl. Polym. Sci.* **2021**, *138*, 51284. [CrossRef]

37. Castro, J.P.; Nobre, J.R.C.; Napoli, A.; Trugilho, P.F.; Tonoli, G.H.D.; Wood, D.F.; Bianchi, M.L. Pretreatment Affects Activated Carbon from Piassava. *Polymers* **2020**, *12*, 1483. [CrossRef]

38. Borges, T.E.; Almeida, J.H.S.; Amico, S.C.; Amado, F.D.R. Hollow glass microspheres/piassava fiber-reinforced homo- and co-polypropylene composites: Preparation and properties. *Polym. Bull.* **2017**, *74*, 1979–1993. [CrossRef]

39. De Carvalho, J.P.R.G.; Lopes, F.P.D.; Simonassi, N.T.; Monteiro, S.N.; De Carvalho, E.A.; Vieira, C.M.F. Eco-friendly piassava fiber reinforced composite for high performance coating application. *Rev. Cereus* **2020**, *12*, 118–132. [CrossRef]

40. *ASTM C580*; Standard Test Method for Flexural Strength and Modulus of Elasticity of Chemical-Resistant Mortars, Grouts, Monolithic Surfacings, and Polymer Concretes. ASTM International: West Conshohocken, PA, USA, 2011.

41. *ASTM D256*; Standard Test Methods for Determining the Izod Pendulum Impact Resistance of Plastics. ASTM International: West Conshohocken, PA, USA, 2018.
42. *ABNT NBR 12042*; Inorganic Materials—Determination of the Resistance to Abrasion. Brazilian National Standards Organization: Sao Paulo, Brazil, 1992.
43. Miranda, C.S.; Fiuza, R.P.; Ricardo, F.; Carvalho, R.F.; José, N.M. Effect of surface treatment on properties of bagasse piassava fiber. *Quim. Nova* **2015**, *38*, 161–165. [CrossRef]
44. Garcia Filho, F.D.C.; Monteiro, S.N. Piassava fiber as an epoxy matrix composite reinforcement for ballistic armor applications. *JOM* **2019**, *71*, 801–808. [CrossRef]
45. Monteiro, S.N.; Lopes, F.P.D.; Ferreira, A.S.; Nascimento, D.C.O. Natural-fiber polymer-matrix composites: Cheaper, tougher, and environmentally friendly. *JOM* **2009**, *61*, 17–22. [CrossRef]
46. D'Almeida, R.M.; Aquino, R.C.M.P.; Monteiro, S.N. Tensile mechanical properties, morphological aspects and chemical characterization of piassava (*Attalea funifera*) fibers. *Compos. Part A Appl. Sci. Manuf.* **2006**, *37*, 1473–1479. [CrossRef]
47. Santos, E.B.C.; Moreno, C.G.; Barros, J.J.P.; Moura, D.A.; Fim, F.C.; Ries, A.; Wellen, R.M.R.; Silva, L.B. Effect of alkaline and hot water treatments on the structure and morphology of piassava fibers. *Mater. Res.* **2018**, *21*, 1–11. [CrossRef]
48. Santos, E.B.C.; Barros, J.J.P.; De Moura, D.A.; Moreno, C.G.; Fim, F.C.; Da Silva, L.B. Rheological and thermal behavior of PHB/piassava fiber residue-based green composites modified with warm water. *J. Mater. Res. Technol.* **2019**, *8*, 531–540. [CrossRef]
49. Echeverri, D.A.; Rios, L.A.; Rivas, B.L. Synthesis and copolymerization of thermosetting resins obtained from vegetable oils and biodiesel-derived crude glycerol. *Eur. Polym. J.* **2015**, *67*, 428–438. [CrossRef]
50. Merlini, C.; Soldi, V.; Barra, G.M.O. Influence of fiber surface treatment and length on physico-chemical properties of short random banana fiber-reinforced castor oil polyurethane composites. *Polym. Test.* **2011**, *30*, 833–840. [CrossRef]
51. Borrero-López, A.M.; Wang, L.; Valencia, C.; Franco, J.M.; Rojas, O.J. Lignin effect in castor oil-based elastomers: Reaching new limits in rheological and cushioning behaviors. *Compos. Sci. Technol.* **2020**, *203*, 108602. [CrossRef]
52. Oliveira, F.; Gonçalves, L.P.; Belgacem, M.N.; Frollini, E. Polyurethanes from plant- and fossil-sourced polyols: Properties of neat polymers and their sisal composites. *Ind. Crops Prod.* **2020**, *155*, 112821. [CrossRef]

sustainability

MDPI

Article

Feasibility Analysis of Mortar Development with Ornamental Rock Waste for Coating Application by Mechanized Projection

Ana Luiza Paes [1], Jonas Alexandre [2], Gustavo de Castro Xavier [2], Sérgio Neves Monteiro [3] and Afonso Rangel Garcez de Azevedo [2,*]

[1] Laboratory of Advanced Materials, State University of the Northern Rio de Janeiro, Campos dos Goytacazes 28013-602, Brazil; alcpaes@gmail.com
[2] Civil Engineering Laboratory, State University of the Northern Rio de Janeiro, Campos dos Goytacazes 28013-602, Brazil; jonasuenf@gmail.com (J.A.); gxavier@uenf.br (G.d.C.X.)
[3] Department of Materials Science, IME—Military Institute of Engineering, Square General Tibúrcio, 80, Rio de Janeiro 22290-270, Brazil; snevesmonteiro@gmail.com
* Correspondence: afonso@uenf.br

Abstract: The industrial production of lime generates greenhouse gases, which contributes to increase the global warming. Therefore, the present study evaluated the feasibility of replacing lime by ornamental rock waste (ORW) as a by-product of the related stone industry, and developed a cost-effective mortars. These new low-costing mortars are intended as fresh fluid paste coatings to be applied on walls by the mechanized projection technique. The ornamental rock waste was collected from a marble and granite industry as ground stone. It was finely crushed before mixing with cement, sand, water and superplasticizer in amounts of 1.0% (R01), 1.2% (R02) and 1.3% (R03), to prepare the mortars, which had the mixture, cement: ORW: sand, 1:1:4 in wt.%. These novel mortars were characterized in both fresh, for well projection, and hardened state, to evaluate the properties after curing performance. The results showed that mortar R03, achieved the best results and did not present cracks in the hardened state. Its water retention was found above 30%. Both tensile strength of 0.312 MPa and compressive strength 7.88 MPa, which are above the corresponding minimum required by the standard for external coating. Water absorption by immersion of 19.37% and void content of 20.23% were close to the corresponding values for hydrated lime mortar. Dry shrinkage showed that the new R03 mortar reached more than 90% of their total retraction at 7 days of cure without sign of cracking. These findings revealed the R03/ornamental rock waste -based mortar applied by mechanized projection as a promising sustainable substitute for common lime-based mortar.

Keywords: ornamental rock waste; projected mortar; construction

Citation: Paes, A.L.; Alexandre, J.; Xavier, G.d.C.; Monteiro, S.N.; de Azevedo, A.R.G. Feasibility Analysis of Mortar Development with Ornamental Rock Waste for Coating Application by Mechanized Projection. *Sustainability* **2022**, *14*, 5101. https://doi.org/10.3390/su14095101

Academic Editor: José Ignacio Alvarez

Received: 30 March 2022
Accepted: 20 April 2022
Published: 23 April 2022

Publisher's Note: MDPI stays neutral with regard to jurisdictional claims in published maps and institutional affiliations.

1. Introduction

The civil construction sector accounts for approximately 10% of Brazil's gross domestic product [1]. In order to boost this growth in times of economic uncertainly, development of novel construction materials might be a solution to improve the process production and reduce costs [2]. Coating mortar, largely applied in all reinforced concrete carried out in Brazil, is a construction material with potential to be improved in a sustainable and cost-effective way [3]. Worldwide coating mortars are used, mainly, to protect the masonry, and represent a substantial cost to the total budget.

A conventional coating mortar is usually composed of cement, water, lime and fine aggregate. To optimize the coating mortar process and to achieve high performance in the building coating execution, a modern procedure is to use the mechanized projection method. Indeed, through this system it is possible to cover a substrate at a rate of $60/m^2/day$ per worker, which is practically twice as much as would be achieved by conventional hand application [4]. For this reason, the interest in the use of mortars designed as an application system for these coatings has been increasing [5,6].The mechanized projection method, well established around the world, is an adequate solution to minimize the problems in the execution of internal and external coating of the buildings, by reducing human interference during the execution. So, it can be optimized by increasing productivity, and above all, improving the quality and gives uniformity to the coating [5,6]. There are variations of this method, for supplying the spraying equipment with fresh mortar. One system is known in Brazil as *canequinha* or compressed air spray. In more complex processes, there is a mixer coupled to the pump, so that the mortar is homogenized and sent directly to the pump from where it is projected to the substrate [4,5]. This last method was used in the present work.

According to Ribeiro et al. [5], an industrial company using this method can achieve a production up to 0.75 m^2/h, while the average used to be 0.42 m^2/h. With this in mind, a construction company of Campos dos Goytacazes, Brazil, chose this method and decided to replace the hydrating lime. However, it was observed the occurrence of cracks in the coating after drying and, as a reference regional center of study, the company came with this problem to be solved. The most common method of mortar application in Brazil, and in several countries with less industrialization of civil construction, is manual, which leads to higher costs and great waste. Thus, research involving the validation of mechanized projection is important in order to provide subsidies to the civil construction sector for the application of new technologies, which aim to provide greater rationality to the application of mortars. Currently, numerous companies have been using mechanized projection techniques in Brazil in order to increase their productivity. However there are still gaps related to the post-application condition of these mortars [6].

In order to mitigate such shortcomings, the present work has as its main objective to propose a new mortar mixture to be projected replacing the hydrated lime by ornamental rock waste (ORW), from Cachoeiro do Itapemirim, Brazil. There is a large amount of ornamental rock production worldwide, in the past six years, about 150 million tons were produced yearly, mainly in China, India, Turkey, Iran and Brazil [7,8]. China alone has produced in 2015 around 350,000,000 m^2 of marble, while the estimated export from Egypt is nearly 1,360,500 ton/year of stone, either processed or unprocessed [9,10]. In 2019 Brazil exported 2.16 million/ton [10,11]. The rock beneficiation process is divided into two stages: the blocks extraction and their beneficiation. According to Angelin et al. [12], in this last stage, the waste generation can reach 20 to 30% of the block volume. Furthermore, during the extraction of marble and granite, the loss can reach 60%, and this activity produces a fine dust that can cause health problems and damage the soil [10,11,13]. So, not only to reduce the consumption of hydrated lime, a material with high added value and a polluting production process, but also to decrease the amount of rock waste disposal at the cities, recent researches have incorporated this material into concrete, mortar, tile, pavement, and even used it for soil stabilization [10].

Some authors, such as Arce et al. [14], replaced lime by ornamental rock waste and concluded that, when finely ground, this type of waste gave mortars characteristics similar to those obtained with the lime used in fresh state. For Oliveira et al. [15], its use was interesting, since as a powdery material, there was an improvement in packaging. The choice to incorporate the ornamental rock waste into mortars contribute to mitigate the impact generated by its improper disposal. In this way, giving the ornamental rock waste an appropriate final destination by incorporating into mortars, the civil construction sector becomes more sustainable. In addition, its cost is about 90% cheaper than hydrated lime, being related only to its transport from one city to another [16]. The ornamental rock waste material had already been incorporated into self-compacting mortars and concretes by Corinaldesi et al. [17], who concluded that, owing to the waste high fineness, this material ensured cohesion and workability.

Other authors have studied the ornamental rock waste incorporation into mortar, Singh et al. [18], carried out an economic and environmental study of advantages of replacing cement by marble waste in concrete and produced three types of concrete. One type for reference and the other two with waste, from which they obtained compressive strength 20–25% higher than the reference mixture. The properties of porosity, resistance to abrasion and carbonation, as well as resistance to sulfates and water penetration, were found superior for the modified mixtures, which can be justified by the filler effect due to the use of marble waste. In addition, they also report that the marble waste can be used to replace hydrated lime, since this rock is classified as limestone and the waste has a chemical composition similar to lime. Amaral et al. [19] studied the partial replacement of sand by rock waste. The authors defined the mixture in 1:1:6, and replacement percentages between 9 and 21% in sand weight. They analyzed their properties in the fresh state (density, incorporated air content and consistency) and in the hardened state (density, flexural and compressive strength), and concluded that these mortars showed satisfactory results, being the incorporation of 21% of residue the most recommended one. Leite et al. [20] studied mortars with the substitution of sand for residue generated in the cutting and polishing of rocks, in proportions of 0%, 10%, 20% and 30%. The authors tested their properties in the fresh and hardened state and concluded that based on physical and mechanical properties that there was not significant modification with the different incorporation level of residue. However, the sample composed of 20 wt.% of cutting residue showed the best performance for being the only one with an appropriate consistency index. This amount of residue did not significantly change the other properties, except capillarity.

Based on these preliminary considerations, the objective of this research is to evaluate the potential use of ornamental rock waste to replace hydrated lime, in order to develop a mortar to be applied by mechanized projection. Although many researcher works have already evaluated this incorporation, to our knowledge, no work exists specifically for mortar to be used by this method and on how the mechanized projection modifies the mortar properties. Indeed, modifying one component alters the entire mixture and its properties, such as void index, water absorption by capillarity and immersion, as well as the compressive, flexural and tensile bond strengths. Consequently, affects the durability and applicability of mortars. The investigation on how mortars properties were modified due to the replacing the lime by ornamental rock waste, and the creation of a superplasticizer to be used in mortar with ornamental rock waste, generating a new material to be applied by the mechanized projection method, is for the first time performed. It is also noteworthy that there are limited studies regarding the application of mechanized mortar with the addition of solid waste. Indeed, this collaborates in two ways for sustainability, either in the reduction of waste at the time of application, or in the reuse of other solid waste that would be discarded. This complementary approach will certainly fulfill a gap in the main studies already carried out.

2. Materials and Methods

It is important to highlight that for projection application, some mortars characteristics in the fresh state must be different from those used in the conventional application method. The main one being the workability. This can be defined as the material fluidity, i.e., the ability of the mass to flow or spread over the substrate surface [19]. The fluidity required for the projection will depend on the corresponding equipment used. For this purpose, a sample of mortar used by a construction firm in the region of Campos dos Goytacazes, Brazil, was collected and its workability was defined by the consistency index. This test consists of molding a conical trunk with fresh mortar and, after 30 drops on the consistency table, measure its spread [20].

Once this index was determined, the workability was kept constant using the same consistency index in all mixture. Then, the amount of water and proportion of additive, which, each mortar should have as reference consistency index was defined [21].

In addition to develop a mortar suitable for projection without cracks, a mortar with ornamental rock waste and an additive developed at our laboratory, were tested for the creation of a novel and sustainable product. In order to analyze whether the materials are economically viable substitutes for hydrated lime, a cost-effective study was carried out. For that, Table 1 presents a comparison made in three commercial centers at Campos dos Goytacazes to estimate the hydrated lime CH-III price. The only cost related to the ornamental rock waste is its transportation from the city of Cachoeiro do Itapemirim to Campos dos Goytacazes, both in the southeast of Brazil.

Table 1. Price of hydrated lime and ornamental rock waste.

Material	Building Material	Quantity (ton)	Price (U$)	Average Price
Hydrated lime CH III	1	1	80.00	
	2	1	92.00	90.00
	3	1	98.00	
Ornamental rock waste	-	1	14.70	14.70
		1.5		

The cement used in this work was the same used as the reference mortar, CPIII-40 RS [22], purchased from the same batch so that there was no variation in the constituents percentage (clinker + calcium sulfate, blast furnace slag and carbonate material).

The small aggregates selected for mortars manufacture came was quartz sand extracted from the bed of the South Paraíba River, with a maximum diameter of 2.4 mm [23].

The ornamental rock waste used (Figure 1a) was collected from Decolores, an ornamental rock processing industry, located at Cachoeiro do Itapemirim. To be applied in this study, the ornamental rock waste was first dried naturally in the sun to remove excess moisture and then placed in a greenhouse (Sterilife brand, Brazil) at 60 °C during 72 h. Completely dry, the waste was crushed and passed through the ASTM 50 sieve (Figure 1b). Corinaldesi et al. [17] characterized Italian marble waste and observed that these were mainly composed of calcite (60% calcium carbonate—$CaCO_3$), while Vardhan et al. [24] studied Indian rocks, and concluded that they were composed of 40.73% CaO and 15.21% MgO. This indicates that the waste chemical composition will depend on the region of origin of the rock from which it was extracted [25].

(a)   (b)

Figure 1. Ornamental Rock Waste (**a**) in the ornamental rock industry with humidity, (**b**) dry and passed at #50.

Figure 2 shows a flowchart that demonstrates the experimental steps of this research.

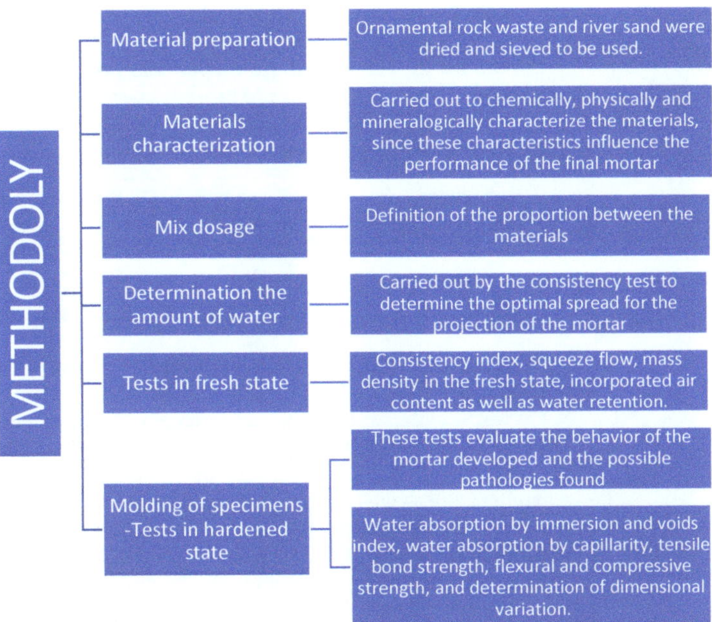

Figure 2. Experimental flowchart of the steps developed.

To increase the mortars workability without adding large proportions of water, and thereby decrease the likelihood of cracking, a superplasticizer additive was developed at our Laboratory of Non-Conventional Materials of the State University of the North Fluminense, in Campos dos Goytacazs), to be added in ornamental rock waste mortars, in order to create an entirely new material. Due to the percentage of fine particle size, which allows the waste to act as a filler and improve packing by reducing the voids left by the other two components, the requirement of water increases as the amount of ornamental rock waste in samples is enhanced [26]. Therefore, superplasticizers are used to achieve the required workability [23], since it is a projected mortar, and workability needs to be superior to the conventional ones.

First of all, the ornamental rock waste characterization was carried out using granulometric analysis [27], dispersive energy X-ray fluorescence (EDX), X-ray diffraction (DRX) and scanning electron microscopy (SEM). The EDX aims to determine the oxides percentages present in the analyzed materials using a Shimadzu EDX-700 equipment. The DRX determines the crystal atomic and molecular structure, in samples using a Rigaki MiniFlex 600. The SEM reveals the particle morphological analysis in a Shimadzu SSX-50 Superscan equipment, Japan. These tests were made concurrently with the consistency test [28], to determine the ideal amount of water for the desired workability.

Once the components have been characterized, their dosage was set for the reference mortar, with design mix of 1:1:4, obtained at the aforementioned construction site, have shrinkage cracks after drying. It contributes to find the consistency index determination to be used to maintain a fixed workability of 310 ± 5 mm. As mentioned before, this is one of the most important characteristics in a mechanized projected mortar, as it allows the material to flow through the projection pump hose without clogging or disaggregation between the paste and aggregates.

With this in mind, the reference mortar was subjected to the consistency test [28], where it was molded in a conical trunk. After performing 30 drops the spreading was measured. The more fluid the material, the greater this spread and consequently the consistency index. The value found was 310 ± 5 mm, which is higher than the limit established by NBR

13276 [28]. However, the standard value must be disregarded, considering that it applies only to conventional mortars.

With the known spread, it was possible to determine the amount of water and the percentage of additive that would be used in each mix, as shown in Table 2.

Table 2. Consistency index and water/dry materials ratio.

Sample	Consistency Index	Water/Dry Materials	Additive (%)
Hydrated lime	310	0.189	1.0
R01	314	0.199	1.0
R02	310	0.197	1.2
R03	315	0.192	1.3

As the aim of this work was to develop a mortar with ornamental rock waste to be applied in a mechanized projected way and within the parameters of the Brazilian standard, it was tested in the fresh and hardened state. With the amount of water and superplasticizer known, the mortars were molded following the NBR 13276 [28] procedures and tested by the fresh state for squeeze flow [29], mass density in the fresh state and the incorporated air content [30] as well as water retention [31]. Like the consistency index, the squeeze flow test is used to determine workability. It was carried out in a universal EMIC press, and consists of compressing the sample between an upper plate, which will apply load to the material and has the same diameter as the initial sample, and a lower plate, which has at least twice the sample initial diameter. The sample is placed and spread when the test is started [32,33].

The mass density test in the fresh state is applied to determine the mortar weight at the application moment. This test is important in situations where the mortar is mechanized projected into the substrate and also to determine the incorporated air content. This characteristic affects its workability, and corresponds to the amount of air existing inside the mortar, expressed as a percentage [34]. The water retention property is related to the mortar's ability to retain water in the fresh state, maintaining its workability or consistency when subjected to conditions that cause water loss [35]. Properties such as mechanical strength, adhesion and durability are influenced by this characteristic [36].

For hardened state parameters, the mortars were molded according to NBR 13279 [37] and their parameters analyzed. These were water absorption by immersion and voids index [38] as well as water absorption by capillarity [39], tensile bond strength [40] and determination of dimensional variation [41]. In addition the flexural and compressive strength, were both obtained in the universal EMIC press. The speeds to perform each test, flexural and compressive strength, was respectively, 50 ± 10 N/s and 500 ± 50 N/s. The compressive strength is statistically associated with the ability to resist surface abrasion, traction, impact and shear, being representative of the others. The water absorption tests by immersion and capillarity are related to coating durability. The first simulates exposure to rain, excessive humidity, and action from coating washing and cleaning. While the capillarity absorption is related to the pores present on the coating surface. The tensile bond strength is the mortar's ability to remain fixed to the substrate when normal and tangential stresses are imposed, without breaking. Its test consists of using a continuous load application equipment to pull the mortar paste from the coating. The shrinkage determination in the mechanized projected mortars is important due to the more fluid consistency of this material and its greater propensity to cracks appearance. The drying shrinkage begins on the coating surface and moves towards its interior, generating tensions and, at the moment when the humidity begins to decrease, the drying rate is not uniform [42].

3. Results and Discussion

3.1. Physical Characterization

Granulometric curve, presented in Figure 3, shows that ornamental rock waste grain size curve, in red, had 95% of its fraction smaller than 0.075 mm, similar to the characteristics of hydrated lime, pink curve, (97% fines content), facilitating the substitution by this material. The specific mass was also defined, using NBR NM 52 [43], resulting in a value of 2.48 g/cm^3 for the ornamental rock waste, approximately equal to that found by Vardhan et al. [24], 2.60 g/cm^3, meanwhile the results found for hydrated lime were 2.27 g/cm^3.

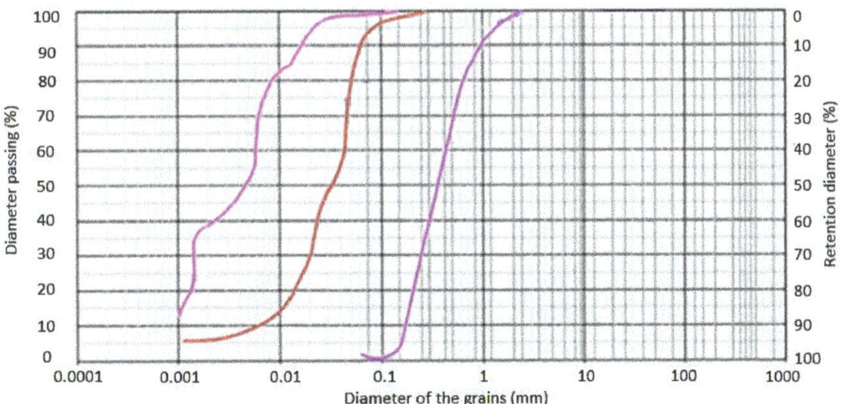

Figure 3. Granulometric curves: hydrated lime in pink, ornamental rock waste in red and sand in purple.

3.2. Chemical and Microstructural Characterization

By the chemical characterization it was observed the presence of 76.34% of SiO$_2$ and 12.49% of CaO in the ornamental rock waste, close to those found in the literature [16,44–46], while lime is mostly composed of CaO, around 93.43%. In the mineralogical analysis of ornamental rock waste, the predominant minerals were quartz and dolomite, showing good correlation between the chemical analysis performed and studies previously carried out on these materials [8,12,18,20,24,45]. Lime is predominantly composed of portlandite and calcite [46]. Thus, based on chemical and mineralogical characterization, the ornamental rock waste might be an effective substitute hydrated lime.

The microstructural characterization, Figure 4, also reveals similarity between hydrated lime and ornamental rock waste. Both ornamental rock waste and hydrated lime have grains of similar texture in terms of particle size and pore size. Although the shape of its grains is a little different, the lime presents rounder and more homogeneous particles while the waste elongated and irregular grains due to the industrial processing [3,47]. Furthermore, the ornamental rock waste presented a rough surface in many of its particles, as also observed by Lozano-Lunar et al. [48]. The fact that sizes and shapes of hydrated lime and ornamental rock waste are similar, it is believed that the replacement of lime by the ornamental rock waste will not impair at the mortars properties, as could be seen in the following tests.

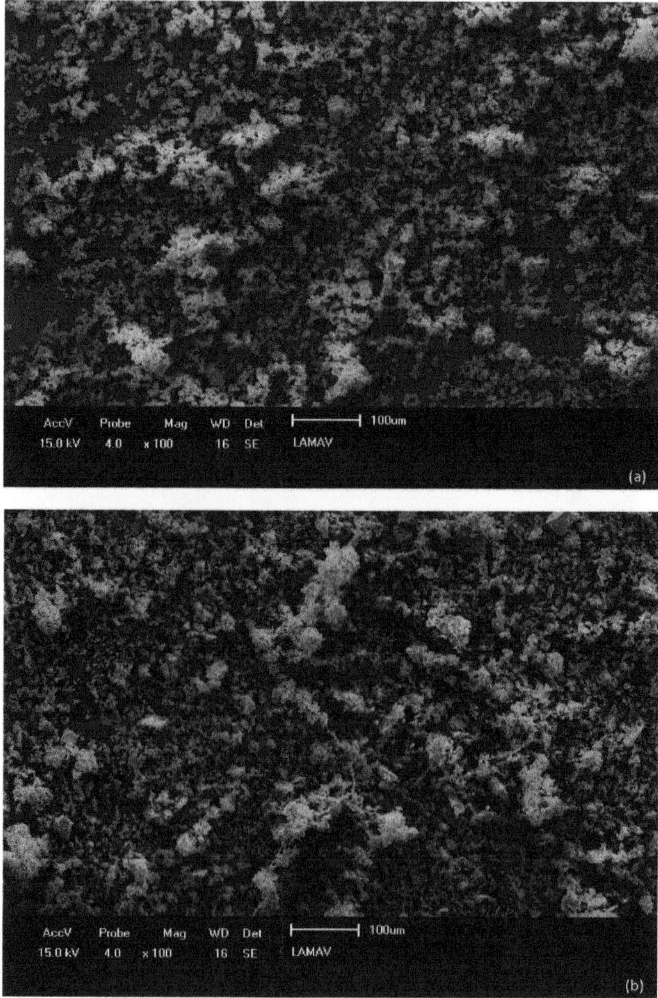

Figure 4. SEM of: (**a**) hydrated lime, (**b**) ornamental rock waste.

3.3. Fresh State Tests

The results obtained in the squeeze flow test [30] are presented in graphs of load x displacement. For this purpose, they are first compared with the reference graph presented in Figure 5. This graph is divided into 3 regions or phases. In region I the material behaves as a solid and has linear elastic deformation. In region II there are intermediate displacements, and the material will flow with plastic or viscous deformation. In phase III there is a significant increase in the load to continue with the material deformation, characterizing the restricted flow of the material, because the frictional forces are predominant, being this stage of great deformations.

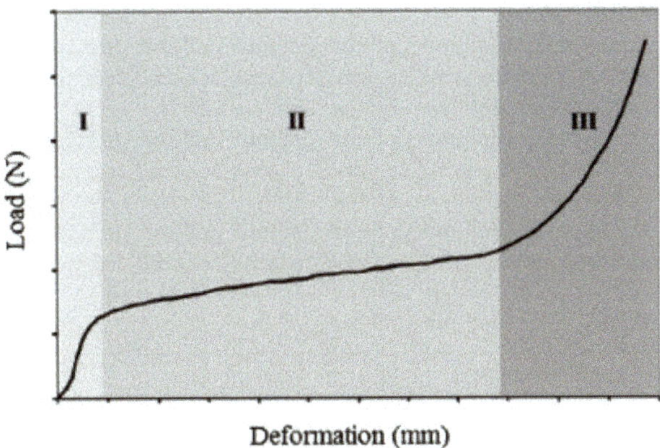

Figure 5. Typical profile load x deformation a squeeze flow test [19].

Each developed mortar in Figure 6 has a specific and unique curve, and comparing them and Figure 5, it is possible to define the most workable ones. The lower the load obtained to reach the same displacement, the less viscous the material is, i.e., the more workable [49]. So, it can be said that these mortars have a predominant stage II. The test was performed at a displacement 3 mm/s speed, with a stop criterion of either 1000 N or 9 mm of displacement, whichever came first. Observing the graphs of Figure 6, for the displacement of 4 mm, the mortar R01 needed a larger load than the others to reach this mark. Therefore, it is considered the least workable among the studied mixtures. All the mortars developed presented curves similar to hydrated lime, concluding that their workability showed also be close. Therefore, it is possible to be replace lime by ornamental rock waste. Of the three regions that are presented in the typical curve of Figure 5, these mortars in Figure 6 have a significant stage II, which means that they will have a higher deformation without a significant increase in strength [50,51].

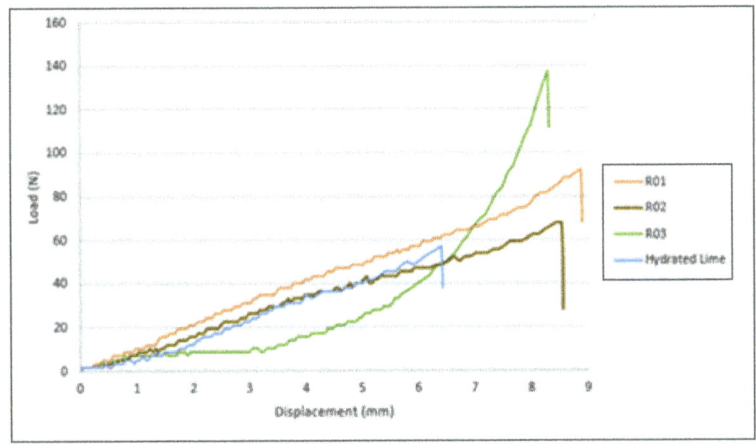

Figure 6. Load x displacement curve of the mixtures developed in this study.

As discussed by Ashish [52] and required by the ASTM standard [53,54], when a material is replaced by another with a larger specific surface, such as cement for ornamental rock waste, the workability of the material decreased, due the specific surface of the waste be higher than that of the cement, generating greater friction between the grains. On the contrary, when a material with a low specific surface is replaced by one with a smaller surface, as sand being replaced by waste, the workability increased, due to fine filler effect of ornamental rock powder. Therefore, the workability was not affected by the incorporation of the ornamental rock waste replacing the hydrated lime, since their grains have similar morphology, and also their specific surface.

Table 3 provides some results of properties of the different mortars in the fresh state. Mortars composed of ornamental rock waste presented a higher mass density when compared to those produced with lime. This result was already expected, considering that the mass density of the residue found was around 2.49 g/cm^3, which is higher than the hydrated lime, 2.27 g/cm^3, according to results mentioned in the literature [7,8,30,36].

Table 3. Fresh state parameters for the developed mortars and hydrated lime.

Parameters	Samples			
	R01	R02	R03	Hydrated lime
Density mass (g/cm^3)	1.954	1.971	1.963	1.366
Incorporated air content (%)	9.35	8.67	8.93	9.50
Water retention (%)	92.259	91.154	91.935	97.000

The incorporated air content is a property with a direct influence on workability. The higher this parameter, the greater the workability and the greater the time the material remains workable, resulting in less effort to handle the mortar, and higher productivity. Therefore, comparing Figure 6 with Table 4, it can be said that the more viscous the mortar the less is the incorporated air content, although the values found are close to each other. Thus it is difficult to say how much this property influenced the others. Knowing that the Brazilian standard does not establish a range of incorporated air content, this result was compared to the literature [53], which indicates that it should be between 2% and 7%, so all the values found are slightly above the recommended by the bibliography. Nonetheless only this result is not enough to validate or discard the mixtures.

Table 4. Results of water absorption by immersion, voids content and capillarity coefficient for the developed mortars and hydrated lime.

Sample	Water Absorption by Immersion (%)	Voids Content (%)	Capillarity Coefficient (g/dm$^2\cdot\sqrt{\text{min}}$)
R01	16.942	17.529	5.452
R02	17.054	17.857	5.846
R03	16.002	16.924	4.767
Hydrated Lime	15.350	16.599	6.800

Regarding water retention, a property that prevents excessive water loss, guaranteeing the workability and complete cement hydration, the Brazilian standard does not stipulate a minimum value. Thus, 90% was adopted according to ASTM C270 [54] and Azevedo et al. [21], reaching the conclusion that the replacement of lime by ornamental rock waste did not influence this characteristic [36].

3.4. Hardened State Tests

A hydrated lime mortar was produced at our laboratory as a reference index's for comparison, and obtained a water absorption by immersion of 15.35%, lower than the three values found in the developed mixtures, although R03 has reached values similar to that of lime [55–57]. From the analysis of this parameter, there was no significant change due to the replacement, a fact explained by the proximity between the granulometries and fine particles found in the rock residue, leading to better packing by the filler effect [17,18,24,58,59].

It is known that the water absorption by immersion and the void index are interconnected. As the absorption tends to increase, the same happens for the void index, evidenced in Table 4. The resulting void indices were compared with that reported by Azevedo et al. [3], and the mixtures developed were below the limit established by these authors. The slight increase in the void ratio was due to the introduction of an inert material replacing one of the mortar's binders. Thus, its contribution is only for the filler effect and not for the development of hydration products [20].

Regarding the water absorption by capillarity, it concerns the capillary pores that are on the coating surface, being its determination important since it is the most effective means of pathological and aggressive agents to penetrate the mortar. Fine particles as ornamental rock waste can be used to reduce void-filling, triggering the reduction of water absorption by capillarity [20]. The capillarity coefficient of R03 mixture in Table 4 displays the best result among the three ornamental rock waste mixtures analyzed. Finally, none of the mortars studied were above the limit established by the literature [1,20,37].

In order to optimize the tests time, the mixtures developed were mechanized projected on the substrate and the cracks appearance was observed, as shown in Figure 7. Despite the appearance or not of cracks being one of the acceptance criteria for the mortars developed, all of them were tested in the hardened state, even those that presented this pathology. Thus, it was possible to analyze whether the mortars with cracks also had other properties affected [55,56].

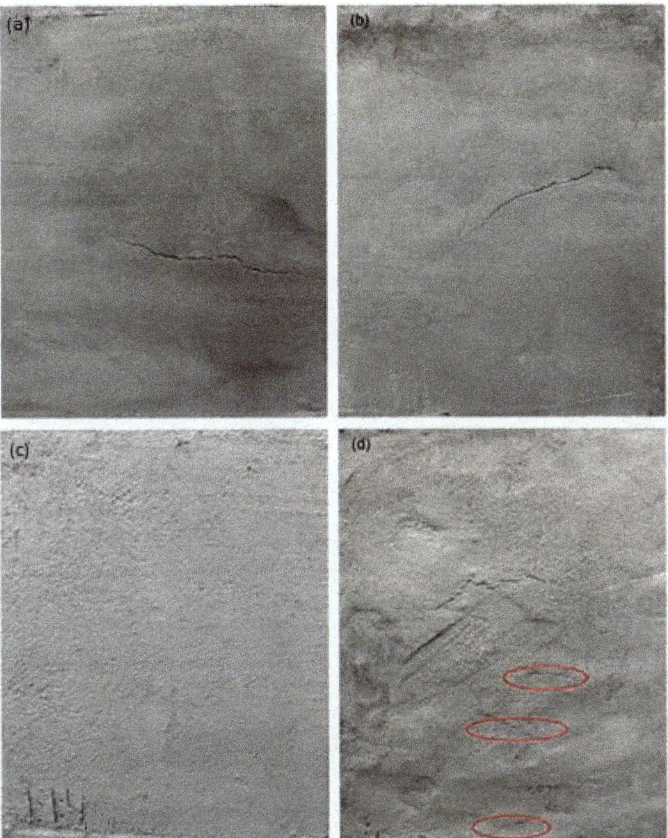

Figure 7. Visualization of cracks presence (red circle). Mixtures: (**a**) R01, (**b**) R02, (**c**) R03, (**d**) Hydrated Lime.

The samples R01 and R02, Figure 7a,b presented a small incidence of cracks, however their amplitude may cause problems. Mortars composed by hydrated lime showed small cracks in greater quantity, while the R03 mixture did not show cracks at all. Although all mortars have the same workability range, sample R03 has a higher percentage of super-plasticizer in its composition, thus reducing the amount of water present. The appearance of most cracks was observed when the coating was not fully hardened, configuring the so-called plastic shrinkage, which occurs by the rapid evaporation of the kneading water before the end of the setting and results in a superficial cracking [57,58].

The values found in flexural and compressive strength, Figure 8, were compared with the results of Kamani et al. [36], since the Brazilian standard does not establish minimum limits for these two parameters. Comparing the results obtained with [36], which indicates a minimum of 1 MPa, all the mortars studied showed results above the minimum established, although the R01 and R02 mortars reached the lowest values, the others compositions presented acceptable results. Analyzing the compressive strength, mortars R02 and R01 showed close averages, followed by the hydrated lime and finally, mortar R03. The mixtures containing 1.3% of additive, R03, are above the expected average [7,36,59].

According to NBR 13529 [41], to be used as an external coating, the minimum bond strength that mortar must have is 0.30 MPa and as internal, 0.20 MPa. Therefore, observing Figure 8, only the mixtures R03 and hydrated lime can be used as an internal and external coating, since the other two are below the value of 0.30 MPa, thus they can be used only as internal coating. In the mortars with ornamental rock waste, the flexural strength remained practically constant, as well as the tensile bond strength, while the compressive strength improved in the R03 mix and decreased in the R01 and R02 mortars. The only variation between the developed mortars is the amount of water in the mixture, thus it was responsible for the variation found in the strength.

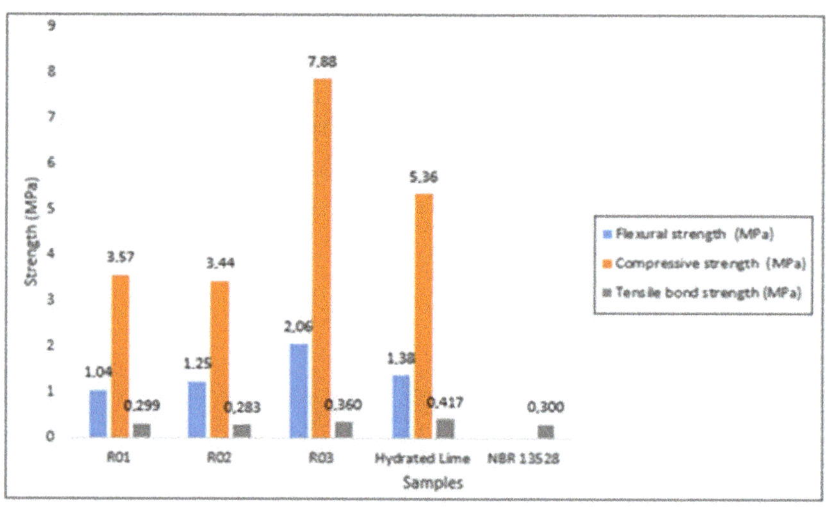

Figure 8. Average flexural strength, compression strength and tensile bond strength.

Due to the existence of cracks, the dimensional variation test [42] was performed, and the curve with its results was plotted in Figure 9. Ornamental rock waste mortars reached 90% of their total retraction at 14 days, while the curve stabilization of hydrated lime occurred around 9 days. The shrinkage found in the dimensional variation test can be correlated to the appearance of cracks in recent aging times observed in the coatings [60].

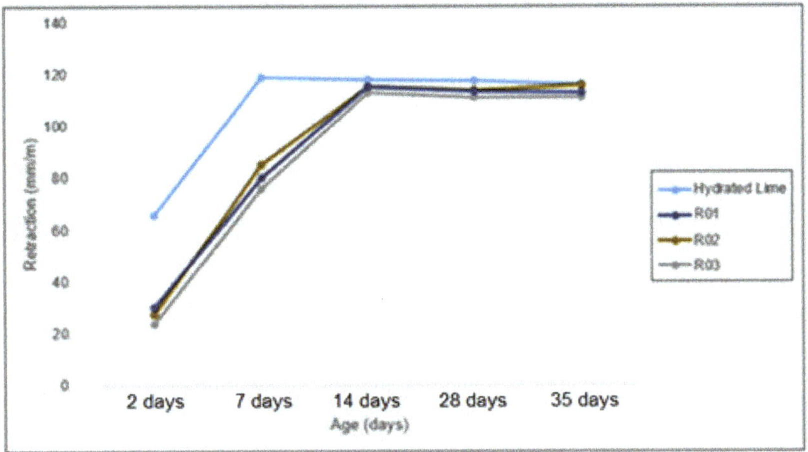

Figure 9. Dry shrinkage (mm/m) x age (days).

4. Conclusions

The material characterization showed that ornamental rock waste () is an advantageous substitute for hydrated lime, since its granulometry is closer to the lime as well as its microstructural composition. Concerning the mechanized projected mortars, workability is a fundamental characteristic for this process to occur correctly, being fixed in this specific case, according to the consistency index, at 310 ± 5 mm. It is influenced by the amount of water and additive in the mixture, incorporated air content and water retention.

The visual cracks analysis showed that among the three mortars developed; only one did not present this pathology, the sample that had the lowest water/dry material ratio. This shows that, although water increases the workability, in excess it affects others properties, such as the coating durability. From the results of dimensional variation, it was observed that all mortars showed a similar process of retraction, although R03 with 1.3% ornamental rock waste had the smallest retraction.

There was an increase in the mass density in the fresh state, of approximately 16%, in all developed mortars with ornamental rock waste in relation to hydrated lime mortars. The incorporated air content was around 9% and the water retention above 90%, following to the adopted values from literature [36].

In relation to the hardened state parameters, the R03 mortar presented a greater performance than the other two mortars, R01 and R02, mainly in the mechanical strength, being approximately 50% above the obtained values. Regarding the capillarity water absorption test, the three mortars, presented good coefficients, close to the values found for the hydrated lime mortar of 6.800 $g/dm^2 \cdot min^{1/2}$. For the void index and water absorption by immersion, the ornamental rock waste incorporated mortar presented values that were approximately the same as lime mortar.

Therefore, it can be said that among the developed mortars, the one that presented the best performances, did not show any cracks, and its workability was in accordance with that required by the projection equipment, was the R03. Thus, with all the arguments presented, the ornamental rock waste studied at this work can replace hydrated lime by mechanized projected mortars in a satisfactory way. These results corroborate other studies on the feasibility of using this waste in mortars for coating. Moreover, they fills an important gap regarding the validation of the condition of mechanized projected application, giving a strong reduction of time, cost and waste in civil construction, contributing to make mortars more sustainable. Thus, this research complements other studies in the literature and advances clearly in this field, aiming at a real and practical application for the industries of the sector.

Author Contributions: Conceptualization: J.A. and G.d.C.X.; methodology: A.L.P.; formal analysis: A.L.P.; data curation: A.L.P.; writing—original draft preparation, A.L.P.; writing—review and editing: S.N.M.; supervision: A.R.G.d.A. All authors have read and agreed to the published version of the manuscript.

Funding: This research was funded by State University of the Northern Fluminense (UENF), partial financed by CAPES (Coordenação de Aperfeiçoamento de Pessoal de Nível Superior—Brazil) and provided additional financial CNPq (Coordenação Nacional de Pesquisa) Code 309428/2020-3. The participation of A.R.G.A. was sponsored by FAPERJ through the research fellowships proc. no: E-26/210.150/2019, E-26/211.194/2021, E-26/211.293/2021, andE-26/201.310/2021 and by CNPq through the research fellowship PQ2 307592/2021-9.

Institutional Review Board Statement: Not applicable.

Informed Consent Statement: Not applicable.

Data Availability Statement: Not applicable.

Acknowledgments: The authors acknowledge the Brazilian governmental research agencies CAPES (Coordenação de Aperfeiçoamento de Pessoal de Nível Superior), CNPq (Conselho Nacional de Desenvolvimento Científico e Tecnológico), and FAPERJ (Fundação de Amparo à Pesquisa do Estadodo Rio de Janeiro).

Conflicts of Interest: The authors declare no conflict of interest.

References

1. de Azevedo, A.R.G.; Alexandre, J.; Xavier, G.; Pedroti, L.G. Recycling paper industry effluent sludge for use in mortars: A sustainability perspective. *J. Clean. Prod.* **2018**, *192*, 335–346. [CrossRef]
2. Hammes, G.; De Souza, E.D.; Rodriguez, C.M.T.; Millan, R.H.R.; Herazo, J.C.M. Evaluation of the reverse logistics performance in civil construction. *J. Clean. Prod.* **2020**, *248*, 119212. [CrossRef]
3. de Azevedo, A.R.G.; Marvila, M.T.; da Silva Barroso, L.; Zanelato, E.B.; Alexandre, J.; de Castro Xavier, G.; Monteiro, S. Effect of Granite Residue Incorporation on the Behavior of Mortars. *Materials* **2019**, *12*, 1449. [CrossRef] [PubMed]
4. Morales, L.; Garzón, E.; Romero, E.; Sánchez-Soto, P. Microbiological induced carbonate ($CaCO_3$) precipitation using clay phyllites to replace chemical stabilizers (cement or lime). *Appl. Clay Sci.* **2019**, *174*, 15–28. [CrossRef]
5. Ribeiro, D.; De Morais, G.A.T.; Júnior, A.C.L. Research on coating technology with wet-process sprayed mortar: Waste and productivity. *Gest. Prod.* **2020**, *27*, 198. [CrossRef]
6. Melo, J.P.; Medina, N.F.; Aguilar, A.S.; Olivares, F.H. Rheological and thermal properties of aerated sprayed mortar. *Constr. Build. Mater.* **2017**, *154*, 275–283. [CrossRef]
7. Montani, C. *Marble and Stones in the World, XXXI Report Italy*; Aldus Casa di Edizioni in Carrara: Carrara, Italy, 2020.
8. Allam, M.E.; Bakhoum, E.S.; Garas, G.L. Re-Use of Granite Sludge in Producing Green Concrete. *J. Eng. Appl. Sci.* **2014**, *9*, 2731–2737.
9. Li, L.G.; Huang, Z.H.; Tan, Y.P.; Kwan, A.K.H.; Liu, F. Use of marble dust as paste replacement for recycling waste and improving durability and dimensional stability of mortar. *Constr. Build. Mater.* **2018**, *166*, 423–432. [CrossRef]
10. Ashish, D.K. Concrete made with waste marble powder and supplementary cementitious material for sustainable development. *J. Clean. Prod.* **2019**, *211*, 716–729. [CrossRef]
11. Mehta, A.; Ashish, D.K. Silica fume and waste glass in cement concrete production: A review. *J. Build. Eng.* **2020**, *29*, 100888. [CrossRef]
12. Angelin, A.F.; Lintz, R.C.C.; Barbosa, L.A.G. Fresh and hardened properties of self-compacting concrete modified with lightweight and recycled aggregates. *Rev. IBRACON Estrut. Mater.* **2018**, *11*, 76–94. [CrossRef]
13. MSME. 2016. Available online: https://www.google.com/url?sa=t&rct=j&q=&esrc=s&source=web&cd=&cad=rja&uact=8&ved=2ahUKEwjn987V0Kb3AhVNmVYBHYQTClwQFnoECA8QAQ&url=https%3A%2F%2Fmsme.gov.in%2Fsites%2Fdefault%2Ffiles%2FMSME_at_a_GLANCE_2016_Final.pdf&usg=AOvVaw0neRgacSfYiNGyl3OzaWTo (accessed on 2 March 2022).
14. Arce, C.; Garzón, E.; Sánchez-Soto, P.J. Phyllite clays as raw materials replacing cement in mortars: Properties of new impermeabilizing mortars. *Constr. Build. Mater.* **2019**, *224*, 348–358. [CrossRef]
15. de Oliveira, T.F.; Beck, M.H.; Escosteguy, P.V.; Bortoluzzi, E.C.; Modolo, M.L. The effect of the substitution of hydrated lime with phyllite on mortar quality. *Appl. Clay Sci.* **2015**, *105–106*, 113–117. [CrossRef]
16. de Azeredo Melo, L.G.; Thaumaturgo, C. Filito: Um material estratégico para fabricação de novos cimentos. *Rev. Mil. Ciência Tecnol. Jan.* **2012**, *29*, 10–24.
17. Corinaldesi, V.; Moriconi, G.; Naik, T.R. Characterization of marble powder for its use in mortar and concrete. *Constr. Build. Mater.* **2010**, *24*, 113–117. [CrossRef]
18. Singh, M.; Choudhary, K.; Srivastava, A.; Sangwan, K.S.; Bhunia, D. A study on environmental and economic impacts of using waste marble powder in concrete. *J. Build. Eng.* **2017**, *13*, 87–95. [CrossRef]

19. Amaral, L.F.; Girondi Delaqua, G.C.; Nicolite, M.; Marvila, M.T.; de Azevedo, A.R.G.; Alexandre, J.; Fontes Vieira, C.M.; Monteiro, S.N. Eco-friendly mortars with addition of ornamental stone waste—A mathematical model approach for granulometric optimization. *J. Clean. Prod.* **2020**, *248*, 119283. [CrossRef]
20. Leite, F.R.; Antunes, M.L.P.; Silva, D.A.L.; Rangel, E.C.; da Cruz, N.C. An ecodesign method application at the experimental stage of construction materials development: A case study in the production of mortar made with ornamental rock wastes. *Constr. Build. Mater.* **2021**, *293*, 123505. [CrossRef]
21. de Azevedo, A.R.G.; Alexandre, J.; Zanelato, E.B.; Marvila, M.T. Influence of incorporation of glass waste on the rheological properties of adhesive mortar. *Constr. Build. Mater.* **2017**, *148*, 359–368. [CrossRef]
22. *ABNT NBR 16697*; Cimento Portland—Requisitos, ABNT. Associação Brasileira de Normas Técnicas: São Paulo, Brazil, 2011.
23. *NBR 7211:2009*; Agregados Para Concreto—Especificação, Rio Janeiro. Associação Brasileira de Normas Técnicas: São Paulo, Brazil, 2009.
24. Vardhan, K.; Goyal, S.; Siddique, R.; Singh, M. Mechanical properties and microstructural analysis of cement mortar incorporating marble powder as partial replacement of cement. *Constr. Build. Mater.* **2015**, *96*, 615–621. [CrossRef]
25. de Azevedo, A.R.G.; Costa, A.M.; Cecchin, D.; Marvila, M.T.; Adesina, A. Economic potential comparative of reusing different industrial solid wastes in cementitious composites: A case study in Brazil. *Environ. Dev. Sust.* **2022**, *24*, 5938–5961. [CrossRef]
26. Garzón, E.G.; Ruíz-Conde, A.; Sánchez-Soto, P.J. Multivariate Statistical Analysis of Phyllite Samples Based on Chemical (XRF) and Mineralogical Data by XRD. *Am. J. Anal. Chem.* **2012**, *3*, 347–363. [CrossRef]
27. *N. NM248, NBR NM 248*; Agregados—Determinação da Composição Granulométrica, Test. ABNT—Associação Brasileira de Normas Técnicas: Rio de Janeiro, Brazil, 2003.
28. *NBR 13276*; Argamassa Para Assentamento e Revestimento de Paredes e Tetos—Determinação do Índice de Consistência. Associação Brasileira de Normas Técnicas: São Paulo, Brazil, 2016.
29. Cardoso, F.A.; Grandes, F.A.; Sakano, V.K.; Rego, A.C.A.; Lofrano, F.C.; John, V.M.; Pileggi, R.G. Experimental Developments of the Squeeze Flow Test for Mortars. In *RILEM Bookseries*; Springer Science and Business Media LLC: Berlin/Heidelberg, Germany, 2019; pp. 182–190.
30. *ABNT, NBR 13278*; Argamassa Para Assentamento e Revestimento de Paredes e Tetos—Determinação da Densidade de Massa e do Teor de ar Incorporado. Associação Brasileira de Normas Técnicas: São Paulo, Brazil, 2005.
31. *ABNT, ABNT NBR 13277*; Argamassa Para Assentamento e Revestimento de Paredes e Tetos—Determinação da Retenção de Água. ABNT—Associação Brasileira de Normas Técnicas: Rio de Janeiro, Brazil, 2005.
32. Engmann, J.; Servais, C.; Burbidge, A. Squeeze flow theory and applications to rheometry: A review. *J. Non-Newton. Fluid Mech.* **2005**, *132*, 1–27. [CrossRef]
33. Campanella, O.H.; Peleg, M. Squeezing Flow Viscosimetry of Peanut Butter. *J. Food Sci.* **1987**, *52*, 180–184. [CrossRef]
34. Piasta, W.; Sikora, H. Effect of air entrainment on shrinkage of blended cements concretes. *Constr. Build. Mater.* **2015**, *99*, 298–307. [CrossRef]
35. Kamani, M.; Ajalloeian, R. The effect of rock crusher and rock type on the aggregate shape. *Constr. Build. Mater.* **2020**, *230*, 117016. [CrossRef]
36. Hendrickx, R.; Roels, S.; Van Balen, K. Measuring the water capacity and transfer properties of fresh mortar. *Cem. Concr. Res.* **2010**, *40*, 1650–1655. [CrossRef]
37. *ABNT, NBR 13279*; Argamassa Para Assentamento e Revestimento de Paredes e Tetos—Determinação da Resistência à Tração na Flexão e à Compressão. Associação Brasileira de Normas Técnicas: São Paulo, Brazil, 2005.
38. *ABNT, NBR 9778*; Argamassa e Concreto Endurecidos—Determinação da Absorção de Água, Índice de Vazios e Massa Específica. Associação Brasileira de Normas Técnicas: São Paulo, Brazil, 2005.
39. *ABNT, NBR 15259*; Argamassa Para Assentamento e Revestimento de Paredes e Tetos—Determinação da Absorção de Água por Capilaridade e do Coeficiente de Capilaridade. Associação Brasileira de Normas Técnicas: São Paulo, Brazil, 2005.
40. *ABNT, NBR 13529*; Revestimento de Paredes e Tetos de Argamassas Inorgânicas—Terminologia. Associação Brasileira de Normas Técnicas: São Paulo, Brazil, 2013.
41. *ABNT NBR 15261:2005*; Argamassa Para Assentamento e Revestimento de Paredes e Tetos—Determinação da Variação Dimensional (Retratação ou Expansão linear) (n.d.). Associação Brasileira de Normas Técnicas: São Paulo, Brazil, 2005.
42. Penev, D.; Kawamura, M. Moisture diffusion in soil-cement mixtures and compacted lean concrete. *Cem. Concr. Res.* **1991**, *21*, 137–146. [CrossRef]
43. *ABNT, NBR NM 52*; Agregados Miúdo—Determinação da Massa Específica e Massa Específica Aparente, Abnt Nbr Mn 52. ABNT—Associação Brasileira de Normas Técnicas: Rio de Janeiro, Brazil, 2003.
44. Romano, R.C.d.O.; Seabra, M.A.; John, V.M.; Pileggi, R.G. Caracterização reológica de suspensões cimentíacute;cias mistas com cales ou filitos TT—Rheological characterization of cementitious suspensions mixed with lime or phyllites. *Ambiente Construído* **2014**, *14*, 75–84. [CrossRef]
45. Amaral, L.F.; Vieira, C.M.F.; Delaqua, G.C.G.; Nicolite, M. Evaluation of Phyllite and Sand in the Heavy Clay Body Composition. In *Materials Science Forum*; Trans Tech Publications Ltd.: Freienbach, Switzerland, 2018; Volume 912, pp. 55–59. [CrossRef]
46. Santos, M.M.A.; Destefani, A.; Holanda, J. Caracterização de resíacute;duos de rochas ornamentais provenientes de diferentes processos de corte e beneficiamento. *Rev. Matéria* **2013**, *18*, 1442–1450. [CrossRef]

47. Cho, J.-S.; Moon, K.-Y.; Choi, M.-K.; Cho, K.-H.; Ahn, J.-W.; Yeon, K.-S. Performance improvement of local Korean natural hydraulic lime-based mortar using inorganic by-products. *Korean J. Chem. Eng.* **2017**, *34*, 1385–1392. [CrossRef]
48. Lozano-Lunar, A.; Dubchenko, I.; Bashynskyi, S.; Rodero, A.; Fernández, J.M.; Jiménez, J.R. Performance of self-compacting mortars with granite sludge as aggregate. *Constr. Build. Mater.* **2020**, *251*, 118998. [CrossRef]
49. Min, B.H.; Erwin, L.; Jennings, H.M. Rheological behaviour of fresh cement paste as measured by squeeze flow. *J. Mater. Sci.* **1994**, *29*, 1374–1381. [CrossRef]
50. Taylor, H.F.W. *Cement Chemistry*; Thomas Telford: London, UK, 1997.
51. García-Cuadrado, J.; Rodríguez, A.; Cuesta, I.; Calderón, V.; Gutiérrez-González, S. Study and analysis by means of surface response to fracture behavior in lime-cement mortars fabricated with steelmaking slags. *Constr. Build. Mater.* **2017**, *138*, 204–213. [CrossRef]
52. Ashish, D.K. Feasibility of waste marble powder in concrete as partial substitution of cement and sand amalgam for sustainable growth. *J. Build. Eng.* **2018**, *15*, 236–242. [CrossRef]
53. Adhikary, S.K.; Rudžionis, Ž.; Tučkutė, S.; Ashish, D.K. Effects of carbon nanotubes on expanded glass and silica aerogel based lightweight concrete. *Sci. Rep.* **2021**, *11*, 2104. [CrossRef]
54. *ASTM C270*; Standard Specification for Mortar for Unit Masonry. ASTM International: West Conshohocken, PA, USA, 2017.
55. Marvila, M.T.; Alexandre, J.; De Azevedo, A.R.G.; Zanelato, E.B. Evaluation of the use of marble waste in hydrated lime cement mortar based. *J. Mater. Cycles Waste Manag.* **2019**, *21*, 1250–1261. [CrossRef]
56. Marvila, M.T.; Alexandre, J.; Azevedo, A.R.G.; Zanelato, E.B.; Xavier, G.C.; Monteiro, S.N. Study on the replacement of the hydrated lime by kaolinitic clay in mortars. *Adv. Appl. Ceram.* **2019**, *118*, 373–380. [CrossRef]
57. Barrios, A.M.; Vega, D.F.; Martínez, P.S.; Atanes-Sánchez, E.; Fernández, C.M. Study of the properties of lime and cement mortars made from recycled ceramic aggregate and reinforced with fibers. *J. Build. Eng.* **2021**, *35*, 102097. [CrossRef]
58. Morón, A.; Ferrández, D.; Saiz, P.; Vega, G.; Morón, C. Influence of Recycled Aggregates on the Mechanical Properties of Synthetic Fibers-Reinforced Masonry Mortars. *Infrastructures* **2021**, *6*, 84. [CrossRef]
59. de Azevedo, A.R.G.; Alexandre, J.; Xavier, G.D.C.; França, F.C.C.; Silva, F.D.A.; Monteiro, S.N. Addition of Paper Sludge Waste into Lime for Mortar Production. In *Materials Science Forum*; Trans Tech Publications Ltd.: Freienbach, Switzerland, 2015; Volume 820, pp. 609–614. [CrossRef]
60. Ashish, D.K.; Verma, S.K.; Kumar, R.; Sharma, N. Properties of concrete incorporating sand and cement with waste marble powder. *Adv. Concr. Constr.* **2016**, *4*, 145–160. [CrossRef]

Review

Environmental Impact and Sustainability of Calcium Aluminate Cements

John F. Zapata [1], Afonso Azevedo [2], Carlos Fontes [3], Sergio Neves Monteiro [4] and Henry A. Colorado [5,*,†]

1   Faculty of Engineering, Institución Universitaria de Envigado, Envigado 055422, Colombia; jfzapata@correo.iue.edu.co
2   Materials Engineering, Faculty of Engineering, State University of the Northern Rio de Janeiro, Campos dos Goytacazes 28013-602, Brazil; afonso@uenf.br
3   Materials Engineering, Faculty of Engineering, Universidade Estadual do Norte Fluminense, Campos dos Goytacazes 28013-602, Brazil; vieira@uenf.br
4   Department of Materials Science, Military Institute of Engineering—IME, Praça General Tibúrcio 80, Urca, Rio de Janeiro 22290-270, Brazil; sergio.neves@ime.eb.br
5   Composites Laboratory, Universidad de Antioquia (UdeA), Medellín 050010, Colombia
*   Correspondence: henry.colorado@udea.edu.co
†   Current address: Facultad de Ingeniería, Universidad de Antioquia, Bloque 20, Calle 67 No. 53–108, Medellín 050010, Colombia.

Abstract: This investigation presents a critical analysis of calcium aluminate cements (CAC), specifically associated with sustainability and environmental impact, and the potential of these cements to help solve certain worldwide problems. Areas of research include cements as recycling holding materials, sustainability, circular economy, production costs, and energy. This investigation summarizes the current trends, perspectives, and the main concerns regarding CAC. Detailed information about the materials and processes involved in CAC is also presented. First, a general search was made using the Carrot2 Workbench metasearch engine to identify possible thematic groups correlated with CAC, then a more in-depth and specialized search was done using the Scopus database. The results revealed that these materials have a lot of potential to help solve problems in the circular economy and suggest several exciting areas for conducting future research.

Keywords: cements; materials; sustainability; construction; civil engineering; phases

Citation: Zapata, J.F.; Azevedo, A.; Fontes, C.; Monteiro, S.N.; Colorado, H.A. Environmental Impact and Sustainability of Calcium Aluminate Cements. *Sustainability* **2022**, *14*, 2751. https://doi.org/10.3390/su14052751

Academic Editors: Mazen Alshaaer, Slávka Andrejkovičová and Asterios Bakolas

Received: 14 January 2022
Accepted: 22 February 2022
Published: 26 February 2022

Publisher's Note: MDPI stays neutral with regard to jurisdictional claims in published maps and institutional affiliations.

1. Introduction

Calcium aluminate cements (CAC) are classified as hydraulic cements and are mostly used in applications involving extreme environments, which include refractories [1], acid resistant requirements, and fast setting cements. These materials are currently the subject of research on topics like property improvement, durability and more friendly manufacturing processes. CAC are manufactured industrially from mixtures of limestone and materials with a high content of Al_2O_3 (bauxites, laterites and alumina obtained via the Bayer process, among other materials) [2]. CAC are considered eco-cements due to the reduced carbon emissions created during their production. They also have diverse applications due to their impressive early age strength and enhanced durability in harsh environments [3].

The main mineral phases present in CACs are calcium mono aluminate (CA), calcium dialuminate (CA_2), dodecalcium heptaaluminate ($C_{12}A_7$) and alpha-alumina (α-Al_2O_3). All CAC properties depend only on their mineralogical phase composition [2,4].

The specific gravity of CACs varies between 3000 and 3250 kg/m^3 (3 and 3.3 g/cm^3), and is largely controlled by the iron content. The apparent density is typically between 1.12 and 1.74 g/cm^3. The specific surface area measured by the British Standard method can vary between 2.500 and 4.000 cm^2/gr. The properties of the constituent minerals of CACs as well as their typical chemical composition are shown in Table 1.

Table 1. Properties of the constituent minerals of CAC.

Mineral	Chemical Composition (wt%)				Tm (°C)	Density (g/cm³)	Crystal Structure	Formation Enthalpy (kJ/mol)
	CaO	Al₂O₃	Fe₂O₃	SiO₂				
C	99.8	-	-	-	2570	3.32	Cubic	-
$C_{12}A_7$	48.6	51.4	-	-	1405–1495	2.69	Cubic	-
CA	35.4	64.6	-	-	1600	2.98	Monoclinic	−2323
CA_2	21.7	78.3	-	-	1750–1765	2.91	Monoclinic	−4023
C_2S	65.1	-	-	34.9	2066	3.27	Monoclinic	-
C_4AF	46.2	20.9	32.9	-	1415	3.77	Orthorhombic	-
C_2AS	40.9	37.2	-	21.9	1590	3.04	Tetragonal	-
CA_6	8.4	91.6	-	-	1830	3.38	Hexagonal	-
Al_2O_3	-	99.8	-	-	2051	3.98	Rhombohedral	

The main hydraulic phase in all CACs is the CA phase. Of all the influencing factors, temperature is the most important since hydration products depend heavily on the curing temperature [5–11]. According to several studies, the initial hydration product that emerges within the temperature range of 0 to 15 °C is predominantly CAH_{10} (Monocalcium aluminate decahydrate), while between 15 °C and 35 °C the predominant phases are CAH_{10}, C_2AH_8 (Dicalcium aluminate octahydrate) of hexagonal morphology, and amorphous AH_3 (Gibbsite). Above 35 °C, the major phase formed is C3AH6 (Katoite) [4,12–15]. CA hydration is primarily responsible for the early development of resistance, while CA2 hydration occurs after the main CAC hydration reaction has already been exceeded [16]. Refs. [17–19] shows some properties of the hydrated phases (Table 2).

Calcium aluminate cements (CAC) are mainly used in the refractory industry. They are also mixed with other cements such as Portland cement to create products in the ceramic industry, especially in construction [4,14,20–24]. Can recycled materials be used for the manufacture of calcium aluminate cements? How do calcium aluminate cement products fit into a circular economy? What relationship is established between the energy sector and the production of calcium aluminate cements? How are calcium aluminate cements being implemented within additive manufacturing? These are important questions in a world where factors that affect the environment and energy contribute to global warming.

Table 2. Densities, molar masses of the hydrated phases, and enthalpy of formation.

Hydrated Phase	Density (g/cm³)	Molar Mass (g/mol)	Formation Enthalpy (kJ/mol)
CAH10	1.72	338.1	-
C2AH8	1.96	358.2	−5433 [25,26]
AH3	2.44	156.0	−2578 [25]
C3AH6	2.52	378.3	-

Figure 1 shows the SEM images of the main hydrated phases of CAC at 20 °C, (a) CAH10, and (b) C2AH8. The laminar structure is very clear, with all thicknesses below 0.5 μm.

Concrete based on Portland cement is the most widely used material in the world and its massive production is responsible for approximately 5 to 8% of global manmade CO_2 emissions [27]. CO_2 emissions are mainly due to energy consumption during the manufacturing process and decarbonization of limestone during clinker manufacturing. These emissions could be reduced by replacing clinker methods with inorganic minerals from industrial wastes such as slag, fly ash, and silica fume, or by using CAC despite its much smaller scale production compared to Portland cement [28]. The study and use of CAC is therefore very valuable to reduce global environmental problems.

Figure 1. SEM of hydrated phases of CAC at 20 °C, (**a**) CAH_{10}, and (**b**) C_2AH_8.

The generation of CO_2 is an important consideration in the manufacture of cement. Thus, it is pertinent to know which mechanisms are involved in the process in order to try to reduce such emissions. 3D printing is a revolutionary new process that impacts many sectors, including medicine and construction. Additive manufacturing (AM) is a process whereby materials are joined to make objects from a 3D model, layer upon layer [29–32], which minimizes the materials wasted. Due to the importance of these latter two areas, they are included in this review to see how much they have been researched already. The value of the other areas included in this review needs no further justification.

This research shows how investigations into CAC are helping to solve its most significant environmental issues. In order to understand and classify all this research, a general search was first carried out using the meta search engine Carrot2 Workbench. Carrot2 Workbench is an open-source search results clustering engine. After that, an exhaustive search was performed using the Scopus database, and the information found there was then analyzed.

2. Materials and Methods

A general search was initially carried out using the Carrot2 Workbench metasearch engine and the keywords Calcium Aluminate Cement. The initial search employed the following sources: etools web search and the K-mean algorithm (base line algorithm, bag-of-words labels). A second search was done using the etools web search and the Lingo algorithm (well-described flat clusters).

After that, a deeper and more specialized search was conducted in the Scopus database, with keywords provided by Carrot2 Workbench and complemented by others that were more specific to CACs and the environment, hydration, high temperature, refractory, Portland cement. The Scopus database search was performed with the keywords shown in Figure 2a.

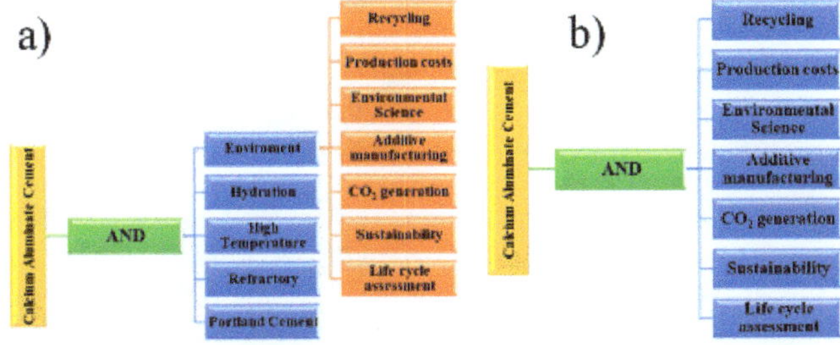

Figure 2. Scopus database search for specific areas. (**a**) CACs and the environment, hydration, high temperature, refractory, Portland cement (**b**) CACs and recycling, productions costs, environmental science, additive manufacturing, CO_2 generation, sustainability, life cycle assessment.

Subsequently, the documents found for specific areas, recycling, productions costs, environmental science, additive manufacturing, CO_2 generation, sustainability and life cycle assessment, related to CACs and the environment were analyzed, as shown in Figure 2b. Finally, the studies found in the last stage were described.

3. Results

The search was conducted using the Carrot2 Workbench with the keyword calcium aluminate cement—Source: etools web search—algorithm: Lingo (well-described flat clusters.). The search found 118 documents in general areas, and 33 documents in clusters (see Figure 3). The few articles found are dispersed over 33 clusters, where 40% (48 documents) focus on clusters such as Portland cement and CAC (see Figure 3). This high dispersion of very few articles means that a common thread in the studies cannot be established and practically every cluster is an area of research to explore.

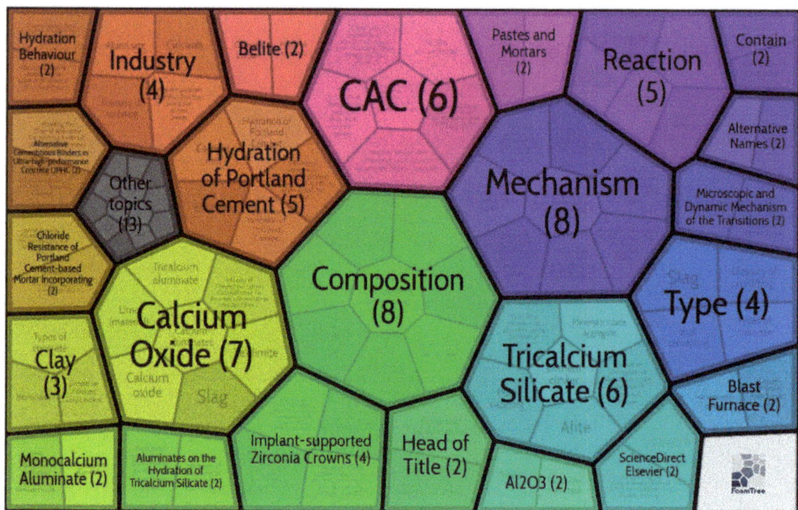

Figure 3. Carrot2 Workbench database search.

The search using Carrot2 Workbench with the keyword calcium aluminate cement-Source [33,34]: etools web search—algorithm: K-mean (base line algorithm, bag-of-words labels), showed 58 documents in related areas and 19 documents in clusters (see Figure 4). This search found more specific clusters but also showed that there are fewer articles and an increased dispersion. According to the metasearch engine, there are not enough studies to establish a common thread of work in a specific area. No cluster is presented as a developed area. On the contrary, each cluster is open to new research, as seen in Figure 4 [33,34].

A deeper and more specialized search was performed in the Scopus database, with the keywords given in Figure 2a. The search in Scopus found that the most commonly investigated factors associated with CAC are the hydration process, refractoriness, high temperature behavior, and the mixture of CAC + Portland cement. A total of 2832 documents were found. It can be established from this review that the factors associated with the most investigated CACs are the hydration process, high temperature behavior, refractoriness, and mixing with Portland cements. A quantification of the type and number of studies carried out in relation to these factors is presented in Figure 5 and in Table 3.

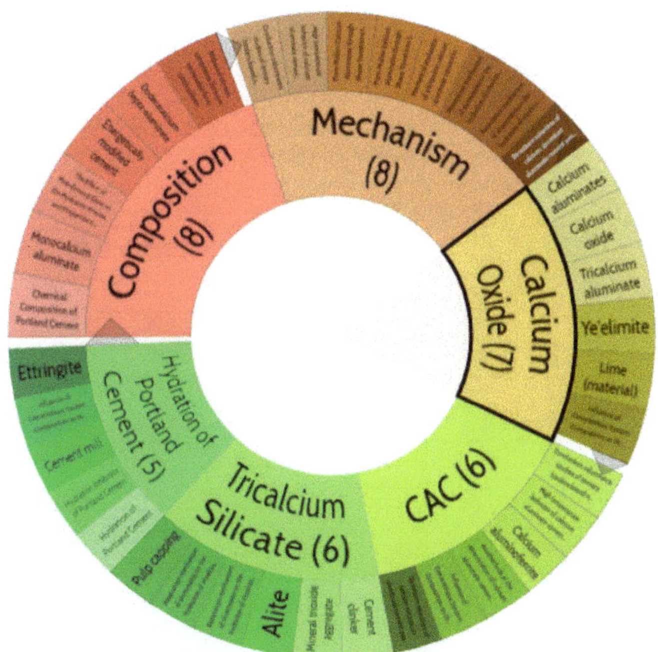

Figure 4. Carrot2 Workbench database search for specific areas.

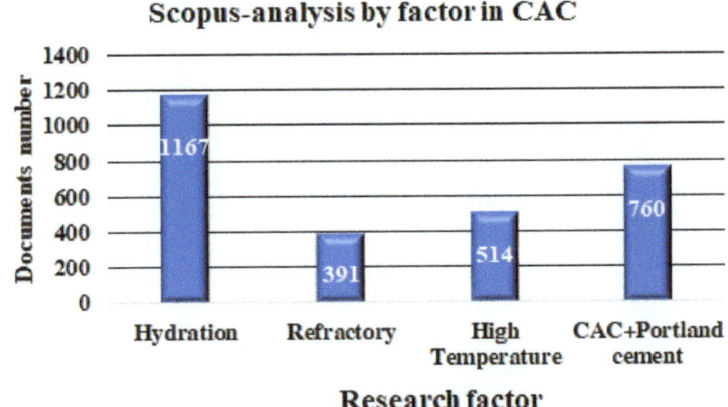

Figure 5. Documents organized by research factor in the Scopus database.

Table 3. Types of documents organized by CAC study factor.

	Article	Conference Paper	Review	Conference Review	Book Chapter	Production Interval (In Years)
Hydration	939	189	27	10	2	1957–2022
High Temperature	412	78	11	8	5	1946–2022
Refractory	298	69	16	5	3	1946–2022
Portland Cement + CAC	579	137	16	10	18	1975–2022

It is clear that the hydration process is the most widely investigated topic in CAC, followed by topics related to refractories, behavior at high temperature, and the CAC + Portland Cement mixtures (see Figure 5).

The bibliographic search was performed using the Scopus database until November 2022, in which 112 documents were found for the environmental areas shown in Figure 6. The search in the specialized Scopus database shows very few studies. It can be divided into 7 clusters, which indicates a high dispersion, and a common thread or an area of research cannot be defined. Therefore, each cluster shown in Figure 3 and Table 4 represent an area of investigation to explore.

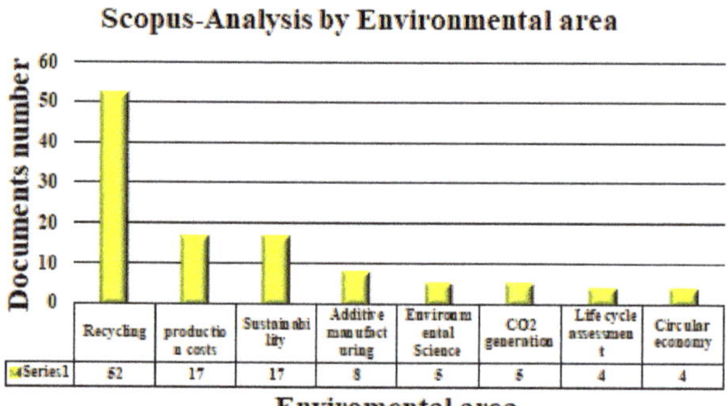

Figure 6. Scopus database search for environmental area.

Table 4. Types of documents organized by environmental cluster.

	Article	Conference Paper	Review	Conference Review	Book Chapter	Production Interval (In Years)
Recycling	37	7	1	4	3	1995–2022
Production costs	13	3	1	0	0	1995–2021
Environmental Science	3	2	0	0	0	2001–2020
Additive manufacturing	3	1	1	1	2	2010–2021
CO$_2$ generation	4	1	0	0	0	2011–2021
Sustainability	10	4	1	1	1	2007–2021
Circular economy	4	0	0	0	0	2017–2021
Life cycle assessment	1	2	0	1	0	2014–2021

Table 4 shows the type of document organized according to environmental clusters, like the years the articles were published. As shown in Table 4, from 112 documents, only 70 articles were found in all areas. In Circular Economy, four articles were registered, and in Life Cycle Assessment, only one was found. In Environmental Science and Additive Manufacturing, three articles per area were found. In Sustainability, 10 articles were found, and in CO$_2$ Generation, four articles were found. The most widely investigated areas are Production Costs and Recycling.

4. Discussion

The revision of the state of the art shows incipient development in the literature related to calcium aluminate cements and the environment, such as CO$_2$ Generation, Sustainability, Additive Manufacturing, Environmental Science, Production Costs, Recycling, and Circular Economy. There is therefore a well-defined research area that can be developed for each topic. Furthermore, this review suggests five emerging research areas for CACs, as shown in Figure 7.

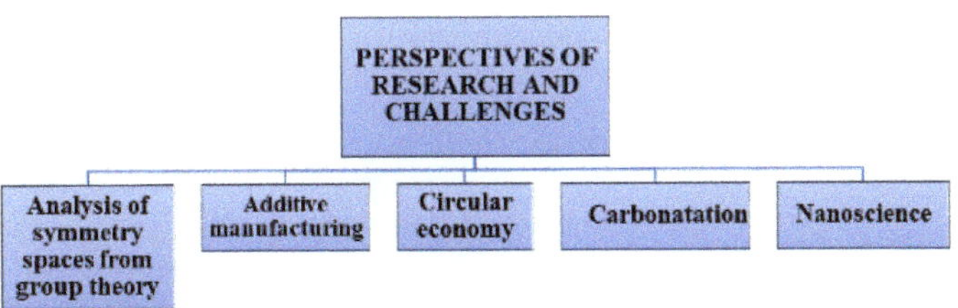

Figure 7. Perspectives of research and challenges.

4.1. Analysis of Symmetry Spaces from Group Theory—Nanoscience

The fundamental chemical interactions that control the structure and performance of cements have been investigated in detail. However, the complex and crystallographic nature of the phases that form in hardened cements makes it difficult to obtain detailed information on the local structure, reaction mechanisms, and kinetics [35–38]. Solid state nuclear magnetic resonance spectroscopy could solve the key atomic structural issues within these materials in combination with the proper use of X-ray diffraction. The X-ray diffraction should focus on unveiling the crystalline symmetry spaces, in order to set out promising lines of research. The research could generate models using highly advanced designs of new materials using knowledge of their quantum behavior and their proper combination with various suitable nano materials suggested by the theoretical analyzes.

4.2. Additive Manufacturing

Additive manufacturing technologies (also known as 3D printing) have expanded rapidly in various industrial sectors, including aerospace, automotive, medical, architecture, arts and design, food and construction. Transition from visualization and prototyping stages to functional and real parts replacement opens up more design possibilities. Among the various AM applications, building and construction is a very promising large-scale area for development and will be included in the coming manufacturing revolution. However, there are obvious challenges and risks to integrating AM into large-scale construction, such as construction and reinforcement regulation. Progress in commercialization is advancing at a slow pace since only a few large-scale 3DP trials for construction can be found in the literature [39–41]. The use of CACs to generate AM-friendly mixtures is therefore a virgin and unexplored research area with great prospects for success.

4.3. Circular Economy

Although CACs can potentially be used to benefit the environment, including the use of waste as aggregates in concrete manufacturing, neutralization of reactive substances, and the storage of hazardous substances, no studies were found that investigate the use of CACs within the circular economy production chain.

Searching in the Scopus database using the keywords calcium aluminate cement + circular economy did not produce any results. This reveals that little work has been carried out on the management of solid waste generated from manufactured products based on CAC. Therefore, new processes using CAC-based waste and the circular economy would be of great interest to industry, and would potentially have a significant environmental impact.

4.4. Carbonatation Process

CAC paste is known to react with atmospheric carbon dioxide. Carbonation causes numerous chemical-mechanical changes in the cement paste, but can also improve properties of the cement material [42–45]. For these reasons, the active use of carbonation as a tool to manipulate the properties of CAC-based materials is a very promising research area that could potentially improve fiber-reinforced cementitious compounds and promote concrete recycling and immobilization of waste.

The carbonation of CACs has been investigated in order to find adequate conditions to improve the mechanical properties of CACs. However, the very specific manufacturing conditions of CAC paste, such as water/cement ratios greater than 0.4, hydration in the presence of alkalis and hydration in a high-humidity atmosphere [44,46], produce a paste with very poor mechanical performance for construction and refractories. However, these pastes do have a high CO_2 absorption capacity. Therefore, hydrating CAC in conditions that are unfavorable to its mechanical properties could drastically increase its CO_2 absorption capacity, which could be used to reduce CO_2 contamination elsewhere. After these pastes are saturated with CO_2, they could be included in the circular economy as a construction material. While no research has been found in this area, it would be an excellent topic to review. The research presented here describes advances in sustainability and many other specific and important topics related to the circular economy of CAC. However, the lack of progress in these areas is quite evident considering the social demands on modern construction materials. Moreover, the required approaches demand a rethinking of the material and process in the circular economy. The process needs to be not only reused but also optimized to reduce energy, water and material waste. From an academic, innovation and production point of view, this technological lag opens up new opportunities to produce a CAC that fits in with the advances of the 21st century. Such research is important because it would pave the way for not only significant progress towards a more sustainable CAC, but also reveal the limitations and opportunities in the area for a real modern CAC material.

5. Analysis by Specific Area

5.1. Analysis by Environmental Area

After refining the search and restricting only to CAC + environment, the Scopus database shows 112 articles distributed by cluster as shown in Table 4. Below is a description of each article found.

5.1.1. CAC + Recycling

A search with the keywords, calcium aluminate cement and recycling (CAC + recycling) in the Scopus database yielded 52 documents only 37 articles were found. The number of documents produced per year is presented in Figure 8a. It can be seen that the highest production was in the last 3 years, with 28 works presented.

The use of CAC can effectively mitigate the expansion of the alkali-silica reaction (ASR) of alkali activated cement mortars (AAC) incorporating glass cullet (GC) as fine aggregates, but the mechanism is unclear [47]. P. He, et al. [47]. present a systematic study on the exploration of the underlying mechanism and to determine the safe use of GC in AAC to maximize the possibility of recycling glass waste in construction materials.

They found that the ASR expansion of the alkali activated glass mortars was more than 1000 $\mu\varepsilon$ after 1 day of alkaline immersion, when using GC and the incorporation of CAC decreases the expansion to less than 1000 $\mu\varepsilon$, even after 14 days of immersion alkaline, decreased by 20%. This result occurs because the aluminium in CAC was incorporated into the AAC matrix with tetrahedral and octahedral units. Increasing the CAC content increases the tetrahedral unit that required more Na ions to balance the excess charge, which caused the alkalinity of the pore solution to decrease as did the expansion of ASR.

B. Zhang, et al. [48] feature a way to incorporate recycled glass materials into AAC-based pastes/mortars. The GC waste was used to replace natural aggregates and the glass waste powder (GP) was used to partially replace conventional precursors such as blast furnace slag (GGBS). Using the developed technology, recycled AAC-based mortars can be produced with a 28-day compressive strength of approximately 40 MPa. The AAC mortar mixes showed good resistance to high temperatures, retaining more than 50% of the original resistance after 2 h of exposure at 800 °C. The experimental results also showed that the damaging expansion caused by the alkaline silica (ASR) reaction between alkali and residual GC in AAC mortars could be successfully controlled by using additional sources of aluminum such as CAC.

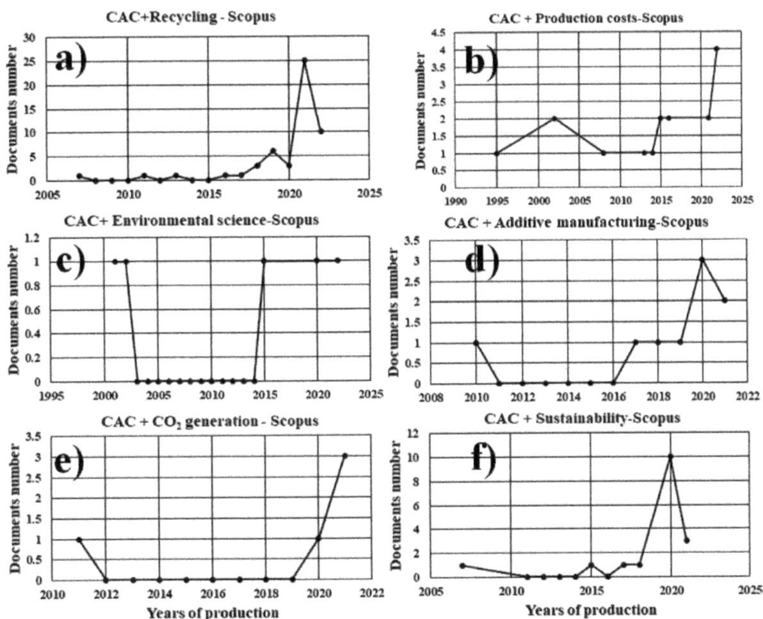

Figure 8. Scopus database search for calcium aluminate cement and recycling (CAC + recycling). (a) CAC + Recycling (b) CAC + Production Costs (c) CAC + Environmental Science (d) CAC + Additive Manufacturing (e) CAC + CO$_2$ Generation (f) CAC + Sustainability.

Chang, et al. [48] also investigated the compressive strength, drying shrinkage and ASR expansion of AAC mortars using GC as aggregates and GP to partially replace GGBS as the precursor. They noted that mortars using GC as an aggregate, replacing GGBS with GP decreased compressive strength, and mortars using sand as an aggregate, showed severe drying shrinkage, while replacement of sand by GC could significantly decrease drying shrinkage.

The alkali activated GGBS mortars using GC as an aggregate showed great expansion after alkaline immersion. When GP is used to partially replace GGBS, the expansion was significantly reduced. The replacement of 15% of GGBS by CAC could further decrease the expansion. The optimal proportion of the AAC mix developed in this study was that the mortars incorporated 15% CAC, 10% GGBS, and 75% GP, could meet stipulated mechanical requirements and durability requirements for bulkhead applications.

W. Panpa, et al. [49] produced a visible light active compound, Ag$_2$O-Ag/CAC/SiO$_2$, by loading an aqueous solution of silver salt onto the hydrated CAC coating layer in a porous SiO$_2$ sphere. The slight hydration condition of the cement induced the formation of Ag$_2$O precipitates in situ on the cement surface which partially decomposed into metallic Ag after drying and appropriate heat treatment at 200 °C to improve both Ag$_2$O decomposition and crystallinity.

The photo catalytic activity of the compound Ag$_2$O-Ag/CAC/SiO$_2$ (calcined at 200 °C), evaluated under irradiation with UV light and visible light through the photo decomposition of CHP in water, quantified by HPLC, was two times greater than that Ag$_2$O-Ag/CAC/SiO$_2$ compound (dried at 45 °C), and capable of completely decomposing CHP in 5 h. Furthermore, the photo stability of the Ag$_2$O-Ag/CAC/SiO$_2$ photocatalyst remained unchanged during five recycling tests.

S.K.S. Hossain and P.K. Roy [50] formulated a nano-lakargiite system (NL) [CaZrO$_3$] for formless refractories. The NL powder formulation is made from a solution mixture, which is easily capable of recycling by-products. Scrap shells were used as a source of CaO for the preparation of NL. They reached at 1100 °C, the single-phase orthorhombic crystal structure with the Pnma space group of nano $CaZrO_3$ (average crystallite size ~19 nm). The formulation involved replaced CAC with NL was heat treated at 1600 °C, resulting in properties that matched by the different advanced bonding systems of high alumina refractory molds, therefore, achieving good densification, heat resistance, and thermal shock resistance for NL bonded refractory molds.

H. Al-musawi, et al. [51] studied the time-dependent transport properties and shrinkage performance of smooth and fiber-reinforced rapid-setting mortars for repair applications, in two single mixes of Steel-Fiber-Reinforced Concrete (SFRC). They found that CSA cement mixes have much lower shrinkage values (about 220 and 365 microspheres) compared to CAC mixes (about 2690 and 2530 microspheres), but most of the shrinkage in these formulations were autogenous. However, the fibers reduced the drying shrinkage of the CAC mixes by approximately 12%.

H. Al-musawi, et al. [52] investigated the flexing performance of fast-setting mortar mixes made with two types of commercial cement, calcium sulfonate cement and CAC, for thin concrete repair applications. They performed three-point bending tests on samples of smooth steel-fiber reinforced (FRC) concrete containing 45 kg/m^3 of recycled clean steel fibers to characterize the bending performance of notched and notched prisms at different ages, ranging from one hour to one year. They found that recycled fibers improve both the flexural strength and toughness of FRC prisms.

One of the best methods of recycling waste in the construction sector is to use it in the preparation of concrete or mortar. S.T. Yildirim, et al. [53] used recycled concrete aggregates and bottom ash as aggregates in the mortar. They used CAC as a binder to improve fire resistance. They investigated flexural and compressive strength, dry unit weight, water absorption, capillary action, thermal conductivity, thermal resistance, and costs, following a Taguchi method for internal cure (IC), the amount of cement, and the aggregate ratio as parameters. They found that IC does not have enough positive effect on mortars. Choosing the appropriate type of cement is believed to be effective. The process of obtaining resistance for CAC is quite fast and water had a negative rather than a positive effect. It is believed that IC water should have increased the amount of water mix. The increase in the amount of CAC has a positive effect because the dosage of cement in the mortars is kept low and the durability with respect to temperature is maintained.

P. Kulu, et al. [54] recycled the niobium slag, separating the metallic Nb and the mineral ballast—calcium aluminate. Preliminary tests of cement to use the main product as a binder or substitute for cement confirmed its potential for use in the production of construction materials. P. Ogrodnik, et al. [55] investigated sanitary ceramic waste as an addition to concrete. Six concrete mixes on Portland cement and CAC were designed with various aeration mix contents. They found that the compressive strength is adequate for concrete containing ceramic aggregate and CAC at high temperatures.

M. Nematzadeh, et al. [56] investigated the compressive behaviour of concrete containing fine recycled refractory brick aggregate, CAC fibers, and polyvinyl alcohol (PVA) in an acidic environment, motivated because earthquake-induced structural debris and other factors cause destruction and problems related to the damaging effects of acidic environments. They found that samples containing CAC along with PVA fibers have an adequate corrosion response against acid attacks, while samples containing fine refractory brick aggregate showed quite unsatisfactory performance in this regard.

M. Nematzadeh and A. Baradaran-Nasiri [57] investigated the compression stress-strain behaviour of recycled aggregate concrete. Different levels of replacement of conventional fine aggregate by recycled particles from used refractory bricks were used (0, 25, 50, 75 and 100% volume replacement level) in two groups, one containing PC and the other containing CAC. After exposure to elevated temperatures (110, 200, 400, 600, 800 and 1000 °C), significant degradation occurred for most of the mechanical properties for concrete containing ordinary cement at 400 °C and for concrete containing aluminate cement at 110 °C. Higher fine brick refractory aggregate contents improved concrete compression behaviour at higher temperatures.

A. Baradaran-Nasiri and M. Nematzadeh [58] investigated the use of recycled aggregate produced by crushing firebricks. Samples were separated based on CAC and OPC with replacement ratios of 0, 25, 50, 75 and 100% of fine aggregate of refractory brick instead of natural sand. After exposing the samples to temperatures of 110, 200, 400, 600, 800, and 1000 °C, it was found that the addition of refractory brick and the CAC improve the residual strength of the concrete up to twice, at the temperature of 800 °C.

R. Stonys, et al. [59] studied the waste from mineral wool (dome dust—CD) and its possible reuse in the production of CAC-based refractory concrete, replacing the silica addition with CD. It was found that CD can be used for the production of refractory concrete. Í. Navarro-Blasco, et al. [60] studied waste foundry sand (WFS) in CAC mortars with a replacement level of 50%. Compared to OPC mortars, the use of CAC showed several advantages, improving compressive strength and retention of toxic metals.

L.J. Fernández, et al. [61] studied the recycling of crystalline solar cells incorporated into cement matrices. The hydration process of a mixture of CAC and photovoltaic solar cell waste was analyzed, founding that the presence of up to 5% solar cell residue in cement matrices does not result in new hydration products that are different from those derived from normal CAC hydration. Furthermore, the developed material can be considered an expansive cement mix because it releases H2 in the early stages. The presence of residues causes a decrease in mechanical resistance and an increase in the total porosity of this material, although, it could be used for applications such as thermal insulation.

Chen et al. [62], showed that CAC had an excellent immobilization efficiency of potentially toxic elements in fly ash from municipal solid waste incineration.

5.1.2. CAC + Production Costs

A search with the keywords, calcium aluminate cement, and production costs (CAC + production costs) in the Scopus database yielded 17 results. The number of documents produced per year is presented in Figure 8b. The low production in this topic can be observed in the last 25 years.

M. Erans, et al. [63] oriented their research to the production of improved sorbents, motivated by the fact that the calcium loop (CaL) is a CO_2 capture technology and a fundamental problem for the commercialization of these technologies is to maintain a high level of sorbent reactivity during the long-term cycle, to mitigate the decrease in carrying capacity. A great strength of CaL compared to other carbon capture technologies is the synergy with the cement industry, due to the use of spent sorbent as Clinker raw material.

M. Erans, et al. [63] investigated limestone doped with HBr through a particle surface impregnation technique and granules prepared from limestone and CAC. These were tested in a 25 kWth dual fluidized bed pilot scale reactor to understand their capture performance and mechanical stability under realistic CaL conditions and the spent sorbent was subsequently used as a raw material to make cement. It was found that HBr-doped limestone showed better performance in terms of mechanical strength and stability of CO_2 absorption compared to that of granules prepared from limestone and CAC, and that the cement produced has characteristics and performance similar to that of commercial CEM I cement. This demonstrates the advantages of using spent sorbent as a raw material for cement manufacturing and shows the benefits of synthetic sorbents in CaL and the end-use suitability of spent sorbents for the cement industry. It can be concluded from the work of M. Erans, et al. [63], the feasibility of using various practical techniques to improve the performance of CaL on a pilot scale, and more importantly, it shows that commercial grade cement can be manufactured at from the lime product of this technology.

Abolhasani et al. [3] discusses the results of a comprehensive study incorporating rice husk ash (RHA) in CAC concrete to limit the phase transition of CAC hydration product and stabilize its long-term properties. The findings indicate that, at 90 days, the mechanical strengths of the mixes containing RHA were higher than those of the control mix, with the maximum improvement occurring at the substitution percentage of 5%.

5.1.3. CAC + Environmental Science

A search with the keywords, calcium aluminate cement and Environmental science (CAC + Environmental science) in the Scopus database yielded 5 results. The number of documents produced per year is presented in Figure 8c. This reveals almost no production in the last 20 years.

L. Xu, et al. [64] investigated setting times, mechanical strength and drying shrinkage ratio of mixed OPC and CAC systems (less than 25%). They found that with CAC, the set times of combined systems are shortened; the compressive strength first increases slightly (peaks with 6% calcium aluminate cement) and then decreases significantly. T.T. Akiti Jr, et al. [65,66] developed a calcium-based sorbent to desulfurize the hot carbon gas, this led to the development of a spherical granule-like material, which has a limestone-based core enclosed in a strong support layer. Strong granules are prepared by incorporating some CAC into the core and a larger amount into the shell, along with the limestone.

The granules are prepared by a two-step granulation method, followed by steam curing and heat treatment. It was found that the granules are capable of absorbing relatively large amounts of hydrogen sulfidic at high temperature (e.g., 1150–1200 K) that can be regenerated by a cyclical oxidation and reduction process.

Secondary aluminum dross (SAD) is a dangerous pollutant as well as a valuable resource. About 95% SAD is disposed by stockpiling on the spot due to its complex composition and technical limitations, causing severe ecological damage and public health threat [67]. Mingzhuang et al. [67] developed a new green process for the preparation of ultrafine and high-whiteness $Al(OH)_3$ from SAD. This phase can then be used as a high-strength cementitious material by mixing it with CAC.

5.1.4. CAC + Additive Manufacturing

A search with the keywords, calcium aluminate cement and Additive manufacturing (CAC + Additive manufacturing) in the Scopus database returned 10 results. The number of documents produced per year is presented in Figure 8d. This reveals almost zero production in the last 10 years.

P. Shakor, et al. [68] They discuss a methodology for replacing the typical powders currently used in 3D printing to make it possible to use printed samples in construction applications. They found that the highest compressive strength (14.68 MPa) is obtained for samples that were first cured in water and then oven dried for one hour at 40 °C, compared to samples that were cured without drying at 40 °C (4.81 MPa). Therefore, the post-processing technique has a significant and effective impact on the resistance of printed samples. P. Shakor, et al. [69] developed a mixture of CAC through a 150 μm sieve and OPC for the Z-Corporation 3D Printing Process (3DP). This cement mix was mixed, and the resulting compound powders were printed with a water-based binder using a Z-Corporation 3D printer. In addition, lithium carbonate was added to some samples to reduce the setting time of the cement mix. The maximum compressive strength of the cubic samples for cementitious 3DP was 8.26 MPa at the 170% saturation level for both the shell and the core. The minimum porosity obtained was 49.28% at the saturation level of 170% and 340% for the cover and the core, respectively.

V. Antonovich, et al. [70] carry out a review on the use of nano technologies in the manufacture of refractory concretes and some other cementitious materials, examining the influence of nanostructure formation on the bonding material on the properties of refractory concretes. In one case, investigations were carried out using two-component bonding materials (sodium silicate solution mixed with dicalcium silicate) and three-components (sodium silicate solution mixed with dicalcium silicate plus calcium aluminate cement). Cement refractory concretes with mullite aggregates, micro silica, and hybrid and simple deflocculant additives (Castament FS20 polycarboxylate ether and sodium tripolyphosphate) was studied, finding that the three-component bond material hardens as opposed to the two-component material, since one of the bonding components (combination of sodium silicate and dicalcium silicate solution) hardens very quickly and affects the hydration process of the other component, CAC. This has a powerful impact on the entire structure of the material. The application of nanotechnology in the manufacture of refractory concretes has increased the compressive strength three times, from 55 MPa to 165 MPa. CACs can be effectively used as an accelerator for PC hydration for the purpose of layered extrusion. Das, et al. [71] established that CAC mixed with a source of calcium sulfate could be a better alternative than using CAC alone.

5.1.5. CAC + CO_2 Generation

The search with the keywords, calcium aluminate cement and CO_2 generation (CAC + CO_2 generation) in the Scopus database returned five results. The number of documents produced per year is presented in Figure 8e. This reveals almost no production in the last 9 years.

S. Bharati, et al. [72] developed, with a sustainable approach, a low cost mouldable for low temperature applications using steelmaking slag as raw materials. The ladle slag generated after the secondary steelmaking process was used in the 50–70% range as a complete source of CaO to develop a CAC. The slag was used as a 100% replacement for the limestone and partial replacement of the Al_2O_3 source in the cement. In this way, it was possible to eliminate the CO_2 emission associated with the dissociation of the limestone during the clinker manufacturing process. The castable was prepared using primary steel slag as an aggregate and slag cement was developed as a binder. The CA and CA2 phases were detected as primary phases in the slag cement and gehlenite and mayenite as secondary phases.

The slag was used as a 100% replacement for the limestone and a partial replacement for the Al_2O_3 source in the cement. Therefore, it was possible to eliminate the CO_2 emission associated with the dissociation of limestone during the clinker manufacturing process. The castable was prepared using primary steel slag as an aggregate, and slag cement was developed as a binder. CA and CA2 were formed as the primary phases in the slag cement, with the Gehlenite phases. Slag cement with a slag content of 60% exhibited superior strength compared to commercial medium purity CAC. In addition, the crush strengths of 110 °C and 1000 °C were found to be better for casting slag cement than commercial cement. Therefore, it was possible to develop a green mold using by-products from the steel manufacturing process that would not only help reduce the carbon footprint, but also reduce manufacturing costs.

5.1.6. CAC + Sustainability

A search with the keywords, calcium aluminate cement and Sustainability (CAC + Sustainability) in the Scopus database returned 17 results. The number of documents produced per year is presented in Figure 8f. This reveals almost zero production in the last 8 years.

M. Giroudon, et al. [73] analyze the deterioration mechanisms of different binders focusing on the impact of the binding nature on the medium (biochemical composition) during digestion. Binders with a favorable composition for chemically aggressive media were tested: slag cement (CEM III/B), CAC and alkaline metakaolin-based material (MKAA), and a reference binder: OPC (CEM I). Under the explored conditions, biodeterioration mainly led to carbonation of the cement pastes.

Alternative cementitious materials (MCAs) are receiving increasing attention worldwide, but there is a lack of knowledge about the resistance of these materials against the intrusion of harmful ions. Furthermore, current accelerated test methods that measure ion diffusion under an electric field are not reliable when comparing binders with very different pore solution chemistry. M.K. Moradllo and M.T. Law [74] solve these problems by using laboratory transmission X-ray microscopy (TXM) and micro-X-ray fluorescence imaging (μXRF) to perform real-time measurements of ion diffusion in paste samples and mortar for five commercially available ACMs and a OPC. They compared the apparent rate of ion diffusion, quantified the change in the rate of ion diffusion over time, and gave an idea of ionic binding. It was found after 42 days of ion exposure that samples made with calcium aluminate cement had the lowest rate of ion penetration, while samples with activated alkali cement and calcium sulfoaluminate had the highest rate of ion penetration. Portland cement had an ion penetration level that was between these two. Furthermore, both alkali activated and calcium sulfoaluminate samples showed a decrease in penetration rate over time. These measurements are important to quantify the sustainability of MCAs, better justifying their uses where durability is a concern, and guiding to future durability tests for these promising materials.

It is well-known that cement production represents 1.65 billion metric tons of annual global CO_2 emissions [75], making it one of the largest contributors to global CO_2 emissions. One way to reduce CO_2 emissions associated with concrete construction is using alternative cementitious materials and binders (ACMs), such as calcium sulfoaluminate, calcium aluminate, and alkali activated binders. These materials often require lower production temperatures than OPCs and have lower calcium contents, reducing the emissions associated with the CO_2 released by calcium carbonate during calcination. Most ACMs are not new materials, but the past uses have been primarily limited to small-scale applications, such as pavement repairs, and there is little field experience regarding their long-term durability in heavily travelled structures, such as pavements and bridge decks. L.E. Burris, et al. [76] present promising results after the first year, of an effort by the U.S. Department of Transportation.

5.1.7. CAC + Life Cycle Assessment

The search with the keywords, calcium aluminate cement and Life cycle assessment (CAC + Life cycle assessment) in the Scopus database returned four results.

Henry-Lanier, et al. [77] evaluated the environmental footprint for the CAC, calculated according to the guidelines defined in the ISO 14,040 and 14,044 standards for Life Cycle Analysis (LCA). After focusing on the methodology and data used, summarizes the intrinsic carbon footprint and energy consumption of CAC production from the cradle to the factory gate. Other data drawn from the LCA literature and databases are provided in this document to give a comparative view of the data for CAC. This document presents results to compare different refractory systems (monolithic and shaped) used for a selected application, showing the intrinsic opportunity to reduce the carbon footprint of specific refractory applications by using monolithic refractories containing calcium aluminate. The conclusions suggest that the use of LCA can be a powerful tool to optimize the types of refractory products and reduce the overall environmental impact of refractories throughout the life cycle of use.

6. Conclusions

- The products of CAC that are recycled as sanitary waste, among other wastes, serve as an aggregate for Portland cement mortars, improving their mechanical properties and behavior at high temperature.
- CAC maximizes the possibility of recycling glass waste in the manufacture of construction materials. Research on CAC waste in relation to its application within a circular economic structure is incipient.
- The manufacture of CAC mortars using mineral wool waste has shown improved resistance to compression and resulted in the stabilization and neutralization of toxic metals.
- CAC proved to be a good additive to Portland cement for the manufacturing of 3D printing mixes with very good mechanical properties.
- There is great potential in CAC research to help solve global issues, such as the use of CAC in 3D printing, the immobilization of solid waste and the manufacture of new

nanostructured materials, as well as the carbonation of CAC pastes at low and high temperature.

Author Contributions: Conceptualization, H.A.C. and J.F.Z.; methodology, A.A.; validation, J.F.Z., C.F., S.N.M. and H.A.C.; investigation, J.F.Z., S.N.M. and A.A.; writing—original draft preparation, H.A.C., J.F.Z., S.N.M., C.F.; writing—review and editing, A.A. and J.F.Z.; supervision, H.A.C.; project administration, A.A. and C.F.; funding acquisition, J.F.Z., C.F. and H.A.C. All authors have read and agreed to the published version of the manuscript.

Funding: The APC was funded by Military Institute of Engineering—IME.

Data Availability Statement: All data will be available upon request of the reviewers and editors.

Conflicts of Interest: Conflicts of interest/Competing interests (None). Availability of data and material (Total). Consent for publication (All authors agree).

References

1. Abolhasani, A.; Samali, B.; Aslani, F. Physicochemical, mineralogical, and mechanical properties of calcium aluminate cement concrete exposed to elevated temperatures. *Materials* **2021**, *14*, 3855. [CrossRef]
2. Scrivener, K. Calcium aluminate. *Adv. Concr. Technol. Set* **2003**, *1*, 2.
3. Abolhasani, A.; Samali, B.; Dehestani, M.; Libre, N.A. Effect of rice husk ash on mechanical properties, fracture energy, brittleness and aging of calcium aluminate cement concrete. *Structures* **2022**, *36*, 140–152.
4. Zapata, J.F.; Gomez, M.; Colorado, H.A. Structure-property relation and Weibull analysis of calcium aluminate cement pastes. *Mater. Charact.* **2017**, *134*, 9–17. [CrossRef]
5. Zapata, J.F.; Gomez, M.; Colorado, H.A. High Temperature Cracking Damage of Calcium Aluminate Cements. In *Proceedings of the TMS Annual Meeting & Exhibition*; Springer: Berlin/Heidelberg, Germany, 2018; pp. 553–563.
6. Zapata, J.F.; Gomezc, M.; Colorado, H.A. Characterization of two calcium aluminate cement pastes. In *Advances in High Temperature Ceramic Matrix Composites and Materials for Sustainable Development*; Wiley-American Ceramic Society: Hoboken, NJ, USA, 2017; pp. 491–503.
7. Zapata, J.F.; Gomez, M.; Colorado, H.A. Calcium Aluminate Cements Subject to High Temperature. *Adv. Mater. Sci. Environ. Energy Technol. VI* **2017**, *262*, 97.
8. Zapata, J.F.; Gomez, M.; Colorado, H.A. Cracking in Calcium Aluminate Cement Pastes Induced at Different Exposure Temperatures. *J. Mater. Eng. Perform.* **2019**, *28*, 7502–7513. [CrossRef]
9. Ukrainczyk, N.; Matusinovic, T.; Kurajica, S.; Zimmermann, B.; Sipusic, J. Dehydration of a layered double hydroxide—C2AH8. *Thermochim. Acta* **2007**, *464*, 7–15. [CrossRef]
10. Cardoso, F.A.; Innocentini, M.D.M.; Akiyoshi, M.M.; Pandolfelli, V.C. Effect of curing time on the properties of CAC bonded refractory castables. *J. Eur. Ceram. Soc.* **2004**, *24*, 2073–2078. [CrossRef]
11. Capmas, A.; Menetrier-Sorrentino, D.; Damidot, D. Effect of temperature on setting time of calcium aluminate cements. In *Calcium Aluminate Cements*; Taylor & Francis: New York, NY, USA, 1990; pp. 65–80.
12. Bushnell-Watson, S.M.; Sharp, J.H. On the cause of the anomalous setting behaviour with respect to temperature of calcium aluminate cements. *Cem. Concr. Res.* **1990**, *20*, 677–686. [CrossRef]
13. Lee, W.E.; Vieira, W.; Zhang, S.; Ahari, K.G.; Sarpoolaky, H.; Parr, C. Castable refractory concretes. *Int. Mater. Rev.* **2001**, *46*, 145–167. [CrossRef]
14. Barnes, P.; Bensted, J. *Structure and Performance of Cements*; CRC Press: Boca Raton, FL, USA, 2014; ISBN 1482295016.
15. Da Luz, A.P.; Braulio, M.d.A.L.; Pandolfelli, V.C. *Refractory Castable Engineering*; FIRE, Federation for International Refractory Research and Education: Baden-Baden, Germany, 2015; ISBN 3872640046.
16. Mercury, J.M.R.; AZAa, A.; Turrillas, X.; Pena, P. Hidratación de los cementos de aluminatos de calcio (Parte I). *Boletín Soc. Española Cerámica Vidr.* **2003**, *42*, 269–276. [CrossRef]
17. Scrivener, K.L.; Capmas, A. Calcium aluminate cements. In *Lea's Chemistry of Cement and Concrete*; Wiley: Hoboken, NJ, USA, 1998; pp. 713–782.
18. Hewlett, P.; Liska, M. *Lea's Chemistry of Cement and Concrete*; Butterworth-Heinemann: Oxford, UK, 2019; ISBN 978-0-08-100773-0.
19. Midgley, H.G. Quantitative determination of phases in high alumina cement clinkers by X-ray diffraction. *Cem. Concr. Res.* **1976**, *6*, 217–223. [CrossRef]
20. Scrivener, K.L.; Cabiron, J.-L.; Letourneux, R. High-performance concretes from calcium aluminate cements. *Cem. Concr. Res.* **1999**, *29*, 1215–1223. [CrossRef]
21. Bentsen, S. Effect of microsilica on conversion of high alumina cement. In *Calcium Aluminate Cements*; Chapman and Hall: London, UK, 1990; p. 294.
22. Parker, K.M. Refractory calcium aluminate cements. *Trans. Br. Ceram. Soc.* **1982**, *82*, 35–42.
23. Antonovič, V.; Kerienė, J.; Boris, R.; Aleknevičius, M. The effect of temperature on the formation of the hydrated calcium aluminate cement structure. *Procedia Eng.* **2013**, *57*, 99–106. [CrossRef]

24. Zapata, J.F.; Colorado, H.A.; Gomez, M.A. Effect of high temperature and additions of silica on the microstructure and properties of calcium aluminate cement pastes. *J. Sustain. Cem. Mater.* **2020**, 1–27. [CrossRef]

25. Lothenbach, B.; Matschei, T.; Möschner, G.; Glasser, F.P. Thermodynamic modelling of the effect of temperature on the hydration and porosity of Portland cement. *Cem. Concr. Res.* **2008**, *38*, 1–18. [CrossRef]

26. Matschei, T.; Lothenbach, B.; Glasser, F.P. Thermodynamic properties of Portland cement hydrates in the system CaO–Al$_2$O$_3$–SiO$_2$–CaSO$_4$–CaCO$_3$–H$_2$O. *Cem. Concr. Res.* **2007**, *37*, 1379–1410. [CrossRef]

27. Gartner, E. Industrially interesting approaches to "low-CO$_2$" cements. *Cem. Concr. Res.* **2004**, *34*, 1489–1498. [CrossRef]

28. Bizzozero, J. Hydration and Dimensional Stability of Calcium Aluminate Cement Based Systems. Available online: https://infoscience.epfl.ch/record/202031 (accessed on 1 January 2022).

29. Rahimi, M.; Esfahanian, M.; Moradi, M. Effect of reprocessing on shrinkage and mechanical properties of ABS and investigating the proper blend of virgin and recycled ABS in injection molding. *J. Mater. Process. Technol.* **2014**, *214*, 2359–2365. [CrossRef]

30. Prendergast, M.E.; Burdick, J.A. Recent advances in enabling technologies in 3D printing for precision medicine. *Adv. Mater.* **2020**, *32*, 1902516. [CrossRef] [PubMed]

31. Browne, M.P.; Redondo, E.; Pumera, M. 3D printing for electrochemical energy applications. *Chem. Rev.* **2020**, *120*, 2783–2810. [CrossRef] [PubMed]

32. Distler, T.; Boccaccini, A.R. 3D printing of electrically conductive hydrogels for tissue engineering and biosensors–A review. *Acta Biomater.* **2020**, *101*, 1–13. [CrossRef]

33. Stefanowski, J.; Weiss, D. Carrot 2 and language properties in web search results clustering. In Proceedings of the International Atlantic Web Intelligence Conference, Madrid, Spain, 5–6 May 2003; Springer: Berlin, Heidelberg, 2003; pp. 240–249.

34. Weiss, D.; Osinski, S. A concept -driven algorithm for clustering search results. *IEEE Intell. Syst.* **2005**, *20*, 48–54.

35. Walkley, B.; Provis, J.L. Solid-state nuclear magnetic resonance spectroscopy of cements. *Mater. Today Adv.* **2019**, *1*, 100007. [CrossRef]

36. Redaoui, D.; Sahnoune, F.; Heraiz, M.; Raghdi, A. Mechanism and kinetic parameters of the thermal decomposition of gibbsite Al(OH) 3 by thermogravimetric analysis. *Acta Phys. Pol. A* **2017**, *131*, 562–565. [CrossRef]

37. Šesták, J. Ignoring heat inertia impairs accuracy of determination of activation energy in thermal analysis. *Int. J. Chem. Kinet.* **2019**, *51*, 74–80. [CrossRef]

38. Ondro, T.; Húlan, T.; Vitázek, I. Non-isothermal kinetic analysis of the dehydroxylation of kaolinite in dynamic air atmosphere. *Acta Technol. Agric.* **2017**, *20*, 52–56. [CrossRef]

39. Al Rashid, A.; Khan, S.A.; Al-Ghamdi, S.G.; Koç, M. Additive manufacturing: Technology, applications, markets, and opportunities for the built environment. *Autom. Constr.* **2020**, *118*, 103268. [CrossRef]

40. Savolainen, J.; Collan, M. How Additive Manufacturing Technology Changes Business Models?–Review of Literature. *Addit. Manuf.* **2020**, *32*, 101070. [CrossRef]

41. Oliveira, J.P.; Santos, T.G.; Miranda, R.M. Revisiting fundamental welding concepts to improve additive manufacturing: From theory to practice. *Prog. Mater. Sci.* **2020**, *107*, 100590. [CrossRef]

42. Šavija, B.; Luković, M. Carbonation of cement paste: Understanding, challenges, and opportunities. *Constr. Build. Mater.* **2016**, *117*, 285–301. [CrossRef]

43. Ashraf, W. Carbonation of cement-based materials: Challenges and opportunities. *Constr. Build. Mater.* **2016**, *120*, 558–570. [CrossRef]

44. Park, S.M.; Jang, J.G.; Son, H.M.; Lee, H.-K. Stable conversion of metastable hydrates in calcium aluminate cement by early carbonation curing. *J. CO$_2$ Util.* **2017**, *21*, 224–226. [CrossRef]

45. Tonoli, G.H.D.; dos Santos, S.F.; Joaquim, A.P.; Savastano, H., Jr. Effect of accelerated carbonation on cementitious roofing tiles reinforced with lignocellulosic fibre. *Constr. Build. Mater.* **2010**, *24*, 193–201. [CrossRef]

46. Fernández-Carrasco, L.; Rius, J.; Miravitlles, C. Supercritical carbonation of calcium aluminate cement. *Cem. Concr. Res.* **2008**, *38*, 1033–1037. [CrossRef]

47. He, P.; Zhang, B.; Lu, J.-X.; Poon, C.S. ASR expansion of alkali-activated cement glass aggregate mortars. *Constr. Build. Mater.* **2020**, *261*, 119925. [CrossRef]

48. Zhang, B.; He, P.; Poon, C.S. Optimizing the use of recycled glass materials in alkali activated cement (AAC) based mortars. *J. Clean. Prod.* **2020**, *255*, 120228. [CrossRef]

49. Panpa, W.; Jinawath, S.; Kashima, D.P. Ag$_2$O-Ag/CAC/SiO$_2$ composite for visible light photocatalytic degradation of cumene hydroperoxide in water. *J. Mater. Res. Technol.* **2019**, *8*, 5180–5193. [CrossRef]

50. Hossain, S.K.S.; Roy, P.K. Development of waste derived nano-lakargiite bonded high alumina refractory castable for high temperature applications. *Ceram. Int.* **2019**, *45*, 16202–16213. [CrossRef]

51. Al-musawi, H.; Figueiredo, F.P.; Guadagnini, M.; Pilakoutas, K. Shrinkage properties of plain and recycled steel–fibre-reinforced rapid hardening mortars for repairs. *Constr. Build. Mater.* **2019**, *197*, 369–384. [CrossRef]

52. Al-musawi, H.; Figueiredo, F.P.; Bernal, S.A.; Guadagnini, M.; Pilakoutas, K. Performance of rapid hardening recycled clean steel fibre materials. *Constr. Build. Mater.* **2019**, *195*, 483–496. [CrossRef]

53. Yıldırım, S.T.; Baynal, K.; Fidan, O. Internal Curing and Temperature Effect on Lightweight and Heat Insulated Mortar with Recycled Concrete Aggregate. *Acta Phys. Pol. A* **2019**, *135*, 865–869. [CrossRef]

54. Kulu, P.; Goljandin, D.; Külaviir, J.; Hain, T.; Kivisto, M. Recycling of Niobium Slag by Disintegrator Milling. In *Key Engineering Materials*; Trans Tech Publications: Baech, Switzerland, 2019; Volume 799, pp. 97–102.

55. Ogrodnik, P.; Szulej, J.; Franus, W. The Wastes of Sanitary Ceramics as Recycling Aggregate to Special Concretes. *Materials* **2018**, *11*, 1275. [CrossRef] [PubMed]

56. Nematzadeh, M.; Dashti, J.; Ganjavi, B. Optimizing compressive behavior of concrete containing fine recycled refractory brick aggregate together with calcium aluminate cement and polyvinyl alcohol fibers exposed to acidic environment. *Constr. Build. Mater.* **2018**, *164*, 837–849. [CrossRef]

57. Nematzadeh, M.; Baradaran-Nasiri, A. Residual properties of concrete containing recycled refractory brick aggregate at elevated temperatures. *J. Mater. Civ. Eng.* **2018**, *30*, 4017255. [CrossRef]

58. Baradaran-Nasiri, A.; Nematzadeh, M. The effect of elevated temperatures on the mechanical properties of concrete with fine recycled refractory brick aggregate and aluminate cement. *Constr. Build. Mater.* **2017**, *147*, 865–875. [CrossRef]

59. Stonys, R.; Kuznetsov, D.; Krasnikovs, A.; Škamat, J.; Baltakys, K.; Antonovič, V.; Černašėjus, O. Reuse of ultrafine mineral wool production waste in the manufacture of refractory concrete. *J. Environ. Manag.* **2016**, *176*, 149–156. [CrossRef]

60. Navarro-Blasco, Í.; Fernández, J.M.; Duran, A.; Sirera, R.; Alvarez, J.I. A novel use of calcium aluminate cements for recycling waste foundry sand (WFS). *Constr. Build. Mater.* **2013**, *48*, 218–228. [CrossRef]

61. Fernández, L.J.; Ferrer, R.; Aponte, D.F.; Fernandez, P. Recycling silicon solar cell waste in cement-based systems. *Sol. Energy Mater. Sol. cells* **2011**, *95*, 1701–1706. [CrossRef]

62. Chen, L.; Wang, Y.-S.; Wang, L.; Zhang, Y.; Li, J.; Tong, L.; Hu, Q.; Dai, J.-G.; Tsang, D.C.W. Stabilisation/solidification of municipal solid waste incineration fly ash by phosphate-enhanced calcium aluminate cement. *J. Hazard. Mater.* **2021**, *408*, 124404. [CrossRef] [PubMed]

63. Erans, M.; Jeremias, M.; Zheng, L.; Yao, J.G.; Blamey, J.; Manovic, V.; Fennell, P.S.; Anthony, E.J. Pilot testing of enhanced sorbents for calcium looping with cement production. *Appl. Energy* **2018**, *225*, 392–401. [CrossRef]

64. Xu, L.; Wang, P.; Wu, G.; Li, N. Effect of calcium aluminate cement on hydration properties and microstructure of portland cement. *Materials* **2015**, *13*, 4000.

65. Akiti, T.T., Jr.; Constant, K.P.; Doraiswamy, L.K.; Wheelock, T.D. An improved core-in-shell sorbent for desulfurizing hot coal gas. *Adv. Environ. Res.* **2002**, *6*, 419–428. [CrossRef]

66. Akiti, T.T., Jr.; Constant, K.P.; Doraiswamy, L.K.; Wheelock, T.D. Development of an advanced calcium-based sorbent for desulfurizing hot coal gas. *Adv. Environ. Res.* **2001**, *5*, 31–38. [CrossRef]

67. Lv, H.; Xie, M.; Shi, L.; Zhao, H.; Wu, Z.; Li, L.; Li, R.; Liu, F. A novel green process for the synthesis of high-whiteness and ultrafine aluminum hydroxide powder from secondary aluminum dross. *Ceram. Int.* **2022**, *48*, 953–962. [CrossRef]

68. Shakor, P.; Nejadi, S.; Paul, G.; Sanjayan, J. A novel methodology of powder-based cementitious materials in 3D inkjet printing for construction applications. In Proceedings of the Sixth International Conference on Durability of Concrete Structures (ICDCS 2018), Leeds, UK, 18–20 July 2018.

69. Shakor, P.; Sanjayan, J.; Nazari, A.; Nejadi, S. Modified 3D printed powder to cement-based material and mechanical properties of cement scaffold used in 3D printing. *Constr. Build. Mater.* **2017**, *138*, 398–409. [CrossRef]

70. Antonovič, V.; Pundiene, I.; Stonys, R.; Česniene, J.; Keriene, J. A review of the possible applications of nanotechnology in refractory concrete. *J. Civ. Eng. Manag.* **2010**, *16*, 595–602. [CrossRef]

71. Das, A.; Reiter, L.; Mantellato, S.; Flatt, R.J. Blended calcium aluminate cements for digital fabrication with concrete. In Proceedings of the 5th International Conference on Calcium Aluminates (2022), Cambridge, UK, 1–3 June 2020; ETH Zurich-Institute of Building Materials: Zurich, Switzerland, 2022. [CrossRef]

72. Bharati, S.; Sah, R.; Sambandam, M. Green Castable Using Steelmaking Slags: A Sustainable Product for Refractory Applications. *J. Sustain. Metall.* **2020**, *6*, 113–120. [CrossRef]

73. Giroudon, M.; Lavigne, M.P.; Patapy, C.; Bertron, A. Biodeterioration mechanisms and kinetics of SCM and aluminate based cements and AAM in the liquid phase of an anaerobic digestion. *MATEC Web Conf.* **2018**, *199*, 2003.

74. Moradllo, M.K.; Ley, M.T. Comparing ion diffusion in alternative cementitious materials in real time by using non-destructive X-ray imaging. *Cem. Concr. Compos.* **2017**, *82*, 67–79. [CrossRef]

75. Boden, T.; Andres, B.; Marland, G. *Global, Regional, and National Fossil-Fuel CO_2 Emissions (1751–2010)*; OSTI. GOV: Oak Ridge, TN, USA, 2013. [CrossRef]

76. Burris, L.E.; Alapati, P.; Moser, R.D.; Ley, M.T.; Berke, N.; Kurtis, K.E. Alternative cementitious materials: Challenges and opportunities. In Proceedings of the International Workshop on Durability and Sustainability of Concrete Structures, Bologna, Italy, 1–3 October 2015.

77. Henry-Lanier, E.; Szepizdyn, M.; Parr, C. Optimisation of the Environmental Footprint of Calcium-Aluminate-Cement Containing Castables. *Refract. Worldforum* **2015**, *8*, 81–86.

 *sustainability*

Communication

Ubim Fiber (*Geonoma baculífera*): A Less Known Brazilian Amazon Natural Fiber for Engineering Applications

Belayne Zanini Marchi [1,*], Michelle Souza Oliveira [1], Wendell Bruno Almeida Bezerra [1], Talita Gama de Sousa [1], Verônica Scarpini Candido [2], Alisson Clay Rios da Silva [2] and Sergio Neves Monteiro [1]

[1] Department of Materials Science, Military Institute of Engineering—IME, Rio de Janeiro 22290-270, Brazil; oliveirasmichelle@ime.eb.br (M.S.O.); wendellbez@ime.eb.br (W.B.A.B.); talitagama@gmail.com (T.G.d.S.); sergio.neves@ime.eb.br (S.N.M.)

[2] Materials Science and Engineering, Federal University of Para-UFPA, Rodovia BR-316, km 7.5-9.0, Ananindeua 67000-000, Brazil; scarpini@ufpa.br (V.S.C.); alissonrios@ufpa.br (A.C.R.d.S.)

[*] Correspondence: belayne@ime.eb.br; Tel.: +55-2198-217-6050

Citation: Marchi, B.Z.; Oliveira, M.S.; Bezerra, W.B.A.; de Sousa, T.G.; Candido, V.S.; da Silva, A.C.R.; Monteiro, S.N. Ubim Fiber (*Geonoma baculífera*): A Less Known Brazilian Amazon Natural Fiber for Engineering Applications. *Sustainability* 2022, 14, 421. https://doi.org/10.3390/su14010421

Academic Editor: Mostafa Ghasemi Baboli

Received: 30 November 2021
Accepted: 29 December 2021
Published: 31 December 2021

Publisher's Note: MDPI stays neutral with regard to jurisdictional claims in published maps and institutional affiliations.

Abstract: The production of synthetic materials generally uses non-renewable forms of energy, which are highly polluting. This is driving the search for natural materials that offer properties similar to synthetic ones. In particular, the use of natural lignocellulosic fibers (NLFs) has been investigated since the end of 20th century, and is emerging strongly as an alternative to replace synthetic components and reinforce composite materials for engineering applications. NLFs stand out in general as they are biodegradable, non-polluting, have comparatively less CO_2 emission and are more economically viable. Furthermore, they are lighter and cheaper than synthetic fibers, and are a possible replacement as composite reinforcement with similar mechanical properties. In the present work, a less known NLF from the Amazon region, the ubim fiber (*Geonoma bacculifera*), was for the first time physically characterized by X-ray diffraction (XRD). Fiber density was statistically analyzed by the Weibull method. Using both the geometric method and the Archimedes' technique, it was found that ubim fiber has one of the lowest densities, 0.70–0.73 g/cm^3, for NLFs already reported in the literature. Excluding the porosity, however, the absolute density measured by pycnometry was relatively higher. In addition, the crystallinity index, of 83%, microfibril angle, of 7.42–7.49°, and ubim fiber microstructure of lumen and channel pores were also characterized by scanning electron microscopy. These preliminary results indicate a promising application of ubim fiber as eco-friendly reinforcement of civil construction composite material.

Keywords: ubim fiber; natural lignocellulosic fiber; X-ray diffraction; density; morphology characterization

1. Introduction

Composites made of synthetic materials have become a class widely applied in various technological sectors [1]. However, their production is associated with a relatively high consumption of energy and non-degradable wastes. Indeed, synthetic materials not only demand greater manufacturing energy but also produce after service life wastes that are highly polluting and cause serious environmental damage [2]. The search for new materials that are environmentally friendly, sustainable and offer similar mechanical properties at a lower cost, has been arousing the interest of industry and researchers since the end of 20th century, to replace synthetic components by cellulose- based ones [3].

Natural fibers and related fabrics are known to be a renewable resource, and have gained considerable interest in different sectors of the industry, from the production of ropes, handcrafts and textiles, to automobile and construction industries [4–7]. In recent decades, the search for new natural lignocellulosic fibers (NLFs) as reinforcement in composite materials has increased, as reported in several review papers [8–12]. Consequently, these NLFs are currently applied to reinforce polymeric composites, used in construction materials, food packaging, and automotive parts [13–15]. Indeed NLFs such as jute, coir, bamboo, hemp and sisal have been reinforcing polymer composites as eco-friendly civil construction materials for panels, doors, roofing tiles and even low cost buildings [15]. In particular, NLFs are currently being considered as part of multilayered ballistic armor [16] for personnel protection against being shot by a rifle with 7.62 mm ammunition.

NLFs stand out in general as they are biodegradable, non-polluting, have a low energy consumption and are more economically viable. In addition to being lighter than synthetic fibers, they present similar chemical and mechanical properties [7]. The properties of NLFs vary according to their chemical composition, diameter, arrangement of constituents in the fiber, degree of polymerization, crystalline fraction of cellulose, part of the plant organism (stem, leaf, root, seed, among others) and cultivation aspects (age, climatic conditions, degradation processes) [17].

The NLFs extracted from the stem have good quality, mainly due to the presence of crystalline cellulose with a high degree of polymerization [18]. The structure of NLFs is basically composed of two different phases: one crystalline and the other amorphous [3,17,18]. Cellulose is the main chemical constituent of these fibers and is responsible for stiffness. A higher cellulose content in the fiber represents better mechanical properties [19].

Most of the crystalline structure of an NLF is restricted to the cellulose. In fact, cellulose is found not to be uniformly crystalline. However, as indicated by Eichhorn et al. [20], the cellulose ordered regions are extensively distributed throughout the material. According to these authors, the amorphous part of an NLF is composed of hemicellulose, as well as lignin, pectin and waxes. As for the crystalline part, the cellulose consists of a threadlike structure called the microfibril. Although the individual densities of cellulose, hemicellulose and lignin might be ~1.60 g/cm^3 [21–23], the density of an NLF measured by the geometric method (mass/volume) or the Archimedes' scale is found to be lower due to the fiber's porosity. Notably, the cellulose films were reported as having densities of 1.49–1.51 g/cm^3 [24].

Cellulose microfibril, as the most crystalline phase in an NLF, is known as the reinforcement phase incorporated into lignin, which is considered to be an amorphous matrix. The cell wall of a fiber is not a homogeneous membrane: it presents an order of cells composed of distinct walls, with different helicoid angles between the fiber axis and the microfibrils. The strength and ductility of cellulosic fibers depend on the angle, at which these microfibrils are curled in relation to the fiber axis, which is also referred to as the microfibrillar angle (MFA) [25]. The primary wall is the first layer where cellulose fibrils are deposited during cell development. Since fibril orientation is more oblique, MFAs are randomly oriented exhibiting varying degrees of alignment. The primary layer involves a secondary layer, which in turn is composed of three layers S1, S2 and S3. The S1 and S3 layers exhibit microfibrils in a smooth helical orientation, whereas the S2 layer is the thickest on the cell wall and has axial orientation. The orientation of these microfibrils is the main responsible for the fiber's tensile strength, it can be said that a low MFA implies superior mechanical properties [26].

Among the most diverse NLFs with potential application as polymer composite reinforcement, are those from the Amazon region [27]. A fiber extracted from the ubim plant found in the Amazon, Brazil, is not well known, has been scientifically denominated *Geonoma bacculifera*, and is a promising reinforcement for composites. In fact, ubim fiber is not mentioned in a study on Brazilian natural fibers with potential for reinforcing polymer composites used in engineering applications [28]. The ubim fiber is extracted from a plant with multiple stems, which is smooth with elongated leaves and unbranched. The plant has a height varying between one to four meters and diameter from one to three centimeters as well as erect or partially crawling stipe, with seven to twelve leaves, little branched inflorescences and globose or ovoid fruits [29]. Ubim fibers have been used in the north of Brazil in hut covers, boat awnings, and rain protection lining. The ubim stem is used in fishing corral walls [30].

Some basic characteristics show that ubim fiber appears as a promising NLF for reinforcing polymeric matrix composites. Thus, the present study aims to investigate the ubim fiber through density, X-ray diffraction (XRD), microfibrillar angle, crystallinity index and scanning electron microscopy. These characteristics may be relevant for composites manufacturing, evaluating their potential for future applications as low cost civil construction materials.

2. Materials and Methods

The ubim fiber (Figure 1a) was extracted from the stem of the plant, scientifically called *Geonoma baculifera*, purchased in Belém do Pará, Brazil. The stem of the ubim plant was divided into longitudinal splints, which are then manually separated into finer fibers.

Figure 1. Ubim (**a**) plant, (**b**) mechanically divided splints form the stalk, and (**c**) bunch of final separated and isolated fibers.

Figure 1 illustrates the ubim plant (a), as well as the split splints of the stalk (b) and final split fibers (c).

The characterization of the ubim fiber, in terms of density, microfibrillar angle and degree of crystallinity was preliminarily evaluated. The fiber was shredded and subsequently measured for the dimensions and length of the fiber cross-section.

2.1. Characteristic Density

2.1.1. Geometric Method

To obtain the average diameter, each fiber was divided into five distinct points uniformly separated, and then rotated by 90°. A new reading was carried out at the same five previous points. A total of 3 measurements were taken at each point, thus obtaining a statistical average. The equipment used was an Olympus microscope, model BX53M with 5x magnification.

The length of each fiber was measured using a 0.01 mm precision caliper, and its mass was weighed on a Gehaka electronic scale, model AG-200, with 0.0001 g precision. The fiber cross-section was considered to be ellipsoidal in shape, and its approximate area calculated according to Equation (1), where a and b represent the major and minor axes of the ellipsoid, respectively:

$$A = \frac{\pi ab}{4} \tag{1}$$

Density results for the five intervals were statistically evaluated using Weibull Analysis (WA) software. This statistical analysis is based on the cumulative Weibull frequency distribution function (F(x)).

$$F(x) = exp\left[-\left(\frac{x}{\theta}\right)^{\beta}\right] \tag{2}$$

where x is measured density, β is the Weibull modulus and θ is the scaling parameter. Weibull modulus (β) is associated with data uniformity, θ has precision adjustment given by R^2. Equation (2) can be modified to fit a linear graph:

$$\ln\left[\ln\left(\frac{1}{F(x)}\right)\right] = \beta \ln x - (\beta \ln \theta) \tag{3}$$

2.1.2. Archimedes' Principle Method

In addition to the density calculated by geometric method, a density evaluation was also carried out using Archimedes' principle (buoyancy). In the Archimedes' test, fiber samples were weighed in air and then subjected to an immersion liquid with a density lower than that of the sample, according to ASTM D3800 [31]. Gasoline was used as the immersion liquid. Density was calculated using the formula indicated in the Equation (4):

$$\rho = \frac{(M_3 - M_1)}{((M_3 - M_1) - (M_4 - M_2))}\rho_l \tag{4}$$

where ρ_l is the density of liquid, as well as M_1 being the weight of the suspension wires in air, M_2 is the weight of the suspension wire in liquid, M_3 is the weight of the suspension wire with fiber in air and M_4 is the weight of the suspension wire with fiber in liquid.

2.1.3. Gas Pycnometry Density

The gas pycnometry was performed by Accupyc 1330 brand Micromeritics, following the procedures established by the ASTM D4892 [32].

2.2. XRD Analysis

X-ray diffraction (XRD) analysis was performed using a PANaytical diffractometer model XPert Pro MRD system, as well as a Pixcel detector and cobald anode with Co K radiation (0.1789 nm). XRD is an effective method of quantifying the cellulose present in the fiber in both its crystalline and amorphous form.

2.2.1. Microfibrillar Angle (MFA)

The microfibril angle (MFA) was determined in association with the cellulose peak (0 0 2), according to the methodology described by Cave [33]. Cave's methodology is the most accurate method for calculating the MFA. Based on XRD, it suggests that the estimation of MFA occurs from the relationship between three curves obtained from the peak associated with the highest intensity, and namely it is Gaussian along with the first-order and second-order derivatives. Through this analysis, a "T" parameter is obtained and applied in the following polynomial Equation (5).

$$MFA = -12.19 * T^3 + 113.67 * T^2 - 348.40 * T + 358.09 \tag{5}$$

2.2.2. Crystallinity Index (CI)

The calculation of the crystallinity index (CI) by XRD was applied using the method described by Segal et al. [34]. The maximum intensity of the peaks (1 0 1) and (0 0 2) was used. These peaks are associated with the amorphous and crystalline phases of the fibers. The CI value was determined using Equation (6) where, I_1 is the intensity of the diffraction minimum (amorphous region) and I_2 is the intensity of the diffraction maximum (crystalline region).

$$CI = 1 - \frac{I_1}{I_2} * 100 \tag{6}$$

2.3. SEM Analysis

The ubim fiber surface morphology was analyzed by scanning electron microscopy (SEM) in a model Quanta FEG 250 Fei microscope (Hillsboro, OR, USA) operating with secondary electrons at 5 kV.

3. Results

3.1. Density Measurements

3.1.1. Geometric Method

The variation in the cross sections, given by the mean values of *a* and *b*, is shown in the histogram of Figure 2, with percentage frequency related to five arbitrary intervals.

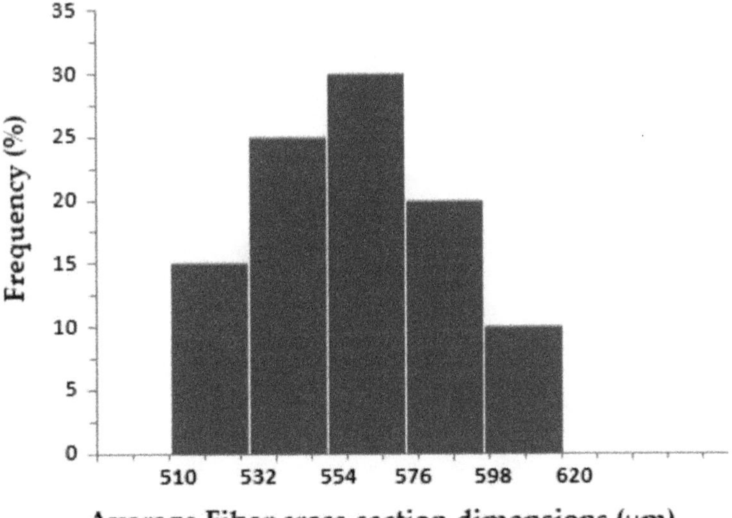

Figure 2. Frequency distribution of the average cross-section dimensions of ubim fibers.

The mean frequency of the cross-sectional dimensions for the investigated ubim fibers (Figure 2) reveals an almost normal distribution within the range of 510 to 620 μm. For each interval, the mean density was measured by dividing the mass by the volume before being statistically analyzed by the Weibull method. Table 1 presents the mean ubim fiber density values for each interval of the mean dimensions of the cross-sections shown in Figure 2. In this table, a trend of greater density for thinner fibers should be noted. Weibull statistical analysis was performed for densities measured by the geometric method. Table 1 presents the Weibull parameters provided by the WA software associated with the shape parameter, scale parameter and precision fit (R^2). The quality of these measurements can be evaluated using the R^2 parameter; the closer to 1, the more reliable the data obtained.

In Table 1 it is important to note that thinner ubim fibers, with lower diameters, display significantly higher density, while the thicker ones are much less dense. As compared to the cellulose, hemicellulose and lignin densities, ~1.6 g/cm^3 [21,23], the ubim fibers revealed an amount of porosity that increases markedly with the fiber diameter.

Table 1. Density of ubim fibers calculated for different intervals of cross-section dimension.

Diameter Range (μm)	ρ (g/cm^3)	Standard Deviation	B	θ	R^2
510–532	0.955	0.06	11.800	0.990	1.000
532–554	0.815	0.02	23.900	0.831	0.960
554–576	0.700	0.03	21.168	0.716	0.990
576–598	0.607	0.03	16.762	0.623	1.000
598–620	0.440	0.02	24.171	0.676	1.000
ρ_{mean}	0.703				

3.1.2. Archimedes' Technique

The density value was also evaluated by the Archimedes' technique and found as 0.73 ± 0.02 g/cm^3. In both cases, ubim fiber density values (0.70 g/cm^3 for geometric and 0.73 g/cm^3 for Archimedes' technique) are among the lowest reported for NLFs and most commonly applied in reinforcing polymer composites such as flax, jute, sisal and cotton [7].

Using the Archimedes' technique it was not possible to precisely identify the difference in densities between thinner and thicker ubim fibers. This was due to the eventual penetration of gasoline (immersion liquid) into the fiber's open porosity. Nevertheless, the obtained value of 0.73 g/cm^3 is close to that found by the geometric method, 0.70 g/cm^3, which in practice attests the mean value of ubim fiber density of 0.72 g/cm^3.

3.1.3. Gas Pycnometry Results

The absolute density obtained by gas pycnometry of the ubim fibers was 1.86 ± 0.26 g/cm^3. This relatively higher density found by the pycnometer is due to the ability of the helium gas to fill all the open pores of the material. The gaseous pycnometry provides the value of the absolute density, including all empty parts of the NLF, and as it is very porous, this higher value is justified. This higher density result by pycnometry compared to other methods was also found by Oliveira et al. [35], as 1.61 g/cm^3.

Higher values for the absolute density, such as that of ubim fiber obtained by pycnometry of the order of ~1.6 g/cm^3, are justified by the density of plain cellulose [21–23]. Most reported densities of NLFs [36,37], including cellulose films [24], are ~1.50 and used for technological application such as polymer composite reinforcement.

The significant amount of porosity in the ubim fiber is responsible for a lower apparent density, 0.72 g/cm^3, which is good for a lighter composite reinforcement in spite of its relatively higher absolute density of 1.86 ± 0.26 g/cm^3.

3.2. XRD Results

Figure 3 shows XRD patterns of samples associated with different related ubim fiber conditions: (a) vertical stalk (VS); (b) vertical fiber (VF); (c) horizontal stalk (HS); and (d) horizontal fiber (HF). It should be noted that in both conditions (a) and (b), the peak associated with (2 0 0) is the prominent one. This peak is related to the cellulose crystalline part of the fiber and was the peak used to assess MFA [38]. As for condition (c) and (d), the same pattern is not repeated and it was not possible to calculate the corresponding MFA.

A preliminary evaluation of the percentage of crystalline cellulose by Rietveld revealed values of around 66% for conditions (a) and (b).

3.2.1. MFA Results

The calculated MFA for the VS and VF corresponded to 7.49° and 7.42°, respectively. An average value of 7.46° can be considered similar to some MFA values reported for other NLFs such as ramie and hemp [3,28,39,40].

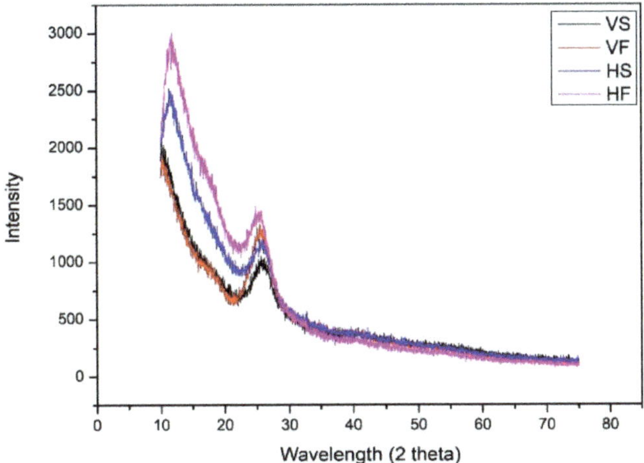

Figure 3. XRD patterns of ubim fiber-related samples vertical stalk, vertical fiber, horizontal stalk and horizontal fiber.

3.2.2. CI Results

The degree of crystallinity of cellulose is an important structural parameter of NLFs. The stiffness of plant fibers is mainly due to the relatively large amount of crystalline regions in the cellulose. Figure 4 shows the convoluted peaks obtained from background free XRD patterns to calculate the crystallinity index (CI) according to Equation (6). In this figure the crystallographic planes (2 0 0) and (1 1 0) are associated with the crystalline and amorphous phases of the ubim stalk (VS) and fiber (VF), respectively. As for VS in Figure 4a, the crystalline peak at $2\theta = 26.3°$ has an intensity $I_2 = 402.1$ a.u, while the amorphous peak at $2\theta = 11.8°$ with $I_1 = 149.7$ a.u allowing to obtain a CI = 63% by Equation (6). Similarly for VF, Figure 4b revealed a crystalline peak at $2\theta = 25.6°$ with $I_2 = 694.5$ a.u and an amorphous peak at $2\theta = 10.4°$ with $I_1 = 116.6$ a.u, giving CI = 83%, which will be considered the proper one for ubim fiber.

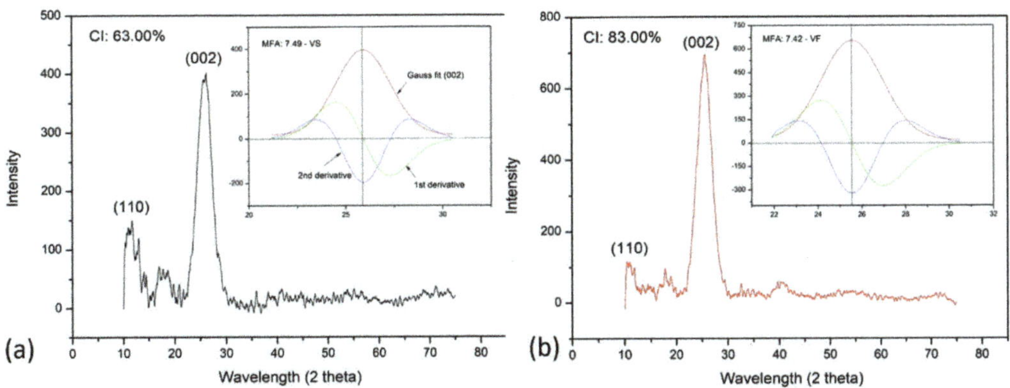

Figure 4. XRD patterns of ubim fiber-related samples of: (**a**) vertical stalk (VS) and (**b**) vertical fiber (VF).

These CIs are within the range of values reported for several common NLFs such as sisal [41], hemp [40], jute [42,43], flax [44], and coir [28], used as reinforcement in polymer composites, which are shown in Table 2. In addition, this table also presents the density, cellulose content, crystallinity index and microfibrillar angle of less known NLFs such as curauá [43,45], banana [7,37,46], pineapple [7,37,38], bamboo [7,37,38] and sugarcane bagasse [7,37,38]. Also included in Table 2 were cotton [7,36,46] and kenaf [37,46], with relatively higher densities and CI.

Table 2. Density, cellulose content, crystallinity index and microfibrillar angle of ubim fibers as compared to other NLFs applied as polymer composite reinforcement.

NLFs	Density (g/cm^3)	Cellulose (%)	Crystallinity Index (%)	Microfibrillar Angle (o)	Reference
Ubim	0.70	66	83.0	7.46	PW
Ramie	1.50	85	75.0	7.50	[17,28,39]
Hemp	1.50	72	87.8	7.50	[7,40]
Sisal	1.50	75	72.2	20.0	[7,17,28,41]
Coir (coconut)	1.40	53	44.0	51.0	[7,28]
Curaua	0.92	71	75.6	18.8	[43,45]
Jute	1.45	60	71.3	8.00	[7,40,42,43]
Flax	1.38	71	63.1	10.0	[7,44]
Banana	1.50	65	39	11.0	[7,37,46]
Pineapple	1.60	83	38	14.0	[7,37,38]
Bamboo	1.21	45	60	6.0	[7,37,38]
Sugarcane Bagasse	0.49	69	45	14.5	[7,37,38]
Kenaf	1.50	72	72	4.1	[37,46]
Cotton	1.60	51	65	25	[7,36,46]

PW: Present Work.

A close look at the values of density, cellulose content, IC and MFA in Table 2 fails to reveal any direct relationship between the density with either the cellulose content or the CI for the presented NLFs. One possible reason is that the reported densities, like the case of ubim fiber in Table 2, are the apparent ones measured by geometric or Archimedes' methods. Indeed, the apparent density considers the amount of porosity, which is strongly influenced by the internal fiber microstructure (lumen and channels), and the characteristics of the fiber species. On the other hand, cellulose content, IC and MFA correspond to another structural hierarchy, which is more associated with the fiber strength. As such, in several cases a lower cellulose content might be related to higher MFA and weaker fiber like in the coir, banana and sugarcane bagasse fibers. To the knowledge of the authors these relationships have never been fully investigated and are the subject of future research.

3.2.3. SEM Results

The SEM analysis disclosed the morphologies of the cross-section and surface of the ubim fiber as shown in Figure 5. The analyzed characteristics reveal the existence of an irregular rough surface, which can contribute to the mechanical interlock between the fiber and the matrix. In Figure 5a,b, one should notice this heterogeneous surface.

The measurement of the diameter of a random fiber is also shown in Figure 5a at a magnification (300×). Figure 5c–e show, with higher magnification, details of the cross section of the ubim fiber.

One should see in Figure 5f a polygonal lumen channel, which is associated with a greater amount of porosity. These characteristics are similar to those observed by other authors [47,48], such as rough surface, defects, flaws and irregularities along the length of the fiber.

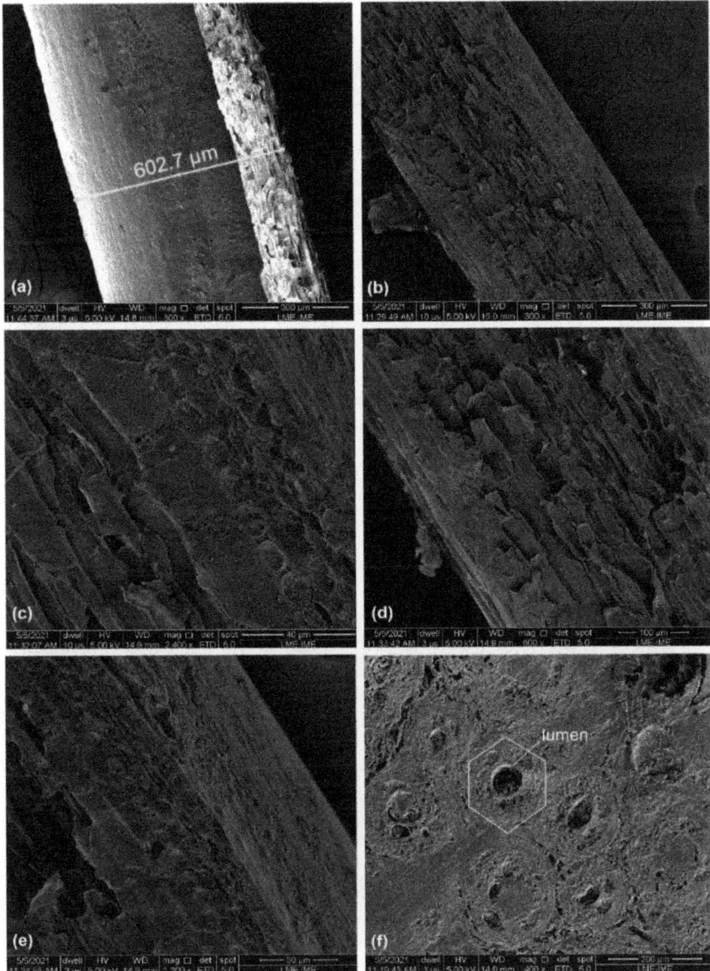

Figure 5. Surface morphology of ubim fiber: (**a**) fiber cross-section; with increasing magnification: (**b**) 300x; (**c**) 2400x; (**d**) 600x; (**e**) 1,200x and (**f**) polygonal lumen channels.

4. Summary and Conclusions

A preliminary investigation on the physical properties of the ubim fibers extracted from the stem of the Brazilian Amazon plant, scientifically known as *Geonoma baculifera*, revealed characteristics supporting a possible reinforcement of novel epoxy composites;

- Densities measured by the geometric method from 0.955 to 0.440 g/cm^3 and analyzed by the Weibull statistical method revealed a tendency to decrease with increasing fiber cross-section dimensions, from 510 to 620 μm. After using the Archimedes' technique, the density showed a mean value of 0.73 g/cm^3, quite similar to the 0.70 g/cm^3 that was found after using the geometric method, which are among the lowest reported so far for natural lignocellulosic fibers. By the gas pycnometry method the density was found as 1.86 ± 0.26 g/cm^3 due to the exclusion of the porosity by the helium gas, but consistent with other NLF and pure cellulose with ~1.60 g/cm^3.

- Microfibril angles from 7.42 to 7.49° were found by X-ray diffraction for different conditions of ubim stalk and fiber. A crystallinity index of 83% was evaluated by XRD in ubim fiber powder. This CI is within the range of reported values for several NLFs.
- These preliminary physical and microstructural characterizations performed for the ubim fiber, in association with a lower cost, indicate a promising application as reinforcement of polymer composites to be used as eco-friendly civil construction materials.

Author Contributions: Conceptualization, S.N.M. and B.Z.M.; methodology, B.Z.M., M.S.O., W.B.A.B., V.S.C.; formal analysis, B.Z.M. and M.S.O.; resources, B.Z.M. and M.S.O.; data curation, M.S.O. and W.B.A.B.; writing—original draft preparation, B.Z.M. and M.S.O.; writing—review and editing, B.Z.M., M.S.O., W.B.A.B., T.G.d.S., A.C.R.d.S. and V.S.C.; supervision, S.N.M.; project administration, S.N.M. All authors have read and agreed to the published version of the manuscript.

Funding: The authors acknowledge the support to this investigation by the brazilian agencies: Conselho Nacional de Desenvolvimento, CNPq; Coordenação de Aperfeiçoamento de Pessoal de Nível Superior, CAPES; and Fundação de Amparo à Pesquisa do Estado do Rio de Janeiro, FAPERJ (process E-26/202.286/2018).

Institutional Review Board Statement: Not applicable.

Informed Consent Statement: Not applicable.

Conflicts of Interest: The authors declare no conflict of interest.

References

1. Chawla, K.K. *Composite Materials: Science and Engineering*, 3rd ed.; Springer Science & Business Media: New York, NY, USA, 2012. [CrossRef]
2. Amcor. Sustainability Review. 2018. Available online: https://www.amcor.com/sustainability/reports (accessed on 10 November 2021).
3. Bledzki, A.K.; Gassan, J. Composites reinforced with cellulose based fibres. *Prog. Polym. Sci.* **1999**, *24*, 221–274. [CrossRef]
4. Li, M.; Pu, Y.; Thomas, V.M.; Yoo, C.G.; Ozcan, S.; Deng, Y.; Nelson, K.; Ragauskas, A.J. Recent advancements of plant-based natural fiber–reinforced composites and their applications. *Compos. Part B Eng.* **2020**, *200*, 108254. [CrossRef]
5. Mohammed, L.; Ansari, M.N.; Pua, G.; Jawaid, M.; Islam, M.S. A Review on natural fiber reinforced polymer composite and its applications. *Int. J. Polym. Sci.* **2015**, *15*, 15. [CrossRef]
6. Shah, D.U. Developing plant fibre composites for structural applications by optimising composite parameters: A critical review. *J. Mater. Sci.* **2013**, *48*, 6083–6107. [CrossRef]
7. Monteiro, S.N.; Lopes, F.P.D.; Barbosa, A.P.; Bevitori, A.B.; Da Silva, I.L.A.; Da Costa, L.L. Natural lignocellulosic fibers as engineering materials: An overview. *Met. Mater. Trans. A* **2011**, *42*, 2963. [CrossRef]
8. Aisyah, H.A.; Paridah, M.T.; Sapuan, S.M.; Ilyas, R.A.; Khalina, A.; Nurazzi, N.M.; Lee, S.H.; Lee, C.H. A comprehensive review on advanced sustainable woven natural fibre polymer composites. *Polymers* **2021**, *13*, 471. [CrossRef]
9. Karimah, A.; Ridho, M.R.; Munawar, S.S.; Adi, D.S.; Damayanti, R.; Subiyanto, B.; Fatriasari, W.; Fudholi, A. A review on natural fibers for development of eco-friendly bio-composite: Characteristics, and utilizations. *J. Mater. Res. Technol.* **2021**, *13*, 2442–2458. [CrossRef]
10. Vigneshwaran, S.; Sundarakannan, R.; John, K.M.; Johnson, R.D.J.; Prasath, K.A.; Ajith, S.; Arumugaprabu, V.; Uthayakumar, M. Recent advancement in the natural fiber polymer composites: A comprehensive review. *J. Clean. Prod.* **2020**, *277*, 124109. [CrossRef]
11. Zhang, Z.; Cai, S.; Li, Y.; Wang, Z.; Long, Y.; Yu, T.; Shen, Y. High performances of plant fiber reinforced composites—A new insight from hierarchical microstructures. *Compos. Sci. Technol.* **2020**, *194*, 108151. [CrossRef]
12. da Luz, F.S.; Garcia Filho, F.D.C.; Del-Rio, M.T.G.; Nascimento, L.F.C.; Pinheiro, W.A.; Monteiro, S.N. Graphene-incorporated natural fiber polymer composites: A first overview. *Polymers* **2020**, *12*, 1601. [CrossRef]
13. Holbery, J.; Houston, D. Natural-fiber-reinforced polymer composites in automotive applications. *JOM* **2006**, *58*, 80–86. [CrossRef]
14. Youssef, A.M.; El-Sayed, S.M. Bionanocomposites materials for food packaging applications: Concepts and future outlook. *Carbohydr. Polym.* **2018**, *193*, 19–27. [CrossRef] [PubMed]
15. Singh, B.; Gupta, M.; Tarannum, H.; Randhawa, A. Natural fiber-based composite building materials. In *Cellulose Fibers: Bio- and Nano-Polymer Composites: Green Chemistry Technology*; Kalia, S., Kaith, B.S., Kaur, I., Eds.; Springer: Berlin/Heidelberg, Germany, 2011. [CrossRef]
16. Oliveira, M.S.; Luz, F.S.D.; Lopera, H.A.C.; Nascimento, L.F.C.; Garcia Filho, F.D.C.; Monteiro, S.N. Energy absorption and limit velocity of epoxy composites incorporated with fique fabric as ballistic armor—A brief report. *Polymers* **2021**, *13*, 2727. [CrossRef] [PubMed]
17. Faruk, O.; Bledzki, A.K.; Fink, H.P.; Sain, M. Biocomposites reinforced with natural fibers: 2000–2010. *Prog. Polym. Sci.* **2012**, *37*, 1552–1596. [CrossRef]

18.	Venugopal, A.; Boominathan, S.K. Physico-chemical, thermal and tensile properties of alkali-treated acacia concinna fiber. *J. Nat. Fibers* **2020**, 1–16. [CrossRef]
19.	Jeyapragash, R.; Srinivasan, V.; Sathiyamurthy, S.J.M.T.P. Mechanical properties of natural fiber/particulate reinforced epoxy composites–A review of the literature. *Mater. Today Proc.* **2020**, *22*, 1223–1227. [CrossRef]
20.	Eichhorn, S.J.; Baillie, C.A.; Zafeiropoulos, N.; Mwaikambo, L.Y.; Ansell, M.P.; Dufresne, A.; Entwistle, K.M.; Herrera-Franco, P.J.; Escamilla, G.C.; Groom, L.; et al. Review: Current international research into cellulosic fibres and composites. *J. Mater. Sci.* **2001**, *36*, 2107–2131. [CrossRef]
21.	Ehrnrooth, E.M. Change in pulp fibre density with acid-chlorite delignification. *J. Wood Chem. Technol.* **2015**, *4*, 91–109. [CrossRef]
22.	Mwaikambo, L.Y.; Ansell, M.P. The determination of porosity and cellulose content of plant fibers by density methods. *J. Mater. Sci. Lett.* **2001**, *20*, 2095–2096. [CrossRef]
23.	Qiu, J.; Khalloufi, S.; Martynenko, A.; Van Dalen, G.; Schutyser, M.; Almeida-Rivera, C. Porosity, bulk density, and volume reduction during drying: Review of measurement methods and coefficient determinations. *Dry. Technol.* **2015**, *33*, 1681–1699. [CrossRef]
24.	Makarov, I.S.; Golova, L.K.; Kuznetsova, L.K.; Antonov, S.V.; Kotsyuk, A.V.; Ignatenko, V.Y.; Kulichikhin, V.G. Influence of Precipitation and Conditioning Baths on the Structure, Morphology, and Properties of Cellulose Films. *Fibre Chem.* **2016**, *48*, 298–305. [CrossRef]
25.	Donaldson, L. Microfibril angle: Measurement, variation and relationships—A review. *Iawa J.* **2008**, *29*, 345–386. [CrossRef]
26.	Meylan, B.A. The influence of microfibril angle on the longitudinal shrinkage-moisture content relationship. *Wood Sci. Technol.* **1972**, *6*, 293–301. [CrossRef]
27.	Marinelli, A.L.; Monteiro, M.R.; Ambrósio, J.D.; Branciforti, M.C.; Kobayashi, M.; Nobre, A.D. Development of poly-meric composites with natural plant fibers from biodiversity: A contribution to Amazon sustainability. *Polimeros* **2008**, *18*, 92–99. [CrossRef]
28.	Satyanarayana, K.G.; Guimarães, J.L.; Wypych, F. Studies on lignocellulosic fibers of Brazil. Part I: Source, production, morphology, properties and applications. *Compos. Part A* **2007**, *38*, 1694–1709. [CrossRef]
29.	Henderson, A.; Galeano, G.; Bernal, R. *Field Guide to the Palms of the Americas*; Princeton University Press: Princeton, NJ, USA, 1995; ISBN 978-0-6916-5612-0.
30.	Oliveira, J.; Potiguara, R.C.V.; Lobato, L.C.B. Vegetal fibers used in artisan fishing in the Salgado region, Pará (In Portu-Guese). Bulletin of the Museu Paraense Emílio Goeldi. *Hum. Sci.* **2006**, *1*, 113–127. [CrossRef]
31.	American Society for Testing and Materials—ASTM. Standard Test Method for Density of High Modulus Fiber. In *D3800-99*; ASTM International: West Conshohoken, PA, USA, 2010.
32.	American Society for Testing and Materials—ASTM. Standard Test Method for Density of Solid Pitch (Helium Pycnometer Method). In *D4892-14*; ASTM International: West Conshohocken, PA, USA, 2019.
33.	Cave, I.D. X-ray measurement of microfibril angle. *For. Prod. J.* **1966**, *16*, 37–42.
34.	Segal, L.E.A. An empirical method for estimating the degree of crystallinity of native cellulose using the x-ray diffractometer. *Text. Res. J.* **1959**, *29*, 786–794. [CrossRef]
35.	Oliveira, M.S.; Luz, F.S.d.; Teixeira Souza, A.; Demosthenes, L.C.d.C.; Pereira, A.C.; Filho, F.d.C.G.; Braga, F.d.O.; Figueiredo, A.B.-H.d.S.; Monteiro, S.N. Tucum fiber from amazon astrocaryum vulgare palm tree: Novel reinforcement for polymer composites. *Polymers* **2020**, *12*, 2259. [CrossRef] [PubMed]
36.	Sanjay, M.R.; Madhu, P.; Jawaid, M.; Senthamaraikannan, P.; Senthil, S.; Pradeep, S. Characterization and properties of natural fiber polymer composites: A comprehensive review. *J. Clean. Prod.* **2018**, *172*, 566–581. [CrossRef]
37.	Pereira, P.H.F.; Rosa, M.D.F.; Cioffi, M.O.H.; Benini, K.C.C.D.C.; Milanese, A.C.; Voorwald, H.J.C.; Mulinari, D.R. Vegetal fibers in polymeric composites: A review. *Polimeros* **2015**, *25*, 9–22. [CrossRef]
38.	Rekha, B.; Nagaraja Ganesh, B. X-ray Diffraction: An Efficient Method to Determine Microfibrillar Angle of Dry and Matured Cellulosic Fibers. *J. Nat. Fibers* **2020**, 1–8. [CrossRef]
39.	Zhou, L.M.; Yeung, K.W.; Yuen, C.W.M.; Zhou, X. Characterization of ramie yarn treated with sodium hydroxide and crosslinked by 1,2,3,4-butanetetracarboxylic acid. *J. Appl. Polym. Sci.* **2004**, *91*, 1857–1864. [CrossRef]
40.	Mwaikambo, L.Y.; Ansell, M.P. Chemical modification of hemp, sisal, jute, and kapok fibers by alkalization. *J. Appl. Polym. Sci.* **2002**, *84*, 2222–2234. [CrossRef]
41.	Gopi Krishna, M.; Kailasanathan, C.; Nagarajaganesh, B. Physico-chemical and morphological characterization of cellulose fibers extracted from *Sansevieria roxburghiana schult. & Schult. F* Leaves. *J. Nat. Fibers* **2020**, *21*, 1–17. [CrossRef]
42.	Raghavendra, G.; Ojha, S.; Acharya, S.K.; Pal, S.K. Jute fiber reinforced epoxy composites and comparison with the glass and neat epoxy composites. *J. Compos. Mater.* **2014**, *48*, 2537–2547. [CrossRef]
43.	Omrani, E.; Menezes, P.L.; Rohatgi, P.K. State of art on tribological behavior of polymer matrix composites reinforced with natural fibers in the green materials world. *Eng. Sci. Technol. Int. J.* **2015**, *19*, 717–736. [CrossRef]
44.	Baley, C.; Bourmaud, A. Average tensile properties of French elementary flax fibers. *Mater. Lett.* **2014**, *122*, 159–161. [CrossRef]
45.	Costa, U.O.; Nascimento, L.F.C.; Garcia, J.M.; Bezerra, W.B.A.; da Luz, F.S.; Pinheiro, W.A.; Monteiro, S.N. Mechanical properties of composites with graphene oxide functionalization of either epoxy matrix or curaua fiber reinforcement. *J. Mater. Res. Technol.* **2020**, *9*, 13390–13401. [CrossRef]

46. Zhan, J.; Li, J.; Wang, G.; Guan, Y.; Zhao, G.; Lin, J.; Coutellier, D. Review on the performances, foaming and injection molding simulation of natural fiber composites. *Polym. Compos.* **2021**, *42*, 1305–1324. [CrossRef]
47. John, M.J.; Thomas, S. Biofibers and Biocomposites. *Carbohydr. Polym.* **2008**, *71*, 343–364. [CrossRef]
48. Junior, H.L.O.; Moraes, A.G.O.; Poletto, M.; Zattera, A.J.; Amico, S.C. Chemical composition, tensile properties and struc-tural characterization of buriti fiber. *Cellul. Chem. Technol.* **2016**, *50*, 15–22. Available online: https://cellulosechemtechnol.ro/pdf/CCT1(2016)/p.15-22.pdf (accessed on 29 November 2021).

 *sustainability*

Article

Composite Soil Made of Rubber Fibers from Waste Tires, Blended Sugar Cane Molasses, and Kaolin Clay

Juan E. Jiménez [1], Carlos Mauricio Fontes Vieira [2] and Henry A. Colorado [1,*]

[1] Composites Laboratory, Universidad de Antioquia UdeA, Medellín 050010, Colombia; juan.jimenezh@udea.edu.co
[2] Materials Engineering, Faculty of Engineering, Universidade Estadual do Norte Fluminense, Rio de Janeiro 28013-602, Brazil; vieira@uenf.br
* Correspondence: henry.colorado@udea.edu.co

Abstract: The use of different chemical and biological admixtures to improve the ground conditions has been a common practice in geotechnical engineering for decades. The use of waste material in these mixtures has received increasing attention in the recent years. This investigation evaluates the effects of using recycled tire polymer fibers (RTPF) and sugar molasses mixed with kaolin clay on the engineering properties of the soil. RTPF were obtained from a tire recycling company, while the molasses were extracted from a sugar cane manufacturer, both located in Colombia. RTPF is a waste and therefore its utilization is the first positive impact of this research, a green solution for this byproduct. Treated kaolin clay is widely used in many industrial processes, such as concrete, paper, paint, and traditional ceramics. The characterization was conducted with scanning electron microscopy, compression strength, particle-size distribution, x-ray diffraction, compressive and density tests. The results showed that the unconfined compressive strength improved from about 1.42 MPa for unstabilized samples, to 2.04 MPa for samples with 0.1 wt% of fibers, and 2.0 wt% molasses with respect to the dry weight of the soil. Furthermore, it was observed that soil microorganisms developed in some of the samples due to the organic nature of the molasses.

Keywords: waste tires; molasses; soil improving

Citation: Jiménez, J.E.; Fontes Vieira, C.M.; Colorado, H.A. Composite Soil Made of Rubber Fibers from Waste Tires, Blended Sugar Cane Molasses, and Kaolin Clay. *Sustainability* **2022**, *14*, 2239. https://doi.org/10.3390/su14042239

Academic Editor: Syed Minhaj Saleem Kazmi

Received: 18 January 2022
Accepted: 11 February 2022
Published: 16 February 2022

Publisher's Note: MDPI stays neutral with regard to jurisdictional claims in published maps and institutional affiliations.

1. Introduction

The accumulation of end-of-life tires has become a worldwide problem with dramatic consequences for sustainability and with a trend to worsen due to the increasing demand of cars, with 1.4 billion tires produced annually across the world, which is equivalent to an amount of 17,000,000 tons/year of tires used [1]. In 2017 in the United States, 287.3 million tires were disposed of, corresponding to 4.7 million tons [2]. In Colombia, an estimated 61,000 tons of tires are annually disposed of [3], from which 18,861 tons are regenerated yearly in the capital city, Bogota [4]. In this city, 15,880 tons of tires were processed and reutilized between 2016 and 2018 [5].

Due to the high stresses in car tires under normal conditions, they are fabricated from rubber matrix reinforced by textile fibers and steel wires. Combining textile fibers and steel wires make tires resistant to biodegradation and give them high strength properties [6]. All these materials can be used after the tire's life cycle. Sienkiewicz et al. [1] summarized five procedures to handle discarded tires: pyrolysis, retreading, product recycling, energy recovery, and material recycling. Retreading deals with the substitution of used parts with new ones [7], although there are serious safety concerns for the vehicle when this is used [8]. Energy process is one of the most common applications for utilizing the used tires as energy source [1]: with a calorific capacity value of 32 MJ/kg, used tires compete with fuel [9]. Pyrolysis is a chemical process that gives flammable gas as byproduct of the tire waste transformation, in addition to carbon black (CBp), liquid fuel, pyrolytic oil, and pyrolysis char [10]. Product recycling involves using the discarded tires in different components, where recycling aims to obtain the original raw materials via processing [1]. Most of these strategies are not massively implemented in many countries representing a significant part of the world, mostly because of the costs involved in the process and other factors including absence of clear environmental laws [11].

In construction materials, perhaps the most known use for waste tires is its use in asphalt pavements to improve the concrete performance [11], although, waste tires are used in multiple applications. One of them is in low-scale retaining structures and architectural furnishing elements [12,13]. Two byproducts of discarded tires have been evaluated in concretes in order to know the mechanical properties: recycled tire polymer fibers (RTPF) and recycled rubber [14–16]. Moreover, steel fibers were evaluated as a potential reinforcement for concrete [17,18]. In another approach, the recycled tire rubber particles were combined with soil containing clay and sand as a potential new material [19–21]. Other composite materials used in construction have been studied and insulation properties were identified [22].

On the other hand, molasses are an organic byproduct of the repeated crystallization of sugar. It is a viscous syrup, with a dark brown color. The main elements in molasses are carbon and oxygen, while others in minor contents are calcium, sulfur, magnesium, chlorine, and potassium. Molasses are commonly used in the agricultural sector to control parasites [23], improve pastures [24], and to create nutritional supplements for cattle [25]. Molasses have also been used in different areas such as an additive for cement and concrete [26], during the cement production [27], stabilizer for expansive clays [28], and as a carbon source for bioremediation of contaminated soils [29].

Kaolin is a white fine-grained clay soil with kaolinite as its main constituent as hydrous aluminum silicate, composed of alumina and silica alternately stacked [30]. Kaolin has commonly been used in specific applications in paper, ceramics, refractories, plastics, rubber, adhesives, and paint industries [30]. Recently, kaolin has been used in additive manufacturing as well [31,32]. Furthermore, it was extensively used as construction material, and in its calcinated form, it has been used in concrete, for improving properties such as strength and durability [33]. Kaolin is also a very important source of alternative cements, geopolymers [34].

Several authors have addressed the use of waste materials from different origins to improve the ground, obtaining considerable improvement in the engineering properties of soils. Plastic wastes have been used to counteract traction stresses in the soil mass [35]. Ground plastic bottles have been used as reinforcement material when mixed with different soil types [36], waste fique natural fibers [37], and ashes from different processes were mixed with soils in order to develop cementing chemical reactions [38].

The present research studied an environmentally friendly alternative to the disposal of RTPF with the addition of molasses. Molasses were used as cementing material and as a matrix for the dispersion of the fibers. Characterization via X-ray diffraction (XRD), scanning electron microscopy (SEM), and shear strength, was used to evaluate the ground improvement. Finally, microorganisms that grew in the soil samples are shown.

2. Materials and Methods

The RTPF were obtained from a tire recycling company located at Rionegro, Colombia. Steel, rubber, and nylon wastes were mechanically separated. The nylon fiber has an average diameter of 22.4 μm. The residual rubber has an average size of 46 μm and a concentration of 60 wt% with respect to the rubber weight content. Molasses were obtained from a sugar cane company, located at Colombia. Table 1 summarizes the chemical composition of molasses. Kaolin soil in natural condition was supplied by Sumicol S.A.

Figure 1a shows polymer fibers on macro-scale image, where fibers look like cotton pads with fine residual rubber particles from the recycling process. Figure 1b shows a higher magnification of these fibers, revealing they are connected to one another due to the recycling process, which destroys the polymer textile and forms polymer fibers. Figure 1c shows the consistency of molasses. Figure 1d shows the supplied raw soil.

Table 1. SEM-EDS chemical composition for molasses.

Element	O	C	K	Cl	Ca	Mg	S	Total
Weight (%)	41.23	46.67	6.86	2.08	2.1	0.64	0.41	99.99
Atomic (%)	51.42	43.7	2.63	0.88	0.78	0.39	0.19	99.99

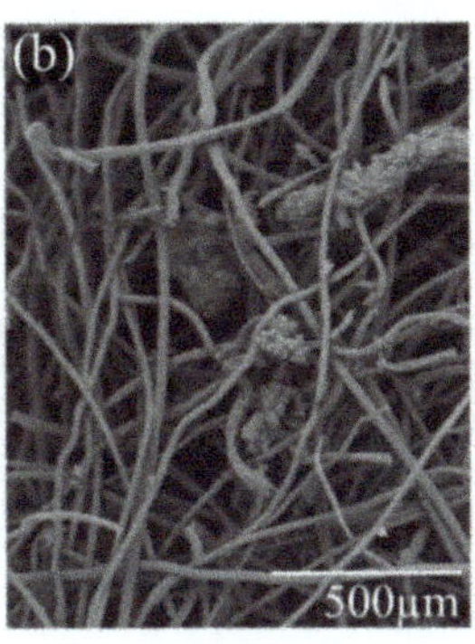

Figure 1. Materials used in this research, (**a**) polymer fibers, (**b**) magnification on polymer fibers, (**c**) molasses, (**d**) soil used in this research.

Figure 2 shows the optimum moisture (water) content showing the optimum moisture content, where soil contains 32% of water and compacted according to ASTM D1557-12e1 achieving 1.42 MPa as its highest dry density.

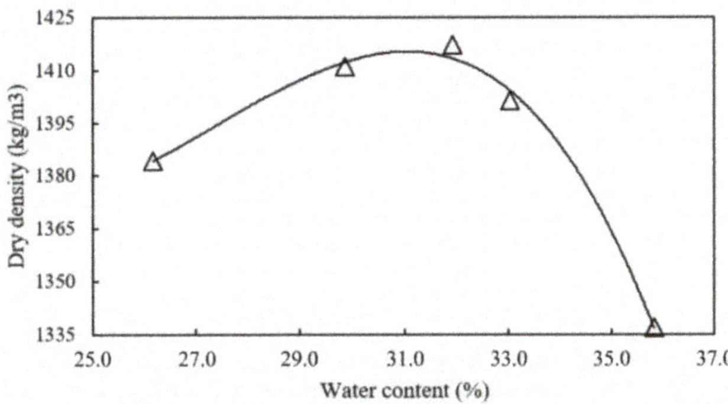

Figure 2. Optimum moisture content.

The dry kaolin clay was crushed using a hand roller in order to decrease agglomeration and have better particle size distribution. Molasses, RTPF, and water were measured by dry soil weight. Molasses and the fiber were hand mixed for three minutes until a homogenous consistency was reached; then the mixture and the water were mechanically mixed using a Hobart N50 apparatus at a speed of 126 rpm for five minutes. For the compaction of the soil, the ASTM D1557-12e1 Method A was used [39]. Three samples were selected from each compacted formulation, obtained by hammering a 2″ diameter aluminum tube, and a total of 21 soil samples. Using a hydraulic jack, soil samples were removed from the tubes. These samples were remolded and cut to 4″ height, and then, covered with plastic membrane in order to maintain the room temperature of 20 ± 3 °C during an air-cured process for 28 days.

The sample fabrication and all tests were conducted at 20 °C, typical room temperature in Medellin. Viscosity of molasses decreases with temperature as shown below [40], from about 8 to 12 Pa.s (8000 to 12,000 cPs). Since molasses change with temperature, the current solution is for areas where the temperature variation is low, such as areas where people do not need AC or heater, like Caracas, Bogota, San Diego, Medellin, etc., [41]. In Medellin, for instance, the temperature variation is between 18 and 28 °C, which in the soil could be just 30% variation (20 to 23 °C), a more acceptable temperature change. Future developments may consider thermal conductivity tests in order to increase insulation via materials science and thus decreasing even more the thermal changes.

For the characterization of the soil, specific gravity, liquid limit, optimum moisture content, plastic limit, and particle-size analysis, tests, have been performed according to respective ASTM test. The soil classification was done according to the ASTM D2487-17 by the Unified Soil Classification System (USCS) [42]. The density of soil samples was estimated based on weight and cylinder volume. Weight was measured using a Mettler Toledo balance, while the cylinder volume was measured using 0.05 mm precision caliper. Soil improvement was tested following the ASTM D2166-16 for the unconfined compression strength (UCS) [43]. For these tests, a universal testing apparatus Shimadzu AG250KN was used at a head speed of 0.68 mm/s.

X-ray diffraction (XRD) characterization was performed in a diffractometer X'Pert PRO, with λ = 1.5406 Å, Cu Kα radiation, voltage of 45 kV, and angles between 5° and 70°. Scanning electron microscopy with energy dispersive spectroscopy (SEM-EDS) was used to understand the microstructure of the samples in a JEOL JSM 6700 R equipment (Medellín, Colombia). For SEM, soil samples were dried at 30 °C for 24 h in a furnace, and coated with gold in a sputtering system at 15 mA AC for 30 s. Using the SEM images, the fiber diameter, rubber grain-size, and the mixed soil size distributions were estimated via image analysis with the Image-J software (version Java 1.8.0_172).

3. Results

3.1. Materials Characterization

The SEM-EDS results of the molasses shows this is mainly an organic compound, see Table 1, with oxygen and carbon with concentrations of 46.67% and 41.23% respectively. It also has potassium, calcium, and chlorine in concentrations of 6.86%, 2.1%, and 2.08% respectively; and magnesium, and sulfur in lower concentrations of 0.64% and 0.41% respectively.

Table 2 summarizes the kaolin´s index properties. According the USCS, the soil is a lean clay with sand (CL). Table 3 presents the experimental design for the formulations prepared. The fibers and molasses contents were estimated in a high precision balance. The soil was mixed with 35% of water, which is three percentage points up from the optimum moisture content to facilitate soil extraction without fracturing the samples (Table 2).

Table 2. Soil index properties.

Parameters	Values
Soil classification	
USCS	CL
Atterberg limits	
Liquid limit (%)	44
Plastic limit (%)	23
Plasticity index (%)	21
Specific gravity	
Specific gravity (20 °C)	2.75
Particle-size distribution	
Gravel (%)	0.1
Sand (%)	20.2
Silt (%)	40.5
Clay (%)	39.2
Parameters of compaction	
Maximum dry density (kg/m^3)	1417
Optimum moisture content (%)	32

Table 3. Soil mixtures.

Designation	Molasses(%)	Fibers (%)	Soil(%)	Total %(By Dry Soil)	% Water Content
M0	0	0	100	100	35
M2	2	0.1	97.9	100	35
M4	4	0.1	95.9	100	35
M6	6	0.1	93.9	100	35
M8	8	0.1	91.9	100	35
M10	10	0.1	89.9	100	35
M12	12	0.1	87.9	100	35

Figure 3a shows the particle-size distribution of the kaolin soil, with grain sizes from 0.2 µm to 4.5 mm, and the thickest particles corresponding to parental rock with less degree of weathering. Figure 3b represents the polymer fibers diameter fitted in a normal probability distribution. Fibers showed a diameter from 10 to 40 µm, with an average of 22.4 µm. Figure 3c shows the rubber grain size in a normal probability distribution. These residual grains showed a diameter from 15 to 105 µm, with an average of 46.03 µm. When comparing these two size distributions for nylon diameters and rubber particles with the sieve analysis for the clay, it is clear that almost all particles are very fine, with very sharp distributions below 100 µm, suggesting a very particular mixture of materials in the microscale.

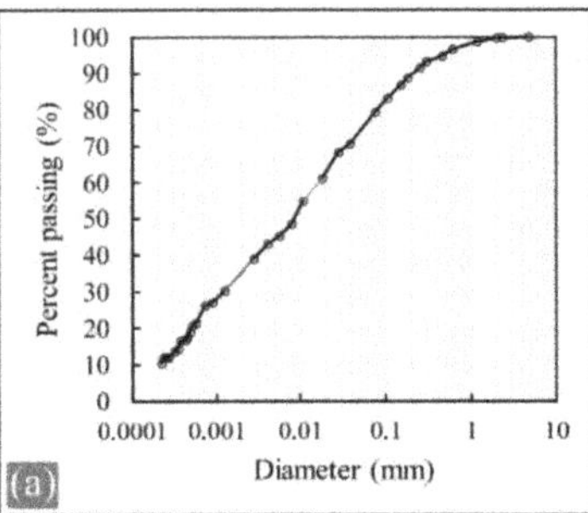
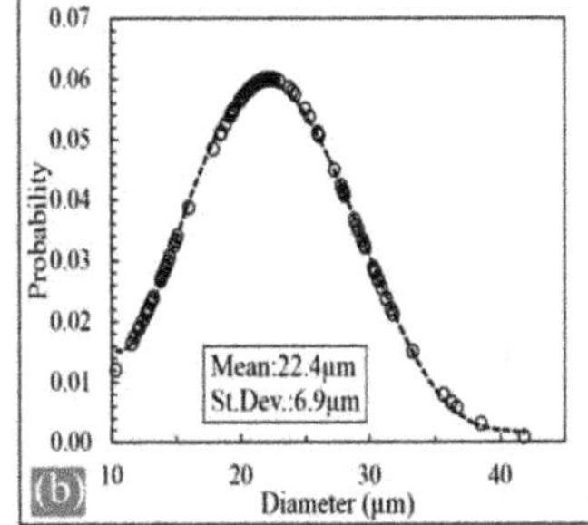

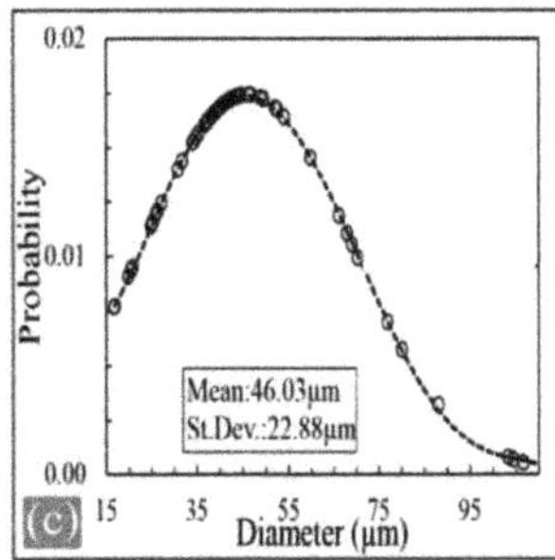

Figure 3. Grain/fiber diameter size distribution, (**a**) shows the particle-size distribution for the kaolin soil used, (**b**) shows the polymer fibers diameter distributions, (**c**) shows the residual rubber grain size distribution.

3.2. Soil Mixtures

Figure 4 shows the optical images representing the soil consistence and grain-sizes of the different mixtures used. These mixtures contain 0.1% fibers and 35% water by dry soil weight. As it can be seen, the molasses concentration has an important role in the grain-size formation, since as its concentration increases the grain-size of the mix becomes larger. Figure 4a–g presents the different mixtures of soil with fibers, water, and molasses as summarized in Table 3.

Figure 5 shows the mixed soil grain-size probability distribution for each mixture prepared. The normal probability distribution represents the random nature of the grain-size forming process, and shows that as molasses content increases the statistical data dispersion and grain-size increase. Samples M0, M2, and M4 showed a distribution in a similar interval, see Figure 5a; while samples M6, M8, M10, and M12 were more diverse in the distribution values, see Figure 5b.

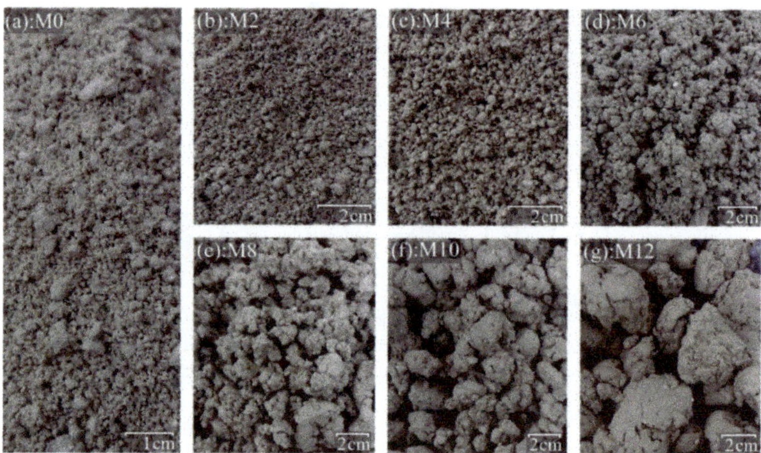

Figure 4. Optical images showing the soil consistence of mixtures.

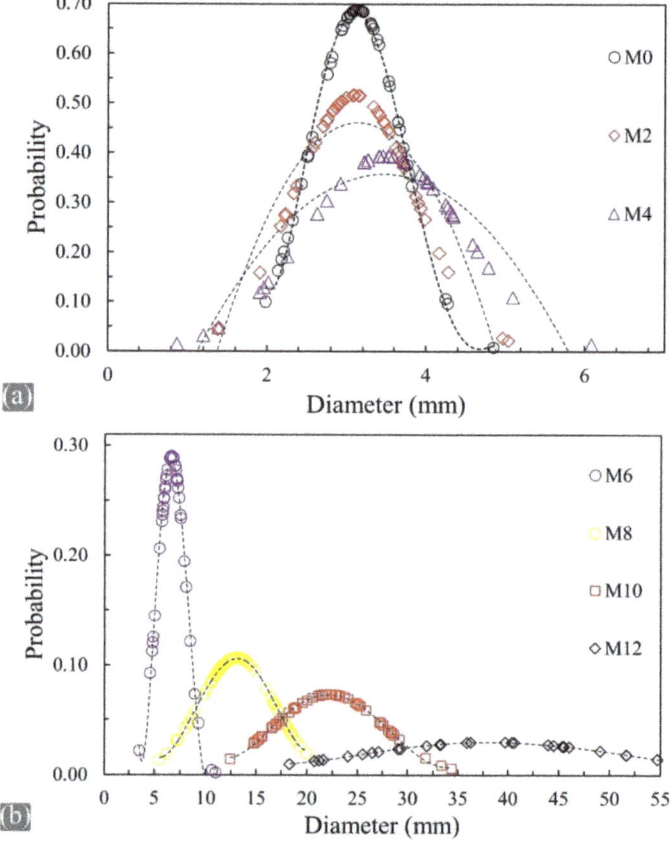

Figure 5. (**a**) Mixed soil grain size probability distribution for samples M0, M2, M4; (**b**) probability distribution for samples M6, M8, M10, M12.

Figure 6 illustrates the XRD spectrum of soil. M0 red line corresponds to soil in its natural condition and the M12 blue line is the chemical composition of the soil after mixing the fibers, 35% of water and 12% of molasses by dry weight of soil. The MINCRYS database was consulted for mineral identification. Halloysite, kaolinite, illite, quartz, dickite, and goethite were identified as the crystalline phases constituting the clay. Kaolinite clay is the main plastic component of this soil, which is important as Colombia has plenty of kaolin-based soils. The particle size distribution via sieve method of the clay is presented before, were more than 80% of the soil is less than 0.1 mm (100 μm) in particle size.

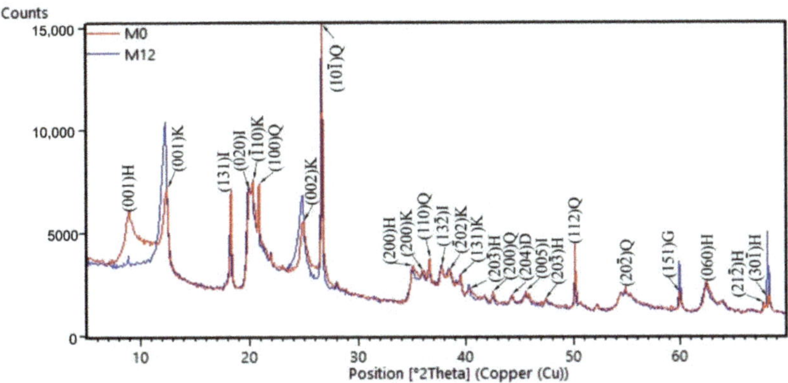

Figure 6. XRD spectrum for the soil. H: halloysite, K: kaolinite, I: illite, Q: quartz, D: dickite, G: goethite peaks.

3.3. UCS Test

Unconfined compression strength test (UCS) was performed using 21 samples, 3 in each batch. UCS and density results are given and compared in Figure 7. First, Figure 7a shows UCS result after 28 days of air curing. Second, Figure 7b presents the UCS and ductility results against the molasses content. Finally, the sample density and the UCS/density ratio are also plotted against the molasses content in Figure 7c. UCS of soil in natural conditions (M0) was 1.34 MPa, for the mixture with optimum molasses content (M2) was 2.14 Mpa, and for the mixture with the maximum molasses content was 1.27 MPa.

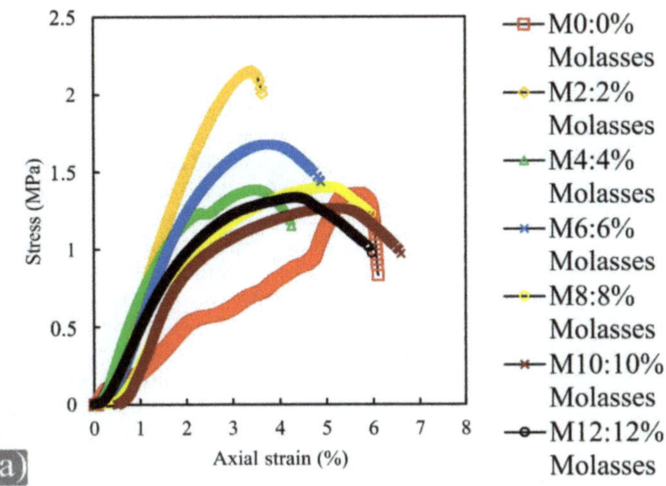

Figure 7. *Cont.*

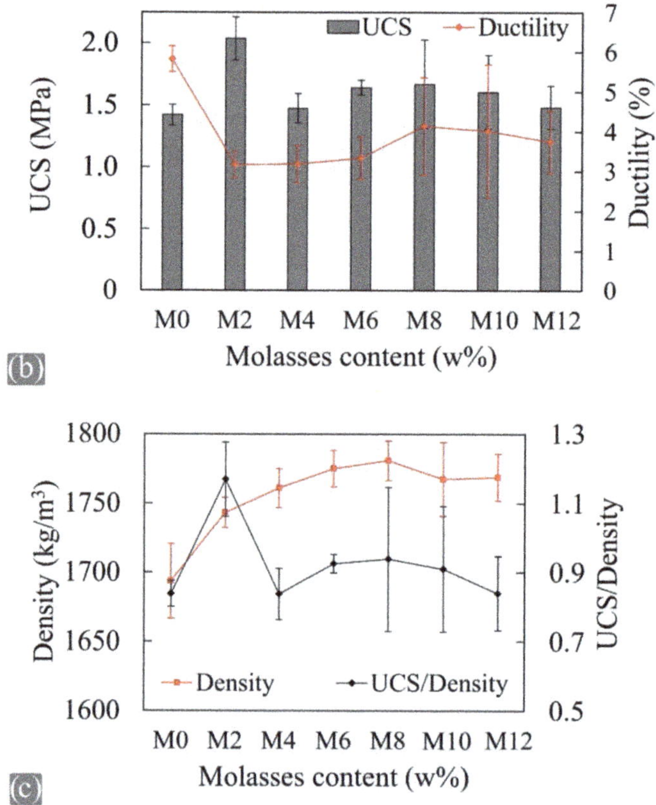

Figure 7. (**a**) UCS test, (**b**) UCS and ductility, (**c**) density and UCS/density.

Figure 8 presents the soil cylinders and failure characteristics of each mixture after compression tests, revealing different failure types, from a brittle failure in soil samples with low molasses content, to a more ductile failure in soils with higher molasses content. Therefore, molasses increases the possibility of a more ductile fracture and behavior.

3.4. SEM Images Results

Figure 9 shows SEM images results for some selected samples. Figure 10a shows the M2 soil sample taken at 500X, showing a fiber coming out from the soil matrix. Moreover, Figure 9b,c correspond to samples M2 and M4, with a fiber still attached to the soil and presenting a good impregnation of the soil to the fiber.

Figure 10 presents SEM-EDS images of kaolin soil. Figure 10a shows a 3×10^{-3} mm^2 spectrum SEM-EDS analysis of the structure and chemical composition of the soil used. Oxygen is the main element present in the material with a concentration of about 48.78%; aluminum and silicon are also significant components with concentrations of about 23.46% and 24.66% respectively. Finally, the soil has low concentrations of thallium and iron, being 1.98% and 1.12% of the respective concentrations. Concentrations are given as percentage of weight. Figure 10b presents a 49 μm^2 spectrum SEM-EDS analysis of the soil showing bacterial growth in the soil under natural conditions. These are rod-like microorganisms with an average 0.5 μm width and 15 μm length and a concentration about 10^4 microorganism/mm^2.

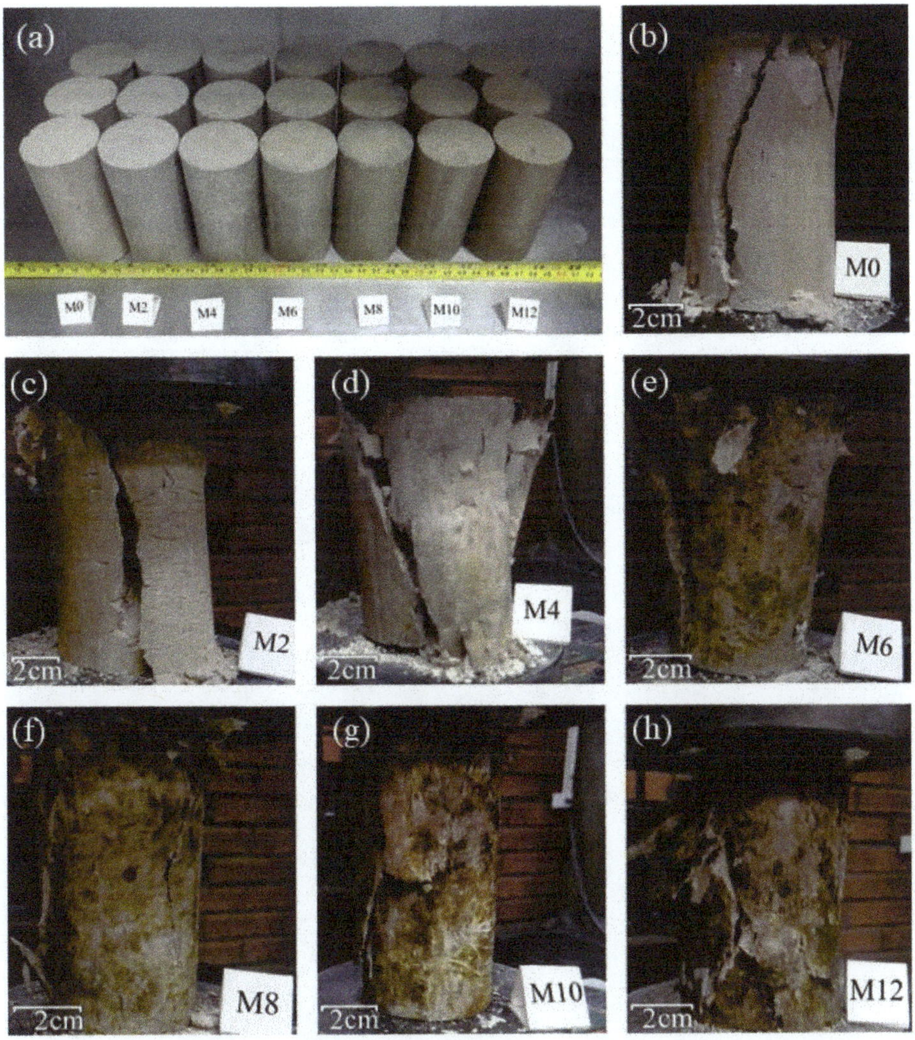

Figure 8. Soil samples after failure in compression tests: (**a**) samples before the test, (**b**) M0, (**c**) M2, (**d**) M4), (**e**) M6, (**f**) M8, (**g**) M10, (**h**) M12.

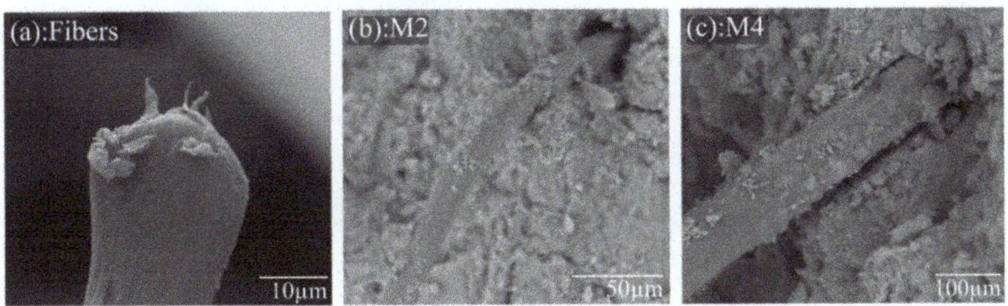

Figure 9. SEM images for (**a**) the end of a fiber; (**b**) the M2 soil sample, and (**c**) the M4 sample.

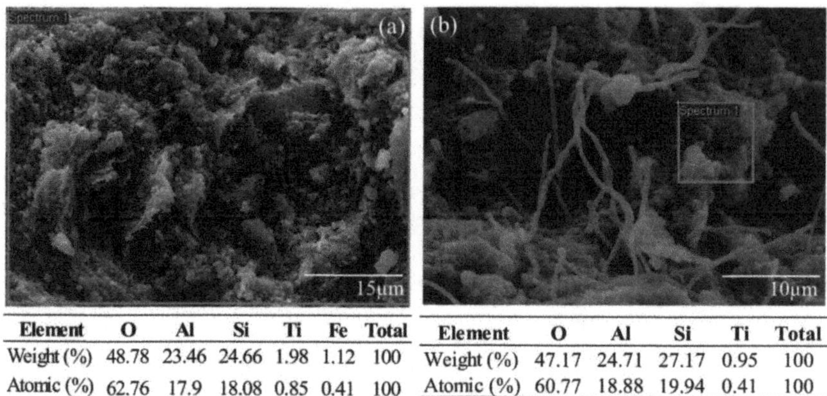

Element	O	Al	Si	Ti	Fe	Total
Weight (%)	48.78	23.46	24.66	1.98	1.12	100
Atomic (%)	62.76	17.9	18.08	0.85	0.41	100

Element	O	Al	Si	Ti	Total
Weight (%)	47.17	24.71	27.17	0.95	100
Atomic (%)	60.77	18.88	19.94	0.41	100

Figure 10. SEM-EDS for kaolin soil images, (**a**) soil used, (**b**) soil with bacterial growth.

Figure 11 shows the biological activity developed in sample M12. Figure 11a displays a microorganism that grew in the soil pores and extended around the micro-channels on the soil matrix with an average width of 2.6 μm, thickness of 0.4 μm, and thin film shape. Microorganism appeared after 4 days in some samples. In Figure 11b, another view of the microorganism described can be seen, as it rises from the bottom and has some particles adhering to it. Figure 11c shows the macro-scale image of microorganism growth in sample M12 at the external layer.

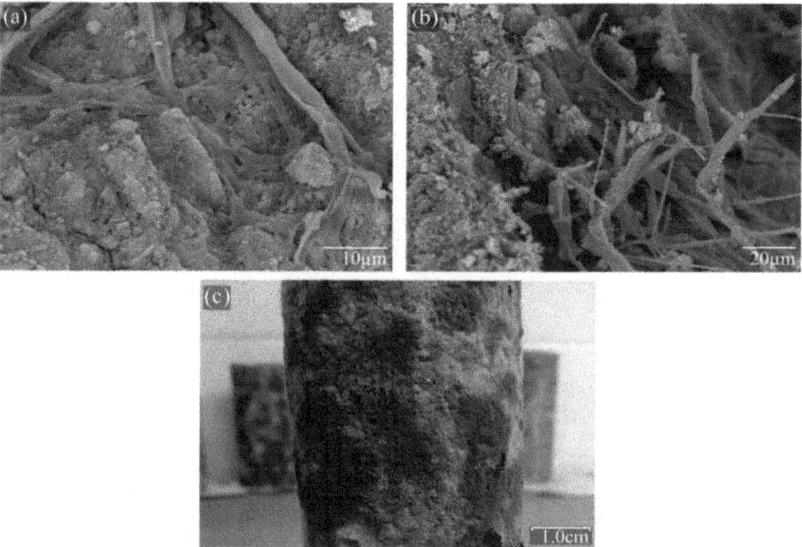

Figure 11. (**a**) Pseudo colored SEM images of M12 internal biological activity; (**b**) another area of the same sample showing a different configuration; (**c**) M12 external biological activity.

4. Discussions

Molasses have several potential uses in engineering applications due to aspects that give it advantages over conventional materials: it is soluble in water and without toxic components, an organic waste material derived from the food industry. Moreover, it is an inexpensive product, easy to acquire and with low dosages addition high maximum resistances can be achieved. For soil stabilization [44], If it is dosed at 2% by dry weight of soil, approximately 20 kg of molasses would be enough to process 1 ton of wet soil.

Saeed et al. [45] studied a kaolin with similar characteristics and engineering properties, treating it with different amounts of cement and curing times. The soil treated with 10% cement and cured for 28 days reached 2.10 MPa in unconfined compressive strength test. Approximately 100 kg of cement were necessary to process 1 ton of wet soil. Despite the research undertaken by Saeed et al. [45] and this work are not strictly equal, they can be compared on the bases of material costs for ground stabilization. In Colombia, molasses costs an average $0.35 USD/kg and cement costs an average $0.17 USD/kg. If 1 ton of soil is treated using cement, the costs would be approximately $17 USD; while if the same amount of soil is treated using molasses the cost would be about $7 USD. Although molasses is an inexpensive material when compared to cement, both as additives to soil, cement is a very competitive material when durability and strength are evaluated. Saeed et al. [45] concluded that 2% of molasses in dry soil matches the compression strength of soil with 10% cement, which is quite an impressive value. The durability on the other hand could be a limitation, as explained before, molasses changes very rapidly with time when compared with cement, although very little is known on this subject, therefore being a good area for research. Durability performance was not studied in this research, and it is possible that molasses does not provide the same durability as cement. However, soil improvement using molasses can be a temporal engineering solution.

Molasses are an organic material that suffers environmental degradation and aging. Therefore, the formulations properties of samples could be significantly affected. Research regarding the molasses stability in the soil have not been conducted yet, although future stability and chemical treatments for overcoming this issue are interesting as future research. Although biological activity was detected, a future systematic study will be conducted on the property's evolution and potential treatments as a function of the microorganisms' growth during the process. The samples were made under controlled fabrication procedure in this research, and the mechanical stability of the samples suggests that samples did not change significantly in a month, mainly because samples were made in a controlled environment, with constant atmosphere conditions. Regarding other construction applications under more severe conditions of temperature and load, the molasses have shown good stability [46] with load and temperature, but poor performance with moisture. Thus, to decrease the aging effect, for now, molasses has to be ideally at constant humidity. The clay used in this research in fact is a perfect material to keep the molasses isolated from water, to seal the soil as other clays [47], which is expected to decrease its natural and working-derived aging effects.

The Ministry of Transport in Colombia classifies tertiary roads as the routes that connect municipalities with rural municipalities [48]. Colombia has 142,284 km tertiary roads, of which 34,148 km do not have any engineering treatment [49]. Some of the aforementioned roads become impassable during the rainy seasons and thus soil improvement is required. Commonly, tertiary roads are stabilized using lime, cement, or a coarse aggregate structure. The analysis of behavior of the soils stabilized using molasses and its environmental impact is an environmentally friendly solution for ground improvement. This could reduce or replace the amount of cement used in tertiary roads, thus, reducing polluting gases generated by cement production. Clearly, this is not only a problem of Colombia, but also an issue in many developing countries.

There is a correlation in agglomerates between the liquid saturation and the growth rate of the agglomerates [50]. Interparticle forces are responsible for the formation of the agglomerates by attaching individual particles to each other [50]. The crushed soil in natural conditions mixed with water and 35% water content, has an average 3.12 mm diameter grain-size. Grain-size and its statistic data dispersion increase as molasses content increases. This occurs as the soil particles are agglomerated by the cohesion of molasses producing coalescence. The bonding forces necessary to obtain coalescence among agglomerates need to be significantly higher than the forces necessary for the primary particles' process of nucleation [51]. In this case, the high viscosity of the molasses increases the bonding forces between the soil particles facilitating coalescence between soil particles.

Addition of molasses has an important effect on the failure mode of the soil. As the molasses content increases the failure mode becomes more ductile. The base-line soil sample (M0) had a brittle failure in which shear cracks fell along a plane to the heading of compression. On the other hand, the soil with the highest content of molasses (M12) showed a plastic failure with multiple fractures extending from top to bottom. Soil samples with the highest UCS (M2 and M8) had a brittle failure with longitudinal splitting. When comparing the failure planes of the unstabilized sample (M0) and the samples with the highest UCS, the change in inclination indirectly shows a cohesion and an angle of internal friction rise.

Biological Activity

This research aims to investigate how molasses and fibers influence the unconfined compressive strength of soils. However, substantial biological activity was evidenced. SEM images showed the development of different microorganism in mixture M12 and after 7 days, microorganism grew on the external layer of the samples. To the naked eye, biological activity was not detected for the mixture with the optimum molasses content (M2). Intensity of the biological activity increases as molasses content increases, which can be used to improve the properties of soils, an area that falls in the bio-mediated soil improvement. This branch of geotechnical engineering has evolved to many methods, which includes bioclogging and biocementation. Bio-mediated methods for soil improving involves the generation of chemical reactions on soil mass that generate byproducts which change the engineering properties of the soil [52].

The use of microorganism to generate the reactions is an attractive and feasible alternative due to the number of microorganisms present in the soil and its resistance to extreme conditions. There are approximately 10^9 to 10^{12}/Kg of organisms of a soil mass near the ground surface [53]. For soil bioclogging or biocementation, the most recommended microorganisms are anaerobic and microaerophilic bacteria [54]. Biological activity is limited by nutrients availability, water availability, and other environmental conditions; carbon and mineral components as energy sources and pores larger than 0.4 μm as environmental conditions are necessary to facilitate microorganism growth [55].

Biological activity was encouraged by conditions under which this research was performed. The molasses used had a composition of 41.2% carbon and 10.01% minerals like Mg, S, K, Ca. Due to its chemical composition, molasses is a considerable source of energy for microorganisms. On the other hand, water and 10-μm pores in the soil stimulated growth of microorganism that are naturally found in the soil.

5. Conclusions

- The RTPF mixed with molasses represent a feasible, environmentally positive, economical, and technical alternative for ground improvement.
- Molasses are a byproduct of the repeated crystallization of sugar with no toxic elements and are environment-friendly. Due to its viscosity, molasses allows the separation of RTPF in soil, thereby producing a good composite solution.
- The improvement in the UCS achieved was of about 43%, this is, from 1.42 MPa for the reference sample, to 2.04 MPa for the mixture sample, which contains 2% of molasses and 0.1% of RTPF.

Author Contributions: Conceptualization, H.A.C., C.M.F.V. and J.E.J.; methodology, H.A.C. and J.E.J.; validation, C.M.F.V.; investigation, J.E.J., C.M.F.V. and H.A.C.; writing—original draft preparation, H.A.C. and C.M.F.V.; writing—review and editing, H.A.C., J.E.J. and C.M.F.V.; supervision, H.A.C.; project administration, H.A.C. and J.E.J.; funding acquisition, H.A.C. and J.E.J. All authors have read and agreed to the published version of the manuscript.

Funding: This investigation has not received external funding.

Data Availability Statement: Datasets generated during the study can be found in: https://drive.google.com/drive/folders/1rjZUQ4tfEI0BoZUOd7LGEUw2A8BZQvmw?usp=sharing (accessed on 30 January 2022).

Acknowledgments: Authors would like to thank Juan Pablo Osorio from Universidad de Antioquia and Sumicol S.A. for supplying soil samples.

Conflicts of Interest: The authors declare no conflict of interest.

References

1. Sienkiewicz, M.; Kucinska-Lipka, J.; Janik, H.; Balas, A. Progress in used tyres management in the European Union: A review. *Waste Manag.* **2012**, *32*, 1742–1751. [CrossRef] [PubMed]
2. Rubber Manufactures Association. *2017 US Scrap tire Management Summary*; Publisher Rubber Manufactures Association: Washington, DC, USA, 2019; pp. 1–20.
3. Ministry of Environment, Housing and Territorial Development. Available online: https://www.minambiente.gov.co/images/AsuntosambientalesySectorialyUrbana/pdf/Programa_posconsumo_existente/RESOLUCION_1457_de_2010_llantas.pdf (accessed on 10 February 2019).
4. District Secretary of the Environment. Available online: https://ambientebogota.gov.co/c/document_library/get_file?uuid=2b1cb194-bfd7-43d9-9350-11dd98b6f426groupId=10157 (accessed on 10 February 2019).
5. Observatorio Ambiental de Bogotá. Available online: https://oab.ambientebogota.gov.co/es/indicadores?id=1057v=1# (accessed on 10 February 2019).
6. Stevenson, K.; Stallwood, B.; Hart, A.G. Tire rubber recycling and bioremediation: A review. *Bioremediat. J.* **2008**, *12*, 1–11. [CrossRef]
7. Lebreton, B.; Tuma, A. A quantitative approach to assessing the profitability of car and truck tire remanufacturing. *Int. J. Prod. Econ.* **2006**, *104*, 639–652. [CrossRef]
8. Zebala, J.; Ciepka, P.; Reza, A.; Janczur, R. Influence of rubber compound and tread pattern of retreaded tyres on vehicle active safety. *Forensic Sci. Int.* **2007**, *167*, 173–180. [CrossRef] [PubMed]
9. Gieré, R.; Smith, K.; Blackford, M. Chemical composition of fuels and emissions from a coal+ tire combustion experiment in a power station. *Fuel* **2006**, *85*, 2278–2285. [CrossRef]
10. Zhang, X.; Wang, T.; Ma, L.; Chang, J. Vacuum pyrolysis of waste tires with basic additives. *Waste Manag.* **2008**, *28*, 2301–2310. [CrossRef]
11. Agudelo, G.; Cifuentes, S.; Colorado, H.A. Ground tire rubber and bitumen with wax and its application in a real highway. *J. Clean. Prod.* **2019**, *228*, 1048–1061. [CrossRef]
12. Shi, S.Q.; Wang, M.; Yin, P.; Luo, X. Experimental Study on a New Type of Waste Tires Combined Stone-blocking Structure. *J. Disaster Prev. Mitig. Eng.* **2011**, *31*, 501–505.
13. Yoon, Y.W.; Heo, S.B.; Kim, K.S. Geotechnical performance of waste tires for soil reinforcement from chamber tests. *Geotext. Geomembr.* **2008**, *26*, 100–107. [CrossRef]
14. Baricevic, A.; Pezer, M.; Rukavina, M.J.; Serdar, M.; Stirmer, N. Effect of polymer fibers recycled from waste tires on properties of wet-sprayed concrete. *Constr. Build. Mater.* **2018**, *176*, 135–144. [CrossRef]
15. Huang, B.; Li, G.; Pang, S.S.; Eggers, J. Investigation into waste tire rubber-filled concrete. *J. Mater. Civil. Eng.* **2004**, *16*, 187–194. [CrossRef]
16. Son, K.S.; Hajirasouliha, I.; Pilakoutas, K. Strength and deformability of waste tyre rubber-filled reinforced concrete columns. *Constr. Build. Mater.* **2011**, *25*, 218–226. [CrossRef]
17. Papakonstantinou, C.G.; Tobolski, M.J. Use of waste tire steel beads in Portland cement concrete. *Cem. Concr. Res.* **2006**, *36*, 1686–1691. [CrossRef]
18. Pilakoutas, K.; Neocleous, K.; Tlemat, H. Reuse of tyre steel fibres as concrete reinforcement. *Proc. ICE-Eng. Sustain.* **2004**, *157*, 131–138. [CrossRef]
19. Hataf, N.; Rahimi, M.M. Experimental investigation of bearing capacity of sand reinforced with randomly distributed tire shreds. *Constr. Build. Mater.* **2006**, *20*, 910–916. [CrossRef]
20. Revelo, F.C.; Colorado, H.A. A green composite material of calcium phosphate cement matrix with additions of car tire waste particles. *Int. J. Appl. Ceram. Technol.* **2021**, *18*, 182–191. [CrossRef]
21. Srivastava, A.; Pandey, S.; Rana, J. Use of shredded tyre waste in improving the geotechnical properties of expansive black cotton soil. *Geomech. Geoengin.* **2014**, *9*, 303–311. [CrossRef]
22. Zhao, J.; Wang, X.M.; Chang, J.M.; Yao, Y.; Cui, Q. Sound insulation property of wood–waste tire rubber composite. *Compos. Sci. Technol.* **2010**, *70*, 2033–2038. [CrossRef]
23. Rodriguez-Kabana, R.; King, P.S. Use of mixtures of urea and blackstrap molasses for control of root-knot nematodes in soil. *Nematropica* **1980**, *10*, 38–44.
24. Yunus, M.; Ohba, N.; Shimojo, M.; Furuse, M.; Masuda, Y. Effects of adding urea and molasses on Napiergrass silage quality. *Asian-Australas. J. Anim. Sci.* **2000**, *13*, 1542–1547. [CrossRef]

25. Tiwari, S.P.; Singh, U.B.; Mehra, U.R. Urea molasses mineral blocks as a feed supplement: Effect on growth and nutrient utilization in buffalo calves. *Anim. Feed. Sci. Technol.* **1990**, *29*, 333–341. [CrossRef]
26. Aalm, A.; Singh, P. Experimental Study on Strength Characteristics of Cement Concrete by Adding Sugar Waste. *Int. J. Enhanc. Res. Sci. Technol. Eng.* **2016**, *5*, 33–34.
27. Zhang, Y.; Fei, A.; Li, D. Utilization of waste glycerin, industry lignin and cane molasses as grinding aids in blended cement. *Constr. Build. Mater.* **2016**, *123*, 785–791. [CrossRef]
28. M'Ndegwa, J.K. The effect of cane molasses on strength of expansive clay soil. *J. Emerg. Trends Eng. Appl. Sci.* **2011**, *2*, 1034–1041.
29. Boopathy, R. Bioremediation of explosives contaminated soil. *Int. Biodeterior. Biodegrad.* **2000**, *46*, 29–36. [CrossRef]
30. Prasada, M.S.; Reid, K.J.; Murray, H.H. Kaolin: Processing, properties and application. *Appl. Clay Sci.* **1991**, *6*, 87–119. [CrossRef]
31. Ordoñez, E.; Gallego, J.M.; Colorado, H.A. 3D printing via the direct ink writing technique of ceramic pastes from typical formulations used in traditional ceramics industry. *Appl. Clay Sci.* **2019**, *182*, 105285. [CrossRef]
32. Revelo, C.F.; Colorado, H.A. 3D printing of kaolinite clay ceramics using the Direct Ink Writing (DIW) technique. *Ceram. Int.* **2018**, *44*, 5673–5682. [CrossRef]
33. Shafiq, N.; Nuruddin, M.F.; Khan, S.U.; Ayub, T. Calcined kaolin as cement replacing material and its use in high strength concrete. *Constr. Build. Mater.* **2015**, *81*, 313–323. [CrossRef]
34. Heah, C.Y.; Kamarudin, H.; Al Bakri, A.M.; Bnhussain, M.; Luqman, M.; Nizar, I.K.; Liew, Y.M. Study on solids-to-liquid and alkaline activator ratios on kaolin-based geopolymers. *Constr. Build. Mater.* **2012**, *35*, 912–922. [CrossRef]
35. Babu, G.S.; Chouksey, S.K. Stress–strain response of plastic waste mixed soil. *Waste Manag.* **2011**, *31*, 481–488. [CrossRef] [PubMed]
36. Sivakumar Babu, G.L.; Chouksey, S.K. Analytical model for stress-strain response of plastic waste mixed soil. *J. Hazard. Toxic Radioact. Waste* **2011**, *16*, 219–228. [CrossRef]
37. García, F.E.; Pérez, A.C.; Colorado, H.A. Kaolinite-based clay ceramics blended with residual fique fibers for potential plastic soil applications. *Int. J. Appl. Ceram. Technol.* **2021**, *18*, 1086–1096. [CrossRef]
38. Sezer, A.; İnan, G.; Yılmaz, H.R.; Ramyar, K. Utilization of a very high lime fly ash for improvement of Izmir clay. *Build. Environ.* **2006**, *41*, 150–155. [CrossRef]
39. ASTM International (n.d). ASTM D1557-12e1 Standard Test Methods for Laboratory Compaction Characteristics of Soil Using Modified Effort (56,000 ft-lbf/ft3 (2700 kN-m/m3)). Available online: https://www.astm.org/d1557-12r21.html (accessed on 10 February 2019).
40. Mišljenović, N.; Schüller, R.B.; Rukke, E.O.; Bringas, C.S. Rheological characterization of liquid raw materials for solid biofuel production. *Annu. Trans. Nord. Rheol. Soc.* **2013**, *21*, 61–68.
41. Cities of the World Where You Don't Need AC or Heat, Mapped. Available online: https://mnolangray.medium.com/cities-of-the-world-where-you-dont-need-ac-or-heat-mapped-2a3d6e018970 (accessed on 10 January 2022).
42. ASTM International (n.d). ASTM D2487-17 Standard Practice for Classification of Soils for Engineering Purposes (Unified Soil Classification System). Available online: https://www.astm.org/d2487-17e01.html (accessed on 10 February 2019).
43. ASTM International (n.d). D2166M-16 Standard Test Method for Unconfined Compressive Strength of Cohesive Soil. Available online: https://www.astm.org/d2166_d2166m-16.html (accessed on 10 February 2019).
44. Rahgozar, M.A.; Saberian, M.; Li, J. Soil stabilization with non-conventional eco-friendly agricultural waste materials: An experimental study. *Transp. Geotech.* **2018**, *14*, 52–60. [CrossRef]
45. Abdulhussein Saeed, K.; Anuar Kassim, K.; Nur, H. Physicochemical characterization of cement treated kaolin clay. *Građevinar* **2014**, *66*, 513–521.
46. Hareru, W.; Ghebrab, T. Rheological Properties and Application of Molasses Modified Bitumen in Hot Mix Asphalt (HMA). *Appl. Sci.* **2020**, *10*, 1931. [CrossRef]
47. Alther, G.R. The qualifications of bentonite as a soil sealant. *Eng. Geol.* **1987**, *23*, 177–191. [CrossRef]
48. Ministerio de Transporte. *Manual de Diseño Geométricos de Carreteas*; Instituto Nacional de Vías: Bogotá, Colombia, 2008.
49. Departamento Nacional de Planeación. Available online: https://colaboracion.dnp.gov.co/CDT/Prensa/Presentaciones/RED%20TERCIARIA%20CCI%20-%20DNP.pdf (accessed on 27 May 2019).
50. Kristensen, H.G. Particle agglomeration in high shear mixers. *Powder Technol.* **1996**, *88*, 197–202. [CrossRef]
51. Lian, G.; Thornton, C.; Adams, M.J. Discrete particle simulation of agglomerate impact coalescence. *Chem. Eng. Sci.* **1998**, *53*, 3381–3391. [CrossRef]
52. DeJong, J.T.; Mortensen, B.M.; Martinez, B.C.; Nelson, D.C. Bio-mediated soil improvement. *Ecol. Eng.* **2010**, *36*, 197–210. [CrossRef]
53. Umar, M.; Kassim, K.A.; Chiet, K.T.P. Biological process of soil improvement in civil engineering: A review. *J. Rock Mech. Geotech. Eng.* **2016**, *8*, 767–774. [CrossRef]
54. Ivanov, V.; Chu, J. Applications of microorganisms to geotechnical engineering for bioclogging and biocementation of soil in situ. *Rev. Environ. Sci. Bio/Technol.* **2008**, *7*, 139–153. [CrossRef]
55. Mitchell, J.K.; Santamarina, J.C. Biological considerations in geotechnical engineering. *J. Geotech. Geoenviron. Eng.* **2005**, *131*, 1222–1233. [CrossRef]

MDPI
St. Alban-Anlage 66
4052 Basel
Switzerland
Tel. +41 61 683 77 34
Fax +41 61 302 89 18
www.mdpi.com

Sustainability Editorial Office
E-mail: sustainability@mdpi.com
www.mdpi.com/journal/sustainability

9 783036 570877